Mechanical Engineering Principles

Second Edition

John O. Bird, BSc(Hons), CEng, CMath, CSci, FIMA, FITE, FCollT
Carl T. F. Ross, BSc(Hons), PhD, DSc, CEng, FRINA, MSNAME

Routledge
Taylor & Francis Group

LONDON AND NEW YORK

Second edition published 2012
by Routledge
2 Park Square, Milton Park, Abingdon, Oxon OX14 4RN

Simultaneously published in the USA and Canada
by Routledge
711 Third Avenue, New York, NY 10017

Routledge is an imprint of the Taylor & Francis Group, an informa business

First edition published by Elsevier in 2002

Trademark notice: Product or corporate names may be trademarks or registered trademarks, and are used only for identification and explanation without intent to infringe.

British Library Cataloguing in Publication Data
A catalogue record for this book is available from the British Library

Library of Congress Cataloguing in Publication Data
A catalog record for this title has been requested

ISBN: 9780415517850 (pbk)
ISBN: 9780203121146 (ebk)

Typeset in Times
by RefineCatch Limited, Bungay, Suffolk

Printed by Bell & Bain Ltd., Glasgow

Mechanical Engineering Principles

Second Edition

Why are competent engine

Engineering is among the most important of all professions. It is the authors' opinions that engineers save more lives than medical doctors (physicians). For example, poor water or the lack of it, is the second largest cause of human death in the world, and if engineers are *given the tools*, they can solve this problem. The largest cause of human death is caused by the malarial mosquito, and even death due to malaria can be decreased by engineers - by providing helicopters for spraying areas infected by the mosquito and making and designing medical syringes and pills to protect people against catching all sorts of diseases. Most medicines are produced by engineers! How does the engineer put 1 mg of 'medicine' precisely and individually into millions of pills, at an affordable price?

Moreover, one of the biggest contributions by humankind was the design of the agricultural tractor, which was designed and built by engineers to increase food production many-fold for a human population which more-or-less quadruples every century! It is also interesting to note that the richest countries in the world are very heavily industrialised. Engineers create wealth! Most other professions don't!

Even in blue sky projects, engineers play a major role. For example, most rocket scientists are chartered engineers or their equivalents and Americans call their Chartered Engineers (and their equivalents), scientists. Astronomers are space scientists and not rocket scientists; they could not design a rocket to conquer outer space. Even modern theoretical physicists are mainly interested in astronomy and cosmology and also nuclear science. In general a theoretical physicist cannot, without special training, design a submarine structure to dive to the bottom of the Mariana Trench, which is 11.52 km or 7.16 miles deep, or design a very long bridge, a tall city skyscraper or a rocket to conquer outer space.

This book presents a solid foundation for the reader in mechanical engineering principles, on which s/he can safely build tall buildings and long bridges that may last for a thousand years or more. It is the authors' experience that it is most unwise to attempt to build such structures on shaky foundations; they may come tumbling down - with disastrous consequences.

John O. Bird is the former Head of Applied Electronics in the Faculty of Technology at Highbury College, Portsmouth, U.K. More recently, he has combined freelance lecturing at the University of Portsmouth, with Examiner responsibilities for Advanced Mathematics with City & Guilds, and examining for the International Baccalaureate Organisation. He is the author of some 120 textbooks on engineering and mathematical subjects with worldwide sales approaching 1 million copies. He is currently a Senior Training Provider at the Royal Naval School of Marine Engineering in the Defence College of Marine and Air Engineering at H.M.S. Sultan, Gosport, Hampshire, U.K.

Carl T. F. Ross gained his first degree in Naval Architecture, from King's College, Durham University; his PhD in Structural Engineering from the Victoria University of Manchester; and was awarded his DSc in Ocean Engineering from the CNAA, London. His research in the field of engineering led to advances in the design of submarine pressure hulls. His publications to date exceed some 270 papers and books and he is Professor of Structural Dynamics at the University of Portsmouth, U.K.

See Carl Ross's website below, which has an enormous content on science, technology and education.
http://homepage.ntlworld.com/carl.ross/page3.htm

Contents

Preface

Mechanical Engineering Principles 2nd Edition aims to broaden the reader's knowledge of the basic principles that are fundamental to mechanical engineering design and the operation of mechanical systems.

Modern engineering systems and products still rely upon static and dynamic principles to make them work. Even systems that appear to be entirely electronic have a physical presence governed by the principles of statics.

In this second edition of Mechanical Engineering Principles, a chapter has been added on revisionary mathematics; it is not possible to progress in engineering studies without a reasonable knowledge of mathematics, a fact that soon becomes obvious to both students and teachers alike. It is therefore hoped that this chapter on basic mathematics revision will be helpful and make the engineering studies more comprehensible. Minor modifications and some further worked problems have also been added throughout the text.

Free Internet downloads of full solutions to the further problems and a **PowerPoint presentation of all the illustrations** contained in the text is available – see page x.

For clarity, the text is divided into **four parts**, these being:

> **Part 1 Revision of Mathematics**
> **Part 2 Statics and strength of materials**
> **Part 3 Dynamics**
> **Part 4 Heat transfer and fluid mechanics**

Mechanical Engineering Principles 2nd Edition is suitable for the following:

(i) National Certificate/Diploma courses in Mechanical Engineering
(ii) Undergraduate courses in Mechanical, Civil, Structural, Aeronautical & Marine Engineering, together with Naval Architecture
(iii) Any introductory/access/foundation course involving Mechanical Engineering Principles at University, and Colleges of Further and Higher education.

Although pre-requisites for the modules covered in this book include Foundation Certificate/ diploma, or similar, in Mathematics and Science, **each topic considered in the text is presented in a way that assumes that the reader has little previous knowledge of that topic.**

Mechanical Engineering Principles 2nd Edition contains over **325 worked problems**, followed by over **550 further problems** (all **with answers**). The further problems are contained within some **140 Exercises**; each Exercise follows on directly from the relevant section of work, every few pages. In addition, the text contains **276 multiple-choice questions** (all **with answers**), and **260 short answer questions**, the answers for which can be determined from the preceding material in that particular chapter. Where at all possible, the problems mirror practical situations found in mechanical engineering. **371 line diagrams** enhance the understanding of the theory.

At regular intervals throughout the text are some **8 Revision Tests** to check understanding. For example, Revision Test 1 covers material contained in Chapter 1, Test 2 covers the material in Chapters 2 to 5, and so on. No answers are given for the questions in the Revision Tests, but a **Lecturer's guide** has been produced giving full solutions and suggested marking scheme. The guide is offered online free to lecturer's/ instructor's – see below.

At the end of the text, a list of relevant **formulae** is included for easy reference.

'Learning by Example' is at the heart of *Mechanical Engineering Principles, 2nd Edition.*

JOHN BIRD
Royal Naval School of Marine Engineering,
HMS Sultan, formerly
University of Portsmouth and Highbury
College, Portsmouth
CARL ROSS Professor, University of Portsmouth

Free web downloads available to lecturers/instructors only at www.routledge.com/cw/bird

Worked Solutions to Exercises

Within the text are some 550 further problems arranged within 140 Exercises; all 550 worked solutions have been prepared.

Instructor's manual

This provides full worked solutions and mark scheme for all 8 Revision Tests in this book.

Illustrations

Lecturers can download electronic files for all 371 illustrations in this second edition.

Part One

Revision of Mathematics

Chapter 1

Revisionary mathematics

Mathematics is a vital tool for professional and chartered engineers. It is used in mechanical & manufacturing engineering, in electrical & electronic engineering, in civil & structural engineering, in naval architecture & marine engineering and in aeronautical & rocket engineering. In these various branches of engineering, it is very often much cheaper and safer to design your artefact with the aid of mathematics - rather than through guesswork. 'Guesswork' may be reasonably satisfactory if you are designing an artefact similar to one that has already proven satisfactory; however, the classification societies will usually require you to provide the calculations proving that the artefact is safe and sound. Moreover, these calculations may not be readily available to you and you may have to provide fresh calculations, to prove that your artefact is 'roadworthy'. For example, if you design a tall building or a long bridge by 'guesswork', and the building or bridge do not prove to be structurally reliable, it could cost you a fortune to rectify the deficiencies. This cost may dwarf the initial estimate you made to construct these artefacts, and cause you to go bankrupt. Thus, without mathematics, the prospective professional or chartered engineer is very severely handicapped.

At the end of this chapter you should be able to:

- convert radians to degrees
- convert degrees to radians
- calculate sine, cosine and tangent for large and small angles
- calculate the sides of a right-angled triangle
- use Pythagoras' theorem
- use the sine and cosine rules for acute-angled triangles
- expand equations containing brackets
- be familiar with summing vulgar fractions

- understand and perform calculations with percentages
- understand and use the laws of indices
- solve simple simultaneous equations

1.1 Introduction

As highlighted above, it is not possible to understand aspects of mechanical engineering without a good knowledge of mathematics. This chapter highlights some areas of mathematics which will make the understanding of the engineering in the following chapters a little easier.

1.2 Radians and degrees

There are 2π radians or $360°$ in a complete circle, thus:

$$\pi \text{ radians} = 180° \qquad \text{from which,}$$

$$1 \text{ rad} = \frac{180°}{\pi} \qquad \text{or} \qquad 1° = \frac{\pi}{180} \text{ rad}$$

where π = 3.14159265358979323846 .... to 20 decimal places!

Problem 1. Convert the following angles to degrees correct to 3 decimal places:

(a) 0.1 rad   (b) 0.2 rad   (c) 0.3 rad

(a) $0.1 \text{ rad} = 0.1 \text{ rad} \times \frac{180°}{\pi \text{ rad}} = \textbf{5.730°}$

(b) $0.2 \text{ rad} = 0.2 \text{ rad} \times \frac{180°}{\pi \text{ rad}} = \textbf{11.459°}$

Mechanical Engineering Principles, Bird and Ross, ISBN 9780415517850

(c) $0.3 \text{ rad} = 0.3 \text{ rad} \times \dfrac{180°}{\pi \text{ rad}} = \textbf{17.189°}$

Problem 2. Convert the following angles to radians correct to 4 decimal places:

(a) 5° (b) 10° (c) 30°

(a) $5° = 5° \times \dfrac{\pi \text{ rad}}{180°} = \dfrac{\pi}{36} \text{ rad} = \textbf{0.0873 rad}$

(b) $10° = 10° \times \dfrac{\pi \text{ rad}}{180°} = \dfrac{\pi}{18} \text{ rad} = \textbf{0.1745 rad}$

(c) $30° = 30° \times \dfrac{\pi \text{ rad}}{180°} = \dfrac{\pi}{6} \text{ rad} = \textbf{0.5236 rad}$

Now try the following Practise Exercise

Practise Exercise 1 Radians and degrees

1. Convert the following angles to degrees correct to 3 decimal places (where necessary):

    (a) 0.6 rad (b) 0.8 rad
    (c) 2 rad (d) 3.14159 rad

    $\begin{bmatrix} \text{(a) } 34.377° & \text{(b) } 45.837° \\ \text{(c) } 114.592° & \text{(d) } 180° \end{bmatrix}$

2. Convert the following angles to radians correct to 4 decimal places:

    (a) 45° (b) 90°
    (c) 120° (d) 180°

    $\left[\begin{array}{l} \text{(a) } \dfrac{\pi}{4} \text{ rad or 0.7854 rad} \\[2mm] \text{(b) } \dfrac{\pi}{2} \text{ rad or 1.5708 rad} \\[2mm] \text{(c) } \dfrac{2\pi}{3} \text{ rad or 2.0944 rad} \\[2mm] \text{(d) } \pi \text{ rad or 3.1416 rad} \end{array}\right]$

1.3 Measurement of angles

Angles are measured starting from the horizontal 'x' axis, in an **anticlockwise direction**, as shown by θ_1 to θ_4 in Figure 1.1. An angle can also be measured in a **clockwise direction**, as shown by θ_5 in Figure 1.1, but in this case the angle has a negative sign before it. If, for example, $\theta_4 = 300°$ then $\theta_5 = -60°$.

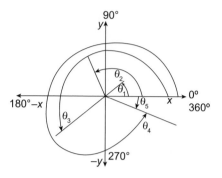

Figure 1.1

Problem 3. Use a calculator to determine the cosine, sine and tangent of the following angles, each measured anticlockwise from the horizontal 'x' axis, each correct to 4 decimal places:

(a) 30° (b) 120° (c) 250°
(d) 320° (e) 390° (f) 480°

(a) $\cos 30° = \textbf{0.8660}$ $\sin 30° = \textbf{0.5000}$
     $\tan 30° = \textbf{0.5774}$

(b) $\cos 120° = \textbf{-0.5000}$ $\sin 120° = \textbf{0.8660}$
     $\tan 120° = \textbf{-1.7321}$

(c) $\cos 250° = \textbf{-0.3420}$ $\sin 250° = \textbf{-0.9397}$
     $\tan 250° = \textbf{2.7475}$

(d) $\cos 320° = \textbf{0.7660}$ $\sin 320° = \textbf{-0.6428}$
     $\tan 320° = \textbf{-0.8391}$

(e) $\cos 390° = \textbf{0.8660}$ $\sin 390° = \textbf{0.5000}$
     $\tan 390° = \textbf{0.5774}$

(f) $\cos 480° = \textbf{-0.5000}$ $\sin 480° = \textbf{0.8660}$
     $\tan 480° = \textbf{-1.7321}$

These angles are now drawn in Figure 1.2. Note that cosine and sine always lie between −1 and +1 but that tangent can be >1 and <1.

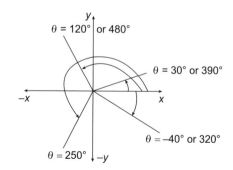

Figure 1.2

Note from Figure 1.2 that $\theta = 30°$ is the same as $\theta = 390°$ and so are their cosines, sines and tangents. Similarly, note that $\theta = 120°$ is the same as $\theta = 480°$ and so are their cosines, sines and tangents. Also, note that $\theta = -40°$ is the same as $\theta = +320°$ and so are their cosines, sines and tangents.

It is noted from above that

- in the **first quadrant**, i.e. where θ varies from $0°$ to $90°$, all (A) values of cosine, sine and tangent are positive
- in the **second quadrant**, i.e. where θ varies from $90°$ to $180°$, only values of sine (S) are positive
- in the **third quadrant**, i.e. where θ varies from $180°$ to $270°$, only values of tangent (T) are positive
- in the **fourth quadrant**, i.e. where θ varies from $270°$ to $360°$, only values of cosine (C) are positive

These positive signs, A, S, T and C are shown in Figure 1.3.

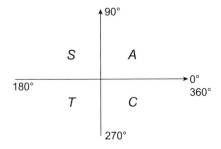

Figure 1.3

Now try the following Practise Exercise

Practise Exercise 2   Measurement of angles

1. Find the cosine, sine and tangent of the following angles, where appropriate each correct to 4 decimal places:

   (a) 60°   (b) 90°   (c) 150°
   (d) 180°   (e) 210°   (f) 270°
   (g) 330°   (h) −30°   (i) 420°
   (j) 450°   (k) 510°

   [(a) 0.5, 0.8660, 1.7321
   (b) 0, 1, ∞
   (c) − 0.8660, 0.5, − 0.5774
   (d) −1, 0, 0
   (e) − 0.8660, − 0.5, 0.5774
   (f) 0, −1, − ∞

(g) 0.8660, − 0.5000, − 0.5774
(h) 0.8660, − 0.5000, − 0.5774
(i) 0.5, 0.8660, 1.7321
(j) 0, 1, ∞
(k) − 0.8660, 0.5, − 0.5774]

1.4   Triangle calculations

(a)   Sine, cosine and tangent

From Figure 1.4, $\sin \theta = \dfrac{bc}{ac}$ $\cos \theta = \dfrac{ab}{ac}$

$$\tan \theta = \frac{bc}{ab}$$

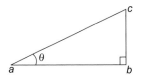

Figure 1.4

Problem 4. In Figure 1.4, if $ab = 2$ and $ac = 3$, determine the angle θ.

It is convenient to use the expression for $\cos \theta$, since 'ab' and 'ac' are given.

Hence, $\cos \theta = \dfrac{ab}{ac} = \dfrac{2}{3} = 0.66667$

from which, $\theta = \cos^{-1}(0.66667) = \mathbf{48.19°}$

Problem 5. In Figure 1.4, if $bc = 1.5$ and $ac = 2.2$, determine the angle θ.

It is convenient to use the expression for $\sin \theta$, since 'bc' and 'ac' are given.

Hence, $\sin \theta = \dfrac{bc}{ac} = \dfrac{1.5}{2.2} = 0.68182$

from which, $\theta = \sin^{-1}(0.68182) = \mathbf{42.99°}$

Problem 6. In Figure 1.4, if $bc = 8$ and $ab = 1.3$, determine the angle θ.

It is convenient to use the expression for $\tan \theta$, since 'bc' and 'ab' are given.

Hence, $\quad\quad\quad \tan\theta = \dfrac{bc}{ab} = \dfrac{8}{1.3} = 6.1538$

from which, $\quad\quad \theta = \tan^{-1}(6.1538) = \mathbf{80.77^{\circ}}$

(b) Pythagoras' theorem

Pythagoras' theorem states that:
(hypotenuse)^2 = (adjacent side)^2 + (opposite side)^2
i.e. in the triangle of Figure 1.5,

$$ac^2 = ab^2 + bc^2$$

Figure 1.5

> **Problem 7.** In Figure 1.5, if $ab = 5.1$ m and $bc = 6.7$ m, determine the length of the hypotenuse, ac.

From Pythagoras, $\quad ac^2 = ab^2 + bc^2$
$$= 5.1^2 + 6.7^2 = 26.01 + 44.89$$
$$= 70.90$$

from which, $\quad\quad\quad ac = \sqrt{70.90} = \mathbf{8.42}$ **m**

Now try the following Practise Exercise

> **Practise Exercise 3** **Sines, cosines and tangents and Pythagoras' theorem**
>
> In problems 1 to 5, refer to Figure 1.5.
>
> 1. If $ab = 2.1$ m and $bc = 1.5$ m, determine angle θ. $\quad\quad\quad\quad\quad\quad$ [35.54°]
>
> 2. If $ab = 2.3$ m and $ac = 5.0$ m, determine angle θ. $\quad\quad\quad\quad\quad\quad$ [62.61°]
>
> 3. If $bc = 3.1$ m and $ac = 6.4$ m, determine angle θ. $\quad\quad\quad\quad\quad\quad$ [28.97°]
>
> 4. If $ab = 5.7$ cm and $bc = 4.2$ cm, determine the length ac $\quad\quad\quad\quad$ [7.08 cm]
>
> 5. If $ab = 4.1$ m and $ac = 6.2$ m, determine length bc. $\quad\quad\quad\quad\quad\quad$ [4.65 m]

(c) The sine and cosine rules

For the triangle ABC shown in Figure 1.6,

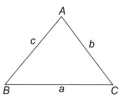

Figure 1.6

the sine rule states: $\quad\quad \dfrac{a}{\sin A} = \dfrac{b}{\sin B} = \dfrac{c}{\sin C}$

and the cosine rule states: $\quad a^2 = b^2 + c^2 - 2bc\cos A$

> **Problem 8.** In Figure 1.6, if $a = 3$ m, $A = 20°$ and $B = 120°$, determine lengths b, c and angle C.

Using the sine rule, $\quad\quad \dfrac{a}{\sin A} = \dfrac{b}{\sin B}$

i.e. $\quad\quad\quad\quad\quad \dfrac{3}{\sin 20°} = \dfrac{b}{\sin 120°}$

from which, $\quad\quad b = \dfrac{3\sin 120°}{\sin 20°} = \dfrac{3 \times 0.8660}{0.3420}$
$$= \mathbf{7.596}\ \mathbf{m}$$

Angle, $C = 180° - 20° - 120° = \mathbf{40°}$

Using the sine rule again gives: $\quad \dfrac{c}{\sin C} = \dfrac{a}{\sin A}$

i.e. $\quad\quad\quad c = \dfrac{a\sin C}{\sin A} = \dfrac{3 \times \sin 40°}{\sin 20°}$
$$= \mathbf{5.638}\ \mathbf{m}$$

> **Problem 9.** In Figure 1.6, if $b = 8.2$ cm. $c = 5.1$ cm and $A = 70°$, determine the length a and angles B and C.

From the cosine rule,

$$a^2 = b^2 + c^2 - 2bc\cos A$$
$$= 8.2^2 + 5.1^2 - 2 \times 8.2 \times 5.1 \times \cos 70°$$
$$= 67.24 + 26.01 - 2(8.2)(5.1)\cos 70°$$
$$= 64.643$$

Hence, **length,** $a = \sqrt{64.643} = \mathbf{8.04}$ **cm**

Using the sine rule: $\quad\quad \dfrac{a}{\sin A} = \dfrac{b}{\sin B}$

i.e.
$$\frac{8.04}{\sin 70°} = \frac{8.2}{\sin B}$$

from which, $8.04 \sin B = 8.2 \sin 70°$

and
$$\sin B = \frac{8.2 \sin 70°}{8.04} = 0.95839$$

and
$$\boldsymbol{B} = \sin^{-1}(0.95839) = \mathbf{73.41°}$$

Since $A + B + C = 180°$, then
$$\boldsymbol{C} = 180° - A - B = 180° - 70° - 73.41° = \mathbf{36.59°}$$

Now try the following Practise Exercise

Practise Exercise 4   Sine and cosine rules

In problems 1 to 4, refer to Figure 1.6.

1. If $b = 6$ m, $c = 4$ m and $B = 100°$, determine angles A and C and length a.
    $$[A = 38.96°, \ C = 41.04°, \ a = 3.83 \text{ m}]$$

2. If $a = 15$ m, $c = 23$ m and $B = 67°$, determine length b and angles A and C.
    $$[b = 22.01 \text{ m}, A = 38.86°, C = 74.14°]$$

3. If $a = 4$ m, $b = 8$ m and $c = 6$ m, determine angle A.
    $$[28.96°]$$

4. If $a = 10.0$ cm, $b = 8.0$ cm and $c = 7.0$ cm, determine angles A, B and C.
    $$[A = 83.33°, B = 52.62°, C = 44.05°]$$

5. In Figure 1.7, PR represents the inclined jib of a crane and is 10.0 m long. PQ is 4.0 m long. Determine the inclination of the jib to the vertical (i.e. angle P) and the length of tie QR.

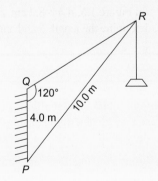

Figure 1.7

$$[P = 39.73°, QR = 7.38 \text{ m}]$$

1.5   Brackets

The use of brackets, which are used in many engineering equations, is explained through the following worked problems.

Problem 10. Expand the bracket to determine A, given $A = a(b + c + d)$

Multiplying each term in the bracket by 'a' gives:

$$A = a(b + c + d) = \boldsymbol{ab + ac + ad}$$

Problem 11. Expand the brackets to determine A, given $A = a[b(c + d) - e(f - g)]$

When there is more than one set of brackets the innermost brackets are multiplied out first. Hence,

$$A = a[b(c + d) - e(f - g)] = a[bc + bd - ef + eg]$$
$$\text{Note that } -e \times -g = +eg$$

Now multiplying each term in the square brackets by 'a' gives:

$$A = \boldsymbol{abc + abd - aef + aeg}$$

Problem 12. Expand the brackets to determine A, given $A = a[b(c + d - e) - f(g - h\{j - k\})]$

The inner brackets are determined first, hence

$$A = a[b(c + d - e) - f(g - h\{j - k\})]$$
$$= a[b(c + d - e) - f(g - hj + hk)]$$
$$= a[bc + bd - be - fg + fhj - fhk]$$
i.e. $$\boldsymbol{A = abc + abd - abe - afg + afhj - afhk}$$

Problem 13. Evaluate A, given $A = 2[3(6 - 1) - 4(7\{2 + 5\} - 6)]$

$$A = 2[3(6 - 1) - 4(7\{2 + 5\} - 6)]$$
$$= 2[3(6 - 1) - 4(7 \times 7 - 6)]$$
$$= 2[3 \times 5 - 4 \times 43]$$
$$= 2[15 - 172] = 2[-157] = \mathbf{-314}$$

Now try the following Practise Exercise

Practise Exercise 5   Brackets

In problems 1 to 2, evaluate A
1. $A = 3(2 + 1 + 4)$ $[21]$
2. $A = 4[5(2 + 1) - 3(6 - 7)]$ $[72]$

Expand the brackets in problems 3 to 7.

3. $2(x - 2y + 3)$ $[2x - 4y + 6]$

4. $(3x - 4y) + 3(y - z) - (z - 4x)$

                                       $[7x - y - 4z]$

5. $2x + [y - (2x + y)]$ $[0]$

6. $24a - [2\{3(5a - b) - 2(a + 2b)\} + 3b]$

                                       $[11b - 2a]$

7. $ab[c + d - e(f - g + h\{i + j\})]$

    $[abc + abd - abef + abeg - abehi - abehj]$

1.6 Fractions

An example of a fraction is $\dfrac{2}{3}$ where the top line, i.e. the 2, is referred to as the **numerator** and the bottom line, i.e. the 3, is referred to as the **denominator**.

A **proper fraction** is one where the numerator is smaller than the denominator, examples being $\dfrac{2}{3}, \dfrac{1}{2}, \dfrac{3}{8}, \dfrac{5}{16}$, and so on.

An **improper fraction** is one where the denominator is smaller than the numerator, examples being $\dfrac{3}{2}, \dfrac{2}{1}, \dfrac{8}{3}, \dfrac{16}{5}$, and so on.

Addition of fractions is demonstrated in the following worked problems.

Problem 14. Evaluate A, given $A = \dfrac{1}{2} + \dfrac{1}{3}$

The lowest common denominator of the two denominators 2 and 3 is 6, i.e. 6 is the lowest number that both 2 and 3 will divide into.

Then $\dfrac{1}{2} = \dfrac{3}{6}$ and $\dfrac{1}{3} = \dfrac{2}{6}$ i.e. both $\dfrac{1}{2}$ and $\dfrac{1}{3}$ have the common denominator, namely 6.

The two fractions can therefore be added as:

$$A = \frac{1}{2} + \frac{1}{3} = \frac{3}{6} + \frac{2}{6} = \frac{3 + 2}{6} = \frac{5}{6}$$

Problem 15. Evaluate A, given $A = \dfrac{2}{3} + \dfrac{3}{4}$

A common denominator can be obtained by multiplying the two denominators together, i.e. the common denominator is $3 \times 4 = 12$.

The two fractions can now be made equivalent, i.e.

$$\frac{2}{3} = \frac{8}{12} \text{ and } \frac{3}{4} = \frac{9}{12}$$

so that they can be easily added together, as follows:

$$A = \frac{2}{3} + \frac{3}{4} = \frac{8}{12} + \frac{9}{12} = \frac{8 + 9}{12} = \frac{17}{12}$$

i.e. $A = \dfrac{2}{3} + \dfrac{3}{4} = 1\dfrac{5}{12}$

Problem 16. Evaluate A, given $A = \dfrac{1}{6} + \dfrac{2}{7} + \dfrac{3}{2}$

A suitable common denominator can be obtained by multiplying $6 \times 7 = 42$, because all three denominators divide exactly into 42.

Thus, $\dfrac{1}{6} = \dfrac{7}{42}$, $\dfrac{2}{7} = \dfrac{12}{42}$ and $\dfrac{3}{2} = \dfrac{63}{42}$

Hence, $A = \dfrac{1}{6} + \dfrac{2}{7} + \dfrac{3}{2}$

$$= \frac{7}{42} + \frac{12}{42} + \frac{63}{42} = \frac{7 + 12 + 63}{42} = \frac{82}{42} = \frac{41}{21}$$

i.e. $A = \dfrac{1}{6} + \dfrac{2}{7} + \dfrac{3}{2} = 1\dfrac{20}{21}$

Problem 17. Determine A as a single fraction, given $A = \dfrac{1}{x} + \dfrac{2}{y}$

A common denominator can be obtained by multiplying the two denominators together, i.e. xy

Thus, $\dfrac{1}{x} = \dfrac{y}{xy}$ and $\dfrac{2}{y} = \dfrac{2x}{xy}$

Hence, $A = \dfrac{1}{x} + \dfrac{2}{y} = \dfrac{y}{xy} + \dfrac{2x}{xy}$

i.e. $A = \dfrac{y + 2x}{xy}$

Note that addition, subtraction, multiplication and division of fractions may be determined using a **calculator** (for example, the CASIO *fx*-83ES or *fx*-991ES).

Locate the $\dfrac{\square}{\square}$ and $\square\dfrac{\square}{\square}$ functions on your calculator (the latter function is a shift function found above the $\dfrac{\square}{\square}$ function) and then check the following worked problems.

Problem 18. Evaluate $\frac{1}{4} + \frac{2}{3}$

(i)   Press $\frac{\square}{\square}$ function

(ii)   Type in 1

(iii) Press ↓ on the cursor key and type in 4

(iv) $\frac{1}{4}$ appears on the screen

(v)   Press → on the cursor key and type in +

(vi)   Press $\frac{\square}{\square}$ function

(vii) Type in 2

(viii) Press ↓ on the cursor key and type in 3

(ix)   Press → on the cursor key

(x)   Press = and the answer $\frac{11}{12}$ appears

(xi)   Press $S \Leftrightarrow D$ function and the fraction changes to a decimal 0.9166666....

Thus, $\frac{1}{4} + \frac{2}{3} = \frac{11}{12} = $ **0.9167** as a decimal, correct to 4 decimal places.

It is also possible to deal with **mixed numbers** on the calculator.

Press Shift then the $\frac{\square}{\square}$ function and $\square\frac{\square}{\square}$ appears

Problem 19. Evaluate $5\frac{1}{5} - 3\frac{3}{4}$

(i)   Press Shift then the $\frac{\square}{\square}$ function and $\square\frac{\square}{\square}$ appears on the screen

(ii)   Type in 5 then → on the cursor key

(iii) Type in 1 and ↓ on the cursor key

(iv)   Type in 5 and $5\frac{1}{5}$ appears on the screen

(v)   Press → on the cursor key

(vi)   Type in - and then press Shift then the $\frac{\square}{\square}$ function and $5\frac{1}{5} - \square\frac{\square}{\square}$ appears on the screen

(vii) Type in 3 then → on the cursor key

(viii) Type in 3 and ↓ on the cursor key

(ix)   Type in 4 and $5\frac{1}{5} - 3\frac{3}{4}$ appears on the screen

(x)   Press = and the answer $\frac{29}{20}$ appears

(xi)   Press $S \Leftrightarrow D$ function and the fraction changes to a decimal 1.45

Thus, $5\frac{1}{5} - 3\frac{3}{4} = \frac{29}{20} = 1\frac{9}{20} = $ **1.45** as a decimal.

Now try the following Practise Exercise

Practise Exercise 6   Fractions

In problems 1 to 3, evaluate the given fractions

1. $\frac{1}{3} + \frac{1}{4}$ $\left[\frac{7}{12}\right]$

2. $\frac{1}{5} + \frac{1}{4}$ $\left[\frac{9}{20}\right]$

3. $\frac{1}{6} + \frac{1}{2} - \frac{1}{5}$ $\left[\frac{7}{15}\right]$

In problems 4 and 5, use a calculator to evaluate the given expressions

4. $\frac{1}{3} - \frac{3}{4} \times \frac{8}{21}$ $\left[\frac{1}{21}\right]$

5. $\frac{3}{4} \times \frac{4}{5} - \frac{2}{3} \div \frac{4}{9}$ $\left[-\frac{9}{10}\right]$

6. Evaluate $\frac{3}{8} + \frac{5}{6} - \frac{1}{2}$ as a decimal, correct to 4 decimal places. $\left[\frac{17}{24} = 0.7083\right]$

7. Evaluate $8\frac{8}{9} \div 2\frac{2}{3}$ as a mixed number. $\left[3\frac{1}{3}\right]$

8. Evaluate $3\frac{1}{5} \times 1\frac{1}{3} - 1\frac{7}{10}$ as a decimal, correct to 3 decimal places. [2.567]

9. Determine $\frac{2}{x} + \frac{3}{y}$ as a single fraction. $\left[\frac{3x + 2y}{xy}\right]$

1.7 Percentages

Percentages are used to give a common standard. The use of percentages is very common in many aspects of commercial life, as well as in engineering. Interest rates, sale reductions, pay rises, exams and VAT are all examples where percentages are used.

Percentages are fractions having 100 as their denominator.

For example, the fraction $\dfrac{40}{100}$ is written as 40% and is read as 'forty per cent'.

The easiest way to understand percentages is to go through some worked examples.

Problem 20. Express 0.275 as a percentage.

$$0.275 = 0.275 \times 100\% = \textbf{27.5\%}$$

Problem 21. Express 17.5% as a decimal number.

$$17.5\% = \frac{17.5}{100} = \textbf{0.175}$$

Problem 22. Express $\dfrac{5}{8}$ as a percentage.

$$\frac{5}{8} = \frac{5}{8} \times 100\% = \frac{500}{8}\% = \textbf{62.5\%}$$

Problem 23. In two successive tests a student gains marks of 57/79 and 49/67. Is the second mark better or worse than the first?

$$57/79 = \frac{57}{79} = \frac{57}{79} \times 100\% = \frac{5700}{79}\%$$
$$= \textbf{72.15\%} \text{ correct to 2 decimal places.}$$
$$49/67 = \frac{49}{67} = \frac{49}{67} \times 100\% = \frac{4900}{67}\%$$
$$= \textbf{73.13\%} \text{ correct to 2 decimal places.}$$

Hence, **the second test is marginally better than the first test.**

This question demonstrates how much easier it is to compare two fractions when they are expressed as percentages.

Problem 24. Express 75% as a fraction.

$$75\% = \frac{75}{100} = \frac{3}{4}$$

The fraction $\dfrac{75}{100}$ is reduced to its simplest form by cancelling, i.e. dividing numerator and denominator by 25.

Problem 25. Express 37.5% as a fraction.

$$37.5\% = \frac{37.5}{100}$$
$$= \frac{375}{1000} \quad \text{by multiplying numerator and denominator by 10}$$
$$= \frac{15}{40} \quad \text{by dividing numerator and denominator by 25}$$
$$= \frac{3}{8} \quad \text{by dividing numerator and denominator by 5}$$

Problem 26. Find 27% of £65.

$$27\% \text{ of } £65 = \frac{27}{100} \times 65 = \textbf{£17.55} \text{ by calculator}$$

Problem 27. A 160 GB iPod is advertised as costing £190 excluding VAT. If VAT is added at 20%, what will be the total cost of the iPod?

$$\text{VAT} = 20\% \text{ of } £190 = \frac{20}{100} \times 190 = £38$$

$$\text{Total cost of iPod} = £190 + £38 = \textbf{£228}$$

A quicker method to determine the total cost is: $1.20 \times £190 = \textbf{£228}$

Problem 28. Express 23 cm as a percentage of 72 cm, correct to the nearest 1%.

$$23 \text{ cm as a percentage of } 72 \text{ cm} = \frac{23}{72} \times 100\%$$
$$= 31.94444...\%$$
$$= \textbf{32\%} \text{ correct to the nearest 1\%}$$

Problem 29. A box of screws increases in price from £45 to £52. Calculate the percentage change in cost, correct to 3 significant figures.

$$\% \text{ change} = \frac{\text{new value} - \text{original value}}{\text{original value}} \times 100\%$$
$$= \frac{52 - 45}{45} \times 100\% = \frac{7}{45} \times 100$$
$$= \textbf{15.6\%} = \textbf{percentage change in cost}$$

Problem 30. A drilling speed should be set to 400 rev/min. The nearest speed available on the machine is 412 rev/min. Calculate the percentage over-speed.

% over-speed

$$= \frac{\text{available speed} - \text{correct speed}}{\text{correct speed}} \times 100\%$$

$$= \frac{412 - 400}{400} \times 100\%$$

$$= \frac{12}{400} \times 100\% = \mathbf{3\%}$$

Now try the following Practise Exercise

Practise Exercise 7   Percentages

In problems 1 and 2, express the given numbers as percentages.

1. 0.057   [5.7%]

2. 0.374   [37.4%]

3. Express 20% as a decimal number
                                               [0.20]

4. Express $\frac{11}{16}$ as a percentage   [68.75%]

5. Express $\frac{5}{13}$ as a percentage, correct to 3 decimal places   [38.462%]

6. Place the following in order of size, the smallest first, expressing each as percentages, correct to 1 decimal place:

    (a) $\frac{12}{21}$ (b) $\frac{9}{17}$ (c) $\frac{5}{9}$ (d) $\frac{6}{11}$

    $$\begin{bmatrix} \text{(b) } 52.9\%, \text{ (d) } 54.5\%, \\ \text{(c) } 55.6\%, \text{ (a) } 57.1\% \end{bmatrix}$$

7. Express 65% as a fraction in its simplest form $\left[\frac{13}{20}\right]$

8. Calculate 43.6% of 50 kg   [21.8 kg]

9. Determine 36% of 27 m   [9.72 m]

10. Calculate correct to 4 significant figures:
    (a) 18% of 2758 tonnes
    (b) 47% of 18.42 grams
    (c) 147% of 14.1 seconds
         [(a) 496.4 t (b) 8.657 g (c) 20.73 s]

11. Express: (a) 140 kg as a percentage of 1 t (b) 47 s as a percentage of 5 min (c) 13.4 cm as a percentage of 2.5 m
                 [(a) 14%   (b) 15.67%   (c) 5.36%]

12. A computer is advertised on the internet at £520, exclusive of VAT. If VAT is payable at 20%, what is the total cost of the computer?   [£624]

13. Express 325 mm as a percentage of 867 mm, correct to 2 decimal places.
                                              [37.49%]

14. When signing a new contract, a Premiership footballer's pay increases from £15,500 to £21,500 per week. Calculate the percentage pay increase, correct 3 significant figures.
                                              [38.7%]

15. A metal rod 1.80 m long is heated and its length expands by 48.6 mm. Calculate the percentage increase in length.
                                              [2.7%]

1.8 Laws of indices

The manipulation of indices, powers and roots is a crucial underlying skill needed in algebra.

Law 1: When multiplying two or more numbers having the same base, the indices are added.

For example, $2^2 \times 2^3 = 2^{2+3} = 2^5$

and $5^4 \times 5^2 \times 5^3 = 5^{4+2+3} = 5^9$

More generally, $\mathbf{a^m \times a^n = a^{m+n}}$

For example, $a^3 \times a^4 = a^{3+4} = a^7$

Law 2: When dividing two numbers having the same base, the index in the denominator is subtracted from the index in the numerator.

For example, $\dfrac{2^5}{2^3} = 2^{5-3} = 2^2$

and $\dfrac{7^8}{7^5} = 7^{8-5} = 7^3$

More generally, $\dfrac{a^m}{a^n} = a^{m-n}$

For example, $\dfrac{c^5}{c^2} = c^{5-2} = c^3$

Law 3: When a number which is raised to a power is raised to a further power, the indices are multiplied.

For example, $\left(2^2\right)^3 = 2^{2\times3} = 2^6$

and $\left(3^4\right)^2 = 3^{4\times2} = 3^8$

More generally, $(a^m)^n = a^{mn}$

For example, $\left(d^2\right)^3 = d^{2\times3} = d^6$

Law 4: When a number has an index of 0, its value is 1.

For example, $3^0 = 1$

and $17^0 = 1$

More generally, $a^0 = 1$

Law 5: A number raised to a negative power is the reciprocal of that number raised to a positive power.

For example, $3^{-4} = \dfrac{1}{3^4}$ and $\dfrac{1}{2^{-3}} = 2^3$

More generally, $a^{-n} = \dfrac{1}{a^n}$

For example, $a^{-2} = \dfrac{1}{a^2}$

Law 6: When a number is raised to a fractional power the denominator of the fraction is the root of the number and the numerator is the power.

For example, $8^{\frac{2}{3}} = \sqrt[3]{8^2} = (2)^2 = 4$

and $25^{\frac{1}{2}} = \sqrt[2]{25^1} = \sqrt{25^1}$

$= \pm 5$ (Note that $\sqrt{\ } \equiv \sqrt[2]{\ }$)

More generally, $a^{\frac{m}{n}} = \sqrt[n]{a^m}$

For example, $x^{\frac{4}{3}} = \sqrt[3]{x^4}$

Problem 31. Evaluate in index form $5^3 \times 5 \times 5^2$

$5^3 \times 5 \times 5^2 = 5^3 \times 5^1 \times 5^2$ (Note that 5 means 5^1)

$= 5^{3+1+2} = \mathbf{5^6}$ from law 1

Problem 32. Evaluate $\dfrac{3^5}{3^4}$

From law 2 $\quad \dfrac{3^5}{3^4} = 3^{5-4} = 3^1 = \mathbf{3}$

Problem 33. Evaluate $\dfrac{2^4}{2^4}$

$\dfrac{2^4}{2^4} = 2^{4-4}$ from law 2

$= 2^0 = \mathbf{1}$ from law 4

Any number raised to the power of zero equals 1

Problem 34. Evaluate $\dfrac{3 \times 3^2}{3^4}$

$\dfrac{3 \times 3^2}{3^4} = \dfrac{3^1 \times 3^2}{3^4} = \dfrac{3^{1+2}}{3^4} = \dfrac{3^3}{3^4}$

$= 3^{3-4} = 3^{-1}$ from laws 1 and 2

$= \dfrac{1}{3}$ from law 5

Problem 35. Evaluate $\dfrac{10^3 \times 10^2}{10^8}$

$\dfrac{10^3 \times 10^2}{10^8} = \dfrac{10^{3+2}}{10^8} = \dfrac{10^5}{10^8}$ from law 1

$= 10^{5-8} = 10^{-3}$ from law 2

$= \dfrac{1}{10^{+3}} = \dfrac{1}{1000}$ from law 5

Hence, $\dfrac{10^3 \times 10^2}{10^8} = \mathbf{10^{-3}} = \dfrac{1}{1000} = \mathbf{0.001}$

Problem 36. Simplify: (a) $(2^3)^4$ (b) $(3^2)^5$ expressing the answers in index form

From law 3: (a) $(2^3)^4 = 2^{3\times4} = \mathbf{2^{12}}$

(b) $(3^2)^5 = 3^{2\times5} = \mathbf{3^{10}}$

Problem 37. Evaluate: $\dfrac{(10^2)^3}{10^4 \times 10^2}$

From laws 1, 2, and 3: $\dfrac{(10^2)^3}{10^4 \times 10^2} = \dfrac{10^{(2\times3)}}{10^{(4+2)}}$

$= \dfrac{10^6}{10^6} = 10^{6-6}$

$= 10^0 = \mathbf{1}$

Problem 38. Evaluate (a) $4^{1/2}$ (b) $16^{3/4}$ (c) $27^{2/3}$ (d) $9^{-1/2}$

(a) $4^{1/2} = \sqrt{4} = \pm\mathbf{2}$

(b) $16^{3/4} = \sqrt[4]{16^3} = (2)^3 = \mathbf{8}$

(Note that it does not matter whether the 4th root of 16 is found first or whether 16 cubed is found first; the same answer will result)

(c) $27^{2/3} = \sqrt[3]{27^2} = (3)^2 = \mathbf{9}$

(d) $9^{-1/2} = \dfrac{1}{9^{1/2}} = \dfrac{1}{\sqrt{9}} = \dfrac{1}{\pm 3} = \pm\dfrac{\mathbf{1}}{\mathbf{3}}$

Problem 39. Simplify $a^2b^3c \times ab^2c^5$

$$a^2b^3c \times ab^2c^5 = a^2 \times b^3 \times c \times a \times b^2 \times c^5$$
$$= a^2 \times b^3 \times c^1 \times a^1 \times b^2 \times c^5$$

Grouping together like terms gives:
$$a^2 \times a^1 \times b^3 \times b^2 \times c^1 \times c^5$$

Using law 1 of indices gives:
$$a^{2+1} \times b^{3+2} \times c^{1+5} = a^3 \times b^5 \times c^6$$

i.e. $\mathbf{a^2b^3c \times ab^2c^5 = a^3b^5c^6}$

Problem 40. Simplify $\dfrac{x^5y^2z}{x^2yz^3}$

$$\frac{x^5y^2z}{x^2yz^3} = \frac{x^5 \times y^2 \times z}{x^2 \times y \times z^3} = \frac{x^5}{x^2} \times \frac{y^2}{y^1} \times \frac{z}{z^3}$$

$$= x^{5-2} \times y^{2-1} \times z^{1-3} \qquad \text{by law 2}$$

$$= x^3 \times y^1 \times z^{-2} = \mathbf{x^3yz^{-2}} \quad \text{or} \quad \mathbf{\dfrac{x^3y}{z^2}}$$

Now try the following Practise Exercise

Practise Exercise 8   Laws of indices

In questions 1 to 18, evaluate without the aid of a calculator

1. Evaluate $2^2 \times 2 \times 2^4$ $[2^7 = 128]$

2. Evaluate $3^5 \times 3^3 \times 3$ in index form $[3^9]$

3. Evaluate $\dfrac{2^7}{2^3}$ $[2^4 = 16]$

4. Evaluate $\dfrac{3^3}{3^5}$ $\left[3^{-2} = \dfrac{1}{3^2} = \dfrac{1}{9}\right]$

5. Evaluate 7^0 $[1]$

6. Evaluate $\dfrac{2^3 \times 2 \times 2^6}{2^7}$ $[2^3 = 8]$

7. Evaluate $\dfrac{10 \times 10^6}{10^5}$ $[10^2 = 100]$

8. Evaluate $10^4 \div 10$ $[10^3 = 1000]$

9. Evaluate $\dfrac{10^3 \times 10^4}{10^9}$

$$\left[10^{-2} = \dfrac{1}{10^2} = \dfrac{1}{100} = 0.01\right]$$

10. Evaluate $5^6 \times 5^2 \div 5^7$ $[5]$

11. Evaluate $(7^2)^3$ in index form $[7^6]$

12. Evaluate $(3^3)^2$ $[3^6 = 729]$

13. Evaluate $\dfrac{3^7 \times 3^4}{3^5}$ in index form $[3^6]$

14. Evaluate $\dfrac{(9 \times 3^2)^3}{(3 \times 27)^2}$ in index form $[3^4]$

15. Evaluate $\dfrac{(16 \times 4)^2}{(2 \times 8)^3}$ $[1]$

16. Evaluate $\dfrac{5^{-2}}{5^{-4}}$ $[5^2 = 25]$

17. Evaluate $\dfrac{3^2 \times 3^{-4}}{3^3}$ $\left[3^{-5} = \dfrac{1}{3^5} = \dfrac{1}{243}\right]$

18. Evaluate $\dfrac{7^2 \times 7^{-3}}{7 \times 7^{-4}}$ $[7^2 = 49]$

In problems 19 to 36, simplify the following, giving each answer as a power

19. $z^2 \times z^6$ $[z^8]$

20. $a \times a^2 \times a^5$ $[a^8]$

21. $n^8 \times n^{-5}$ $[n^3]$

22. $b^4 \times b^7$ $[b^{11}]$

23. $b^2 \div b^5$ $\left[b^{-3} \text{ or } \dfrac{1}{b^3}\right]$

24. $c^5 \times c^3 \div c^4$ $[c^4]$

25. $\dfrac{m^5 \times m^6}{m^4 \times m^3}$ $[m^4]$

26. $\dfrac{(x^2)(x)}{x^6}$ $\left[x^{-3} \text{ or } \dfrac{1}{x^3}\right]$

27. $\left(x^3\right)^4$ $[x^{12}]$

28. $\left(y^2\right)^{-3}$ $\left[y^{-6} \text{ or } \dfrac{1}{y^6}\right]$

29. $\left(t \times t^3\right)^2$ $[t^8]$

30. $\left(c^{-7}\right)^{-2}$ $[c^{14}]$

31. $\left(\dfrac{a^2}{a^5}\right)^3$ $\left[a^{-9} \text{ or } \dfrac{1}{a^9}\right]$

32. $\left(\dfrac{1}{b^3}\right)^4$ $\left[\dfrac{1}{b^{12}} \text{ or } b^{-12}\right]$

33. $\left(\dfrac{b^2}{b^7}\right)^{-2}$ $[b^{10}]$

34. $\dfrac{1}{\left(s^3\right)^3}$ $\left[\dfrac{1}{s^9} \text{ or } s^{-9}\right]$

35. $p^3qr^2 \times p^2q^5r \times pqr^2$ $[p^6q^7r^5]$

36. $\dfrac{x^3y^2z}{x^5yz^3}$ $\left[x^{-2}yz^{-2} \text{ or } \dfrac{y}{x^2z^2}\right]$

1.9 Simultaneous equations

The solution of simultaneous equations is demonstrated in the following worked problems.

Problem 41. If 6 apples and 2 pears cost £1.80 and 8 apples and 6 pears cost £2.90, calculate how much an apple and a pear each cost.

Let an apple = A and a pear = P, then:

$$6A + 2P = 180 \qquad (1)$$
$$8A + 6P = 290 \qquad (2)$$

From equation (1), $6A = 180 - 2P$

and $\quad A = \dfrac{180 - 2P}{6} = 30 - 0.3333P \qquad (3)$

From equation (2), $8A = 290 - 6P$

and $\quad A = \dfrac{290 - 6P}{8} = 36.25 - 0.75P \qquad (4)$

Equating (3) and (4) gives:
$$30 - 0.3333P = 36.25 - 0.75P$$
i.e. $\quad 0.75P - 0.3333P = 36.25 - 30$

and $\quad 0.4167P = 6.25$

and $\quad P = \dfrac{6.25}{0.4167} = 15$

Substituting in (3) gives: $\quad A = 30 - 0.3333(15)$
$$= 30 - 5 = 25$$

Hence, **an apple costs 25p and a pear costs 15p**
The above method of solving simultaneous equations is called the **substitution method**.

Problem 42. If 6 bananas and 5 peaches cost £3.45 and 4 bananas and 8 peaches cost £4.40, calculate how much a banana and a peach each cost.

Let a banana = B and a peach = P, then:

$$6B + 5P = 345 \qquad (1)$$
$$4B + 8P = 440 \qquad (2)$$

Multiplying equation (1) by 2 gives:
$$12B + 10P = 690 \qquad (3)$$

Multiplying equation (2) by 3 gives:
$$12B + 24P = 1320 \qquad (4)$$

Equation (4) – equation (3) gives: $14P = 630$

from which, $\quad P = \dfrac{630}{14} = 45$

Substituting in (1) gives: $\quad 6B + 5(45) = 345$

i.e. $\quad 6B = 345 - 5(45)$

i.e. $\quad 6B = 120$

and $\quad B = \dfrac{120}{6} = 20$

Hence, **a banana costs 20p and a peach costs 45p**
The above method of solving simultaneous equations is called the **elimination method**.

Problem 43. If 20 bolts and 2 spanners cost £10, and 6 spanners and 12 bolts cost £18, how much does a spanner and a bolt cost?

Let s = a spanner   and b = a bolt.

Therefore, $2s + 20b = 10$ (1)

and $6s + 12b = 18$ (2)

Multiplying equation (1) by 3 gives:

$6s + 60b = 30$ (3)

Equation (3) – equation (2) gives: $48b = 12$

from which, $b = \dfrac{12}{48} = 0.25$

Substituting in (1) gives: $2s + 20(0.25) = 10$

i.e. $2s = 10 - 20(0.25)$

i.e. $2s = 5$

and $s = \dfrac{5}{2} = 2.5$

Therefore, **a spanner costs £2.50 and a bolt costs £0.25 or 25p**

Now try the following Practise Exercises

Practise Exercise 9   Simultaneous equations

1. If 5 apples and 3 bananas cost £1.45 and 4 apples and 6 bananas cost £2.42, determine how much an apple and a banana each cost.   [apple = 8p, banana = 35p]

2. If 7 apples and 4 oranges cost £2.64 and 3 apples and 3 oranges cost £1.35, determine how much an apple and an orange each cost.   [apple = 28p, orange = 17p]

3. Three new cars and four new vans supplied to a dealer together cost £93000, and five new cars and two new vans of the same models cost £99000. Find the respective costs of a car and a van.
             [car = £15000, van = £12000]

4. In a system of forces, the relationship between two forces F_1 and F_2 is given by:

$$5F_1 + 3F_2 = -6$$
$$3F_1 + 5F_2 = -18$$

Solve for F_1 and F_2
             $[F_1 = 1.5, F_2 = -4.5]$

5. Solve the simultaneous equations:
    $a + b = 7$
    $a - b = 3$ $[a = 5, b = 2]$

6. Solve the simultaneous equations:
    $8a - 3b = 51$
    $3a + 4b = 14$ $[a = 6, b = -1]$

Practise Exercise 10   Multiple-choice questions on revisionary mathematics

(Answers on page 297)

1. 73° is equivalent to:
    (a) 23.24 rad   (b) 1.274 rad
    (c) 0.406 rad   (d) 4183 rad

2. 0.52 radians is equivalent to:
    (a) 93.6°   (b) 0.0091°
    (c) 1.63°   (d) 29.79°

3. $3\pi/4$ radians is equivalent to:
    (a) 135°   (b) 270°
    (c) 45°   (d) 67.5°

4. In the right-angled triangle ABC shown in Figure 1.8, sine A is given by:
    (a) b/a (b) c/b
    (c) b/c (d) a/b

Figure 1.8

5. In the right-angled triangle ABC shown in Figure 1.8, cosine C is given by:
    (a) a/b (b) c/b
    (c) a/c (d) b/a

6. In the right-angled triangle ABC shown in Figure 1.8, tangent A is given by:
    (a) b/c (b) a/c
    (c) a/b (d) c/a

7. In the right-angled triangle PQR shown in Figure 1.9, angle R is equal to:
    (a) 41.41°   (b) 48.59°
    (c) 36.87°   (d) 53.13°

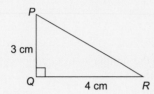

Figure 1.9

8. In the triangle ABC shown in Figure 1.10, side 'a' is equal to:
    (a) 61.27 mm
    (b) 86.58 mm
    (c) 96.41 mm
    (d) 54.58 mm

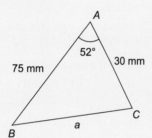

Figure 1.10

9. In the triangle ABC shown in Figure 1.10, angle B is equal to:
    (a) 0.386°   (b) 22.69°
    (c) 74.71°   (d) 23.58°

10. Removing the brackets from the expression: $a[b + 2c - d\{(e-f) - g(m-n)\}]$ gives:
    (a) $ab + 2ac - ade - adf + adgm - adgn$
    (b) $ab + 2ac - ade - adf - adgm - adgn$
    (c) $ab + 2ac - ade + adf + adgm - adgn$
    (d) $ab + 2ac - ade - adf + adgm + adgn$

11. $\dfrac{5}{6} + \dfrac{1}{5} - \dfrac{2}{3}$ is equal to:
    (a) $\dfrac{1}{2}$ (b) $\dfrac{11}{30}$
    (c) $-\dfrac{1}{2}$ (d) $1\dfrac{7}{10}$

12. $1\dfrac{1}{3} + 1\dfrac{2}{3} \div 2\dfrac{2}{3} - \dfrac{1}{3}$ is equal to:
    (a) $1\dfrac{2}{7}$ (b) $\dfrac{19}{24}$
    (c) $2\dfrac{1}{21}$ (d) $1\dfrac{5}{8}$

13. $\dfrac{3}{4} \div 1\dfrac{3}{4}$ is equal to:
    (a) $\dfrac{3}{7}$ (b) $1\dfrac{9}{16}$
    (c) $1\dfrac{5}{16}$ (d) $2\dfrac{1}{2}$

14. 11 mm expressed as a percentage of 41 mm is:
    (a) 2.68, correct to 3 significant figures
    (b) 2.6, correct to 2 significant figures
    (c) 26.83, correct to 2 decimal places
    (d) 0.2682, correct to 4 decimal places

15. The value of $\dfrac{2^{-3}}{2^{-4}} - 1$ is equal to:
    (a) 1   (b) 2
    (c) $-\dfrac{1}{2}$ (d) $\dfrac{1}{2}$

16. In an engineering equation $\dfrac{3^4}{3^r} = \dfrac{1}{9}$. The value of r is:
    (a) –6   (b) 2
    (c) 6   (d) – 2

17. $16^{-\frac{3}{4}}$ is equal to:
    (a) 8   (b) $-\dfrac{1}{2^3}$
    (c) 4   (d) $\dfrac{1}{8}$

18. The engineering expression $\dfrac{(16 \times 4)^2}{(8 \times 2)^4}$ is equal to:
    (a) 4   (b) 2^{-4}
    (c) $\dfrac{1}{2^2}$ (d) 1

19. $(16^{-\frac{1}{4}} - 27^{-\frac{2}{3}})$ is equal to:
    (a) $\dfrac{7}{18}$ (b) – 7
    (c) $1\dfrac{8}{9}$ (d) $-8\dfrac{1}{2}$

20. The solution of the simultaneous equations: $3a - 2b = 13$ and $2a + 5b = -4$ is:

(a) $a = -2, b = 3$

(b) $a = 1, b = -5$

(c) $a = 3, b = -2$

(d) $a = -7, b = 2$

References

There are many aspects of mathematics needed in engineering studies; a few have been covered in this chapter. For further engineering mathematics, see the following references:

[1] BIRD J O *Basic Engineering Mathematics 5th Edition*, Taylor & Francis, 2010

[2] BIRD J O *Engineering Mathematics 6th Edition*, Taylor & Francis, 2010

Revision Test 1   Revisionary mathematics

This Revision Test covers the material contained in Chapter 1. *The marks for each question are shown in brackets at the end of each question.*

1. Convert, correct to 2 decimal places:

   (a) 76.8° to radians

   (b) 1.724 radians to degrees   (4)

2. In triangle *JKL* in Figure RT1.1, find

   (a) length *KJ* correct to 3 significant figures

   (b) sin *L* and tan *K*, each correct to 3 decimal places

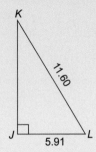

Figure RT1.1

                                          (4)

3. In triangle *PQR* in Figure RT1.2, find angle *P* in decimal form, correct to 2 decimal places

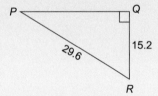

Figure RT1.2

                                          (2)

4. In triangle *ABC* in Figure RT1.3, find lengths *AB* and *AC*, correct to 2 decimal places

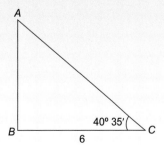

Figure RT1.3

                                          (4)

5. A triangular plot of land *ABC* is shown in Figure RT1.4. Solve the triangle and determine its area

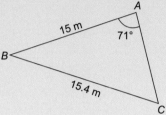

Figure RT1.4

                                          (9)

6. Figure RT1.5 shows a roof truss *PQR* with rafter *PQ* = 3 m. Calculate the length of (a) the roof rise *PP′* (b) rafter PR, and (c) the roof span *QR*. Find also (d) the cross-sectional area of the roof truss

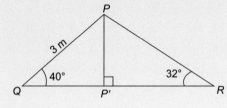

Figure RT1.5

                                          (10)

7. Solve triangle *ABC* given *b* = 10 cm, *c* = 15 cm and ∠*A* = 60°.   (7)

8. Remove the brackets and simplify
   $2(3x - 2y) - (4y - 3x)$ (3)

9. Remove the brackets and simplify
   $10a - [3(2a - b) - 4(b - a) + 5b]$ (4)

10. Determine, correct to 2 decimal places, 57% of 17.64 *g*   (2)

11. Express 54.7 mm as a percentage of 1.15 m, correct to 3 significant figures.   (3)

12. Simplify:

    (a) $\dfrac{3}{4} - \dfrac{7}{15}$ (b) $1\dfrac{5}{8} - 2\dfrac{1}{3} + 3\dfrac{5}{6}$ (8)

13. Use a calculator to evaluate:

   (a) $1\frac{7}{9} \times \frac{3}{8} \times 3\frac{3}{5}$

   (b) $6\frac{2}{3} \div 1\frac{1}{3}$

   (c) $1\frac{1}{3} \times 2\frac{1}{5} \div \frac{2}{5}$ (10)

14. Evaluate:

   (a) $3 \times 2^3 \times 2^2$

   (b) $49^{\frac{1}{2}}$ (4)

15. Evaluate:

   (a) $\dfrac{2^7}{2^2}$ (b) $\dfrac{10^4 \times 10 \times 10^5}{10^6 \times 10^2}$ (4)

16. Evaluate:

   (a) $\dfrac{2^3 \times 2 \times 2^2}{2^4}$

   (b) $\dfrac{\left(2^3 \times 16\right)^2}{\left(8 \times 2\right)^3}$

   (c) $\left(\dfrac{1}{4^2}\right)^{-1}$ (7)

17. Evaluate:

   (a) $(27)^{-\frac{1}{3}}$ (b) $\dfrac{\left(\frac{3}{2}\right)^{-2} - \frac{2}{9}}{\left(\frac{2}{3}\right)^2}$ (5)

18. Solve the simultaneous equations:

   (a) $2x + y = 6$
       $5x - y = 22$

   (b) $4x - 3y = 11$
       $3x + 5y = 30$ (10)

Part Two

Statics and Strength
of Materials

Chapter 2

The effects of forces on materials

A good knowledge of some of the constants used in the study of the properties of materials is vital in most branches of engineering, especially in mechanical, manufacturing, aeronautical and civil and structural engineering. For example, most steels look the same, but steels used for the pressure hull of a submarine are about 5 times stronger than those used in the construction of a small building, and it is very important for the professional and chartered engineer to know what steel to use for what construction; this is because the cost of the high-tensile steel used to construct a submarine pressure hull is considerably higher than the cost of the mild steel, or similar material, used to construct a small building. The engineer must not only take into consideration the ability of the chosen material of construction to do the job, but also its cost. Similar arguments lie in manufacturing engineering, where the engineer must be able to estimate the ability of his/her machines to bend, cut or shape the artefact s/he is trying to produce, and at a competitive price! This chapter provides explanations of the different terms that are used in determining the properties of various materials.

At the end of this chapter you should be able to:

- define force and state its unit
- recognise a tensile force and state relevant practical examples
- recognise a compressive force and state relevant practical examples
- recognise a shear force and state relevant practical examples
- define stress and state its unit
- calculate stress σ from $\sigma = \dfrac{F}{A}$
- define strain
- calculate strain ε from $\varepsilon = \dfrac{x}{L}$
- define elasticity, plasticity, limit of proportionality and elastic limit
- state Hooke's law
- define Young's modulus of elasticity E and stiffness
- appreciate typical values for E
- calculate E from $E = \dfrac{\sigma}{\varepsilon}$
- perform calculations using Hooke's law
- plot a load/extension graph from given data
- define ductility, brittleness and malleability, with examples of each
- define rigidity or shear modulus
- understand thermal stresses and strains
- calculates stresses in compound bars

2.1 Introduction

A **force** exerted on a body can cause a change in either the shape or the motion of the body. The unit of force is the **newton, N**.

No solid body is perfectly rigid and when forces are applied to it, changes in dimensions occur. Such changes are not always perceptible to the human eye since they are so small. For example, the span of a bridge will sag under the weight of a vehicle and a spanner will bend slightly when tightening a nut. It is important for engineers and designers to appreciate the effects of forces on materials, together with their mechanical properties. The three main types of mechanical force that can act on a body are:

(i) tensile (ii) compressive and (iii) shear

2.2 Tensile force

Tension is a force that tends to stretch a material, as shown in Figure 2.1. For example,

(i)   the rope or cable of a crane carrying a load is in tension

Force   Force

Figure 2.1

(ii)   rubber bands, when stretched, are in tension
(iii)   when a nut is tightened, a bolt is under tension

A tensile force, i.e. one producing tension, increases the length of the material on which it acts.

2.3 Compressive force

Compression is a force that tends to squeeze or crush a material, as shown in Figure 2.2. For example,

Force   Force

Figure 2.2

(i)   a pillar supporting a bridge is in compression
(ii)   the sole of a shoe is in compression
(iii)   the jib of a crane is in compression

A compressive force, i.e. one producing compression, will decrease the length of the material on which it acts.

2.4 Shear force

Shear is a force that tends to slide one face of the material over an adjacent face. For example,

(i)   a rivet holding two plates together is in shear if a tensile force is applied between the plates – as shown in Figure 2.3

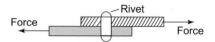

Figure 2.3

(ii)   a guillotine cutting sheet metal, or garden shears, each provide a shear force
(iii)   a horizontal beam is subject to shear force
(iv)   transmission joints on cars are subject to shear forces

A shear force can cause a material to bend, slide or twist.

Problem 1.   Figure 2.4(a) represents a crane and Figure 2.4(b) a transmission joint. State the types of forces acting, labelled A to F.

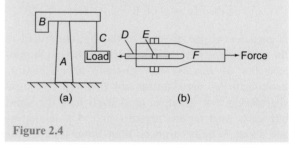

Figure 2.4

(a)   For the crane, A, a supporting member, is in **compression**, B, a horizontal beam, is in **shear**, and C, a rope, is in **tension.**

(b)   For the transmission joint, parts D and F are in **tension**, and E, the rivet or bolt, is in **shear.**

2.5 Stress

Forces acting on a material cause a change in dimensions and the material is said to be in a state of **stress**. Stress is the ratio of the applied force F to cross-sectional area A of the material. The symbol used for tensile and compressive stress is σ (Greek letter sigma). The unit of stress is the **Pascal, Pa**, where $1 \text{ Pa} = 1 \text{ N/m}^2$. Hence

$$\sigma = \frac{F}{A} \text{ Pa}$$

where F is the force in Newtons and A is the cross-sectional area in square metres. For tensile and compressive forces, the cross-sectional area is that which is at right angles to the direction of the force. For a shear force the shear stress is equal to $\frac{F}{A}$, where the cross-sectional area A is that which is parallel to the direction of the force. The symbol used for shear stress is the Greek letter tau, τ.

> **Problem 2.** A rectangular bar having a cross-sectional area of 75 mm² has a tensile force of 15 kN applied to it. Determine the stress in the bar.

Cross-sectional area $A = 75$ mm² $= 75 \times 10^{-6}$ m²

and force $F = 15$ kN $= 15 \times 10^3$ N

Stress in bar, $\quad \sigma = \dfrac{F}{A} = \dfrac{15 \times 10^3 \text{N}}{75 \times 10^{-6} \text{m}^2}$

$$= 0.2 \times 10^9 \text{ Pa} = \textbf{200 MPa}$$

> **Problem 3.** A wire of circular cross-section, has a tensile force of 60.0 N applied to it and this force produces a stress of 3.06 MPa in the wire. Determine the diameter of the wire.

Force $F = 60.0$ N and stress $\sigma = 3.06$ MPa

$$= 3.06 \times 10^6 \text{ Pa}$$

Since $\sigma = \dfrac{F}{A}$ then area, $A = \dfrac{F}{\sigma} = \dfrac{60.0 \text{ N}}{3.06 \times 10^6 \text{ Pa}}$

$$= 19.61 \times 10^{-6} \text{m}^2 = 19.61 \text{ mm}^2$$

Cross-sectional area $\qquad A = \dfrac{\pi d^2}{4}$

hence $\qquad 19.61 = \dfrac{\pi d^2}{4}$

from which, $\qquad d^2 = \dfrac{4 \times 19.61}{\pi}$

and $\qquad d = \sqrt{\left(\dfrac{4 \times 19.61}{\pi}\right)} = 5.0$

i.e. $\qquad$ **diameter of wire = 5.0 mm**

Now try the following Practise Exercise

> **Practise Exercise 11   Further problems on stress**
>
> 1. A rectangular bar having a cross-sectional area of 80 mm² has a tensile force of 20 kN applied to it. Determine the stress in the bar. [250 MPa]
>
> 2. A circular section cable has a tensile force of 1 kN applied to it and the force produces a stress of 7.8 MPa in the cable. Calculate the diameter of the cable. [12.78 mm]

3. A square-sectioned support of side 12 mm is loaded with a compressive force of 10 kN. Determine the compressive stress in the support. [69.44 MPa]

4. A bolt having a diameter of 5 mm is loaded so that the shear stress in it is 120 MPa. Determine the value of the shear force on the bolt. [2.356 kN]

5. A split pin requires a force of 400 N to shear it. The maximum shear stress before shear occurs is 120 MPa. Determine the minimum diameter of the pin. [2.06 mm]

6. A tube of outside diameter 60 mm and inside diameter 40 mm is subjected to a tensile load of 60 kN. Determine the stress in the tube. [38.2 MPa]

2.6 Strain

The fractional change in a dimension of a material produced by a force is called the **strain**. For a tensile or compressive force, strain is the ratio of the change of length to the original length. The symbol used for strain is ε (Greek epsilon). For a material of length L metres which changes in length by an amount x metres when subjected to stress,

$$\varepsilon = \frac{x}{L}$$

Strain is dimension-less and is often expressed as a percentage, i.e.

$$\text{percentage strain} = \frac{x}{L} \times 100$$

For a shear force, strain is denoted by the symbol γ (Greek letter gamma) and, with reference to Figure 2.5, is given by:

$$\gamma = \frac{x}{L}$$

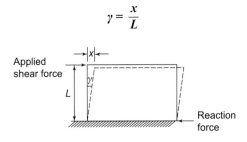

Figure 2.5

Problem 4. A bar 1.60 m long contracts axially by 0.1 mm when a compressive load is applied to it. Determine the strain and the percentage strain.

Strain $\varepsilon = \dfrac{\text{contraction}}{\text{original length}} = \dfrac{0.1\,\text{mm}}{1.60 \times 10^3\,\text{mm}}$

$$= \dfrac{0.1}{1600} = \textbf{0.0000625}$$

Percentage strain $= 0.0000625 \times 100 = \textbf{0.00625\%}$

Problem 5. A wire of length 2.50 m has a percentage strain of 0.012% when loaded with a tensile force. Determine the extension of the wire.

Original length of wire = 2.50 m = 2500 mm

and strain $= \dfrac{0.012}{100} = 0.00012$

Strain $\varepsilon = \dfrac{\text{extension}\,x}{\text{original length}\,L}$ hence,

extension $x = \varepsilon L = (0.00012)(2500) = \textbf{0.30 mm}$

Problem 6. (a) A rectangular section metal bar has a width of 10 mm and can support a maximum compressive stress of 20 MPa; determine the minimum breadth of the bar when loaded with a force of 3 kN. (b) If the bar in (a) is 2 m long and decreases in length by 0.25 mm when the force is applied, determine the strain and the percentage strain.

(a) Since stress, $\sigma = \dfrac{\text{force}\,F}{\text{area}\,A}$

then, area, $A = \dfrac{F}{\sigma} = \dfrac{3000\,\text{N}}{20 \times 10^6\,\text{Pa}}$

$$= 150 \times 10^{-6}\,\text{m}^2 = 150\,\text{mm}^2$$

Cross-sectional area = width × breadth, hence

$$\textbf{breadth} = \dfrac{\text{area}}{\text{width}} = \dfrac{150}{10} = \textbf{15 mm}$$

(b) **Strain,** $\varepsilon = \dfrac{\text{contraction}}{\text{original length}} = \dfrac{0.25}{2000} = \textbf{0.000125}$

Percentage strain $= 0.000125 \times 100 = \textbf{0.0125\%}$

Problem 7. A pipe has an outside diameter of 25 mm, an inside diameter of 15 mm and length 0.40 m and it supports a compressive load of 40 kN. The pipe shortens by 0.5 mm when the load

is applied. Determine (a) the compressive stress (b) the compressive strain in the pipe when supporting this load.

Compressive force $F = 40$ kN = 40000 N,

and cross-sectional area $A = \dfrac{\pi}{4}\left(D^2 - d^2\right),$

where D = outside diameter = 25 mm

and d = inside diameter = 15 mm.

Hence, $A = \dfrac{\pi}{4}(25^2 - 15^2)\,\text{mm}^2$

$$= \dfrac{\pi}{4}(25^2 - 15^2) \times 10^{-6}\,\text{m}^2$$

$$= 3.142 \times 10^{-4}\,\text{m}^2$$

(a) Compressive stress, $\sigma = \dfrac{F}{A} = \dfrac{40000\,\text{N}}{3.142 \times 10^{-4}\,\text{m}^2}$

$$= 12.73 \times 10^7\,\text{Pa}$$

$$= \textbf{127.3 MPa}$$

(b) Contraction of pipe when loaded,
$x = 0.5$ mm = 0.0005 m, and original length
$L = 0.40$ m. Hence, compressive strain,

$$\varepsilon = \dfrac{x}{L} = \dfrac{0.0005}{0.4} = \textbf{0.00125}\ (\text{or } \textbf{0.125\%})$$

Problem 8. A circular hole of diameter 50 mm is to be punched out of a 2 mm thick metal plate. The shear stress needed to cause fracture is 500 MPa. Determine (a) the minimum force to be applied to the punch, and (b) the compressive stress in the punch at this value.

(a) The area of metal to be sheared, A = perimeter of hole × thickness of plate.

Perimeter of hole $= \pi d = \pi(50 \times 10^{-3}) = 0.1571$ m.

Hence, shear area, $A = 0.1571 \times 2 \times 10^{-3}$
$$= 3.142 \times 10^{-4}\,\text{m}^2$$

Since shear stress $= \dfrac{\text{force}}{\text{area}},$

shear force $=$ shear stress × area
$$= (500 \times 10^6 \times 3.142 \times 10^{-4})\,\text{N}$$
$$= \textbf{157.1 kN,}$$

which is the minimum force to be applied to the punch.

(b) Area of punch $= \dfrac{\pi d^2}{4} = \dfrac{\pi(0.050)^2}{4}$

$$= 0.001963\,\text{m}^2$$

Compressive stress

$$= \frac{\text{force}}{\text{area}} = \frac{157.1 \times 10^3 \text{ N}}{0.001963 \text{ m}^2}$$

$$= 8.003 \times 10^7 \text{ Pa} = \mathbf{80.03 \text{ MPa}},$$

which is the compressive stress in the punch.

> **Problem 9.** A rectangular block of plastic material 500 mm long by 20 mm wide by 300 mm high has its lower face glued to a bench and a force of 200 N is applied to the upper face and in line with it. The upper face moves 15 mm relative to the lower face. Determine (a) the shear stress, and (b) the shear strain in the upper face, assuming the deformation is uniform.

(a) Shear stress, $\tau = \dfrac{\text{force}}{\text{area parallel to the force}}$

Area of any face parallel to the force

$$= 500 \text{ mm} \times 20 \text{ mm}$$

$$= (0.5 \times 0.02) \text{ m}^2 = 0.01 \text{ m}^2$$

Hence, **shear stress,** $\boldsymbol{\tau} = \dfrac{200 \text{ N}}{0.01 \text{ m}^2}$

$$= \mathbf{20000 \text{ Pa} \text{ or } 20 \text{ kPa}}$$

(b) Shear strain, $\gamma = \dfrac{x}{L}$ (see side view in Figure 2.6)

$$= \frac{15}{300} = \mathbf{0.05 \text{ (or } 5\%)}$$

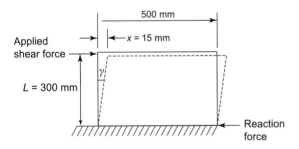

Figure 2.6

Now try the following Practise Exercise

> **Practise Exercise 12 Further problems on strain**
>
> 1. A wire of length 4.5 m has a percentage strain of 0.050% when loaded with a tensile force. Determine the extension in the wire.
>   [2.25 mm]

2. A metal bar 2.5 m long extends by 0.05 mm when a tensile load is applied to it. Determine (a) the strain, (b) the percentage strain.
   [(a) 0.00002 (b) 0.002%]

3. An 80 cm long bar contracts axially by 0.2 mm when a compressive load is applied to it. Determine the strain and the percentage strain. [0.00025, 0.025%]

4. A pipe has an outside diameter of 20 mm, an inside diameter of 10 mm and length 0.30 m and it supports a compressive load of 50 kN. The pipe shortens by 0.6 mm when the load is applied. Determine (a) the compressive stress, (b) the compressive strain in the pipe when supporting this load.
   [(a) 212.2 MPa (b) 0.002 or 0.20%]

5. When a circular hole of diameter 40 mm is punched out of a 1.5 mm thick metal plate, the shear stress needed to cause fracture is 100 MPa. Determine (a) the minimum force to be applied to the punch, and (b) the compressive stress in the punch at this value.
   [(a) 18.85 kN (b) 15.0 MPa]

6. A rectangular block of plastic material 400 mm long by 15 mm wide by 300 mm high has its lower face fixed to a bench and a force of 150 N is applied to the upper face and in line with it. The upper face moves 12 mm relative to the lower face. Determine (a) the shear stress, and (b) the shear strain in the upper face, assuming the deformation is uniform. [(a) 25 kPa (b) 0.04 or 4%]

2.7 Elasticity, limit of proportionality and elastic limit

Elasticity is the ability of a material to return to its original shape and size on the removal of external forces.

Plasticity is the property of a material of being permanently deformed by a force without breaking. Thus if a material does not return to the original shape, it is said to be plastic.

Within certain load limits, mild steel, copper, polythene and rubber are examples of elastic materials; lead and plasticine are examples of plastic materials.

If a tensile force applied to a uniform bar of mild steel is gradually increased and the corresponding extension of the bar is measured, then provided the applied force is not too large, a graph depicting these results is likely to be as shown in Figure 2.7. Since the graph is a straight line, **extension is directly proportional to the applied force**.

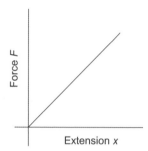

Figure 2.7

The point on the graph where extension is no longer proportional to the applied force is known as the **limit of proportionality**. Just beyond this point the material can behave in a non-linear elastic manner, until the **elastic limit** is reached. If the applied force is large, it is found that the material becomes plastic and no longer returns to its original length when the force is removed. The material is then said to have passed its elastic limit and the resulting graph of force/extension is no longer a straight line. Stress, $\sigma = \dfrac{F}{A}$, from Section 2.5, and since, for a particular bar, area A can be considered as a constant, then $F \propto \sigma$.

Strain $\varepsilon = \dfrac{x}{L}$, from Section 2.6, and since for a particular bar L is constant, then $x \propto \varepsilon$. Hence for stress applied to a material below the limit of proportionality a graph of stress/strain will be as shown in Figure 2.8, and is a similar shape to the force/extension graph of Figure 2.7.

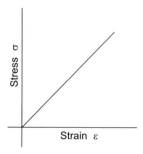

Figure 2.8

2.8 Hooke's law

Hooke's law states:
Within the limit of proportionality, the extension of a material is proportional to the applied force

It follows, from Section 2.7, that:
Within the limit of proportionality of a material, the strain produced is directly proportional to the stress producing it

Young's modulus of elasticity
Within the limit of proportionality, stress α strain, hence

$$\text{stress} = (\text{a constant}) \times \text{strain}$$

This constant of proportionality is called **Young's modulus of elasticity** and is given the symbol E.

The value of E may be determined from the gradient of the straight line portion of the stress/strain graph. The dimensions of E are Pascals (the same as for stress, since strain is dimension-less).

$$E = \frac{\sigma}{\varepsilon} \text{ Pa}$$

Some **typical values** for Young's modulus of elasticity, E, include:

Aluminium alloy 70 GPa (i.e. 70×10^9 Pa), brass 90 GPa, copper 96 GPa, titanium alloy 110 GPa, diamond 1200 GPa, mild steel 210 GPa, lead 18 GPa, tungsten 410 GPa, cast iron 110 GPa, zinc 85 GPa, glass fibre 72 GPa, carbon fibre 300 GPa.

Stiffness
A material having a large value of Young's modulus is said to have a high value of material stiffness, where stiffness is defined as:

$$\text{Stiffness} = \frac{\text{force } F}{\text{extension } x}$$

For example, mild steel is a much stiffer material than lead.

Since $E = \dfrac{\sigma}{\varepsilon}$, $\sigma = \dfrac{F}{A}$ and $\varepsilon = \dfrac{x}{L}$,

then $E = \dfrac{\dfrac{F}{A}}{\dfrac{x}{L}} = \dfrac{FL}{Ax} = \left(\dfrac{F}{x}\right)\left(\dfrac{L}{A}\right)$

i.e. $E = (\text{stiffness}) \times \left(\dfrac{L}{A}\right)$

Stiffness $\left(= \dfrac{F}{x}\right)$ is also the gradient of the force/extension graph, hence

$$E = \text{(gradient of force/extension graph)}\left(\dfrac{L}{A}\right)$$

Since L and A for a particular specimen are constant, the greater Young's modulus the greater the material stiffness.

Problem 10. A wire is stretched 2 mm by a force of 250 N. Determine the force that would stretch the wire 5 mm, assuming that the limit of proportionality is not exceeded.

Hooke's law states that extension x is proportional to force F, provided that the limit of proportionality is not exceeded, i.e. $x \,\alpha\, F$ or $x = kF$ where k is a constant.

When $x = 2$ mm, $F = 250$ N,

thus $2 = k(250)$, from which,

constant $k = \dfrac{2}{250} = \dfrac{1}{125}$

When $x = 5$ mm, then $5 = kF$

i.e. $5 = \left(\dfrac{1}{125}\right)F$

from which, force $F = 5(125) = 625$ N

Thus to stretch the wire 5 mm, a force of 625 N is required.

Problem 11. A tensile force of 10 kN applied to a component produces an extension of 0.1 mm. Determine (a) the force needed to produce an extension of 0.12 mm, and (b) the extension when the applied force is 6 kN, assuming in each case that the limit of proportionality is not exceeded.

From Hooke's law, extension x is proportional to force F within the limit of proportionality, i.e. $x \,\alpha\, F$ or $x = kF$, where k is a constant. If a force of 10 kN produces an extension of 0.1 mm, then $0.1 = k(10)$,

from which, constant $k = \dfrac{0.1}{10} = 0.01$

(a) When an extension $x = 0.12$ mm, then
$0.12 = k(F)$, i.e. $0.12 = 0.01\,F$, from which,

  force $F = \dfrac{0.12}{0.01} = 12$ kN

(b) When force $F = 6$ kN, then
extension $x = k(6) = (0.01)(6) = 0.06$ mm

Problem 12. A copper rod of diameter 20 mm and length 2.0 m has a tensile force of 5 kN applied to it. Determine (a) the stress in the rod (b) by how much the rod extends when the load is applied. Take the modulus of elasticity for copper as 96 GPa.

(a) Force $F = 5$ kN $= 5000$ N and cross-sectional area

$$A = \frac{\pi d^2}{4} = \frac{\pi(0.020)^2}{4} = 0.000314 \text{ m}^2$$

  Stress, $\sigma = \dfrac{F}{A} = \dfrac{5000 \text{ N}}{0.000314 \text{ m}^2}$

$$= 15.92 \times 10^6 \text{ Pa} = \mathbf{15.92 \text{ MPa}}$$

(b) Since $E = \dfrac{\sigma}{\varepsilon}$ then strain $\varepsilon = \dfrac{\sigma}{E} = \dfrac{15.92 \times 10^6 \text{ Pa}}{96 \times 10^9 \text{ Pa}}$

$$= 0.000166$$

  Strain $\varepsilon = \dfrac{x}{L}$, hence extension,

$$x = \varepsilon L = (0.000166)(2.0) = 0.000332 \text{ m}$$

  i.e. **extension of rod is 0.332 mm**

Problem 13. A bar of thickness 15 mm and having a rectangular cross-section carries a load of 120 kN. Determine the minimum width of the bar to limit the maximum stress to 200 MPa. The bar, which is 1.0 m long, extends by 2.5 mm when carrying a load of 120 kN. Determine the modulus of elasticity of the material of the bar.

Force, $F = 120$ kN $= 120000$ N and cross-sectional area $A = (15x)10^{-6}$ m^2, where x is the width of the rectangular bar in millimetres.

Stress $\sigma = \dfrac{F}{A}$, from which,

$$A = \frac{F}{\sigma} = \frac{120000 \text{ N}}{200 \times 10^6 \text{ Pa}} = 6 \times 10^{-4} \text{ m}^2$$

$$= 6 \times 10^{-4} \times 10^6 \text{ mm}^2$$

$$= 6 \times 10^2 \text{ mm}^2 = 600 \text{ mm}^2$$

Hence, $600 = 15x$, from which,

width of bar, $x = \dfrac{600}{15} = 40$ mm

Extension of bar $= 2.5$ mm $= 0.0025$ m

Strain $\varepsilon = \dfrac{x}{L} = \dfrac{0.0025}{1.0} = 0.0025$

Modulus of elasticity, $E = \dfrac{\text{stress}}{\text{strain}} = \dfrac{200 \times 10^6}{0.0025}$

$$= 80 \times 10^9 = \mathbf{80 \text{ GPa}}$$

Problem 14. An aluminium alloy rod has a length of 200 mm and a diameter of 10 mm. When subjected to a compressive force the length of the rod is 199.6 mm. Determine (a) the stress in the rod when loaded, and (b) the magnitude of the force. Take the modulus of elasticity for aluminium alloy as 70 GPa.

(a) Original length of rod, $L = 200$ mm, final length of rod = 199.6 mm, hence contraction, $x = 0.4$ mm.

Thus, strain, $\varepsilon = \dfrac{x}{L} = \dfrac{0.4}{200} = 0.002$

Modulus of elasticity, $E = \dfrac{\text{stress } \sigma}{\text{strain } \varepsilon}$

hence **stress, $\sigma = E\varepsilon = 70 \times 10^9 \times 0.002$**

$$= 140 \times 10^6 \, \text{Pa} = \textbf{140 MPa}$$

(b) Since stress $\sigma = \dfrac{\text{force } F}{\text{area } A}$, then force, $F = \sigma A$

Cross-sectional area, $A = \dfrac{\pi d^2}{4} = \dfrac{\pi (0.010)^2}{4}$

$$= 7.854 \times 10^{-5} \, \text{m}^2.$$

Hence, **compressive force,**
$$F = \sigma A = 140 \times 10^6 \times 7.854 \times 10^{-5} = \textbf{11.0 kN}$$

Problem 15. A brass tube has an internal diameter of 120 mm and an outside diameter of 150 mm and is used to support a load of 5 kN. The tube is 500 mm long before the load is applied. Determine by how much the tube contracts when loaded, taking the modulus of elasticity for brass as 90 GPa.

Force in tube, $F = 5$ kN $= 5000$ N, and

cross-sectional area of tube,

$$A = \frac{\pi}{4}\left(D^2 - d^2\right) = \frac{\pi}{4}\left(0.150^2 - 0.120^2\right)$$

$$= 0.006362 \, \text{m}^2$$

Stress in tube, $\sigma = \dfrac{F}{A} = \dfrac{5000 \, \text{N}}{0.006362 \, \text{m}^2}$

$$= 0.7859 \times 10^6 \, \text{Pa}$$

Since the modulus of elasticity, $E = \dfrac{\text{stress } \sigma}{\text{strain } \varepsilon}$

then strain, $\varepsilon = \dfrac{\sigma}{E} = \dfrac{0.7859 \times 10^6 \, \text{Pa}}{90 \times 10^9 \, \text{Pa}} = 8.732 \times 10^{-6}$

Strain, $\varepsilon = \dfrac{\text{contraction } x}{\text{original length } L}$ thus,

contraction, $x = \varepsilon L = 8.732 \times 10^{-6} \times 0.500$

$$= 4.37 \times 10^{-6} \, \text{m}.$$

Thus, **when loaded, the tube contracts by 4.37 μm**

Problem 16. In an experiment to determine the modulus of elasticity of a sample of mild steel, a wire is loaded and the corresponding extension noted. The results of the experiment are as shown.

Load (N)	0	40	110	160	200	250	290	340
Extension (mm)	0	1.2	3.3	4.8	6.0	7.5	10.0	16.2

Draw the load/extension graph.
The mean diameter of the wire is 1.3 mm and its length is 8.0 m. Determine the modulus of elasticity E of the sample, and the stress at the limit of proportionality.

A graph of load/extension is shown in Figure 2.9

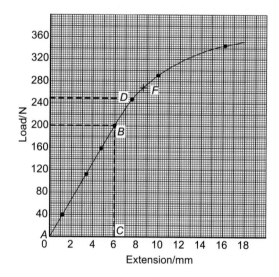

Figure 2.9

$$E = \frac{\sigma}{\varepsilon} = \frac{\dfrac{F}{A}}{\dfrac{x}{L}} = \left(\frac{F}{x}\right)\left(\frac{L}{A}\right)$$

$\dfrac{F}{x}$ is the gradient of the straight line part of the load/extension graph.

Gradient, $\dfrac{F}{x} = \dfrac{BC}{AC} = \dfrac{200 \, \text{N}}{6 \times 10^{-3} \text{m}} = 33.33 \times 10^3 \, \text{N/m}$

Modulus of elasticity = (gradient of graph)$\left(\dfrac{L}{A}\right)$

Length of specimen, $L = 8.0$ m and cross-sectional area

$$A = \frac{\pi d^2}{4} = \frac{\pi(0.0013)^2}{4}$$
$$= 1.327 \times 10^{-6}\,\text{m}^2$$

Hence **modulus of elasticity,**

$$E = (33.33 \times 10^3)\left(\frac{8.0}{1.327 \times 10^{-6}}\right) = \textbf{201 GPa}$$

The limit of proportionality is at point D in Figure 2.9 where the graph no longer follows a straight line. This point corresponds to a load of 250 N as shown.

Stress at the limit of proportionality

$$= \frac{\text{force}}{\text{area}} = \frac{250}{1.327 \times 10^{-6}}$$
$$= 188.4 \times 10^6\,\text{Pa} = \textbf{188.4 MPa}$$

Note that for structural materials the stress at the elastic limit is only fractionally larger than the stress at the limit of proportionality, thus it is reasonable to assume that the stress at the elastic limit is the same as the stress at the limit of proportionality; this assumption is made in the remaining exercises. In Figure 2.9, the elastic limit is shown as point *F*.

Now try the following Practise Exercise

Practise Exercise 13   Further problems on Hooke's law

1. A wire is stretched 1.5 mm by a force of 300 N. Determine the force that would stretch the wire 4 mm, assuming the elastic limit of the wire is not exceeded.
   [800 N]

2. A rubber band extends 50 mm when a force of 300 N is applied to it. Assuming the band is within the elastic limit, determine the extension produced by a force of 60 N.
   [10 mm]

3. A force of 25 kN applied to a piece of steel produces an extension of 2 mm. Assuming the elastic limit is not exceeded, determine (a) the force required to produce an extension of 3.5 mm, (b) the extension when the applied force is 15 kN.
   [(a) 43.75 kN (b) 1.2 mm]

4. A test to determine the load/extension graph for a specimen of copper gave the following results:

Load (kN)	8.5	15.0	23.5	30.0
Extension (mm)	0.04	0.07	0.11	0.14

   Plot the load/extension graph, and from the graph determine (a) the load at an extension of 0.09 mm, and (b) the extension corresponding to a load of 12.0 kN.
   [(a) 19 kN (b) 0.057 mm]

5. A circular section bar is 2.5 m long and has a diameter of 60 mm. When subjected to a compressive load of 30 kN it shortens by 0.20 mm. Determine Young's modulus of elasticity for the material of the bar.
   [132.6 GPa]

6. A bar of thickness 20 mm and having a rectangular cross-section carries a load of 82.5 kN. Determine (a) the minimum width of the bar to limit the maximum stress to 150 MPa, (b) the modulus of elasticity of the material of the bar if the 150 mm long bar extends by 0.8 mm when carrying a load of 200 kN.
   [(a) 27.5 mm (b) 68.2 GPa]

7. A metal rod of cross-sectional area 100 mm^2 carries a maximum tensile load of 20 kN. The modulus of elasticity for the material of the rod is 200 GPa. Determine the percentage strain when the rod is carrying its maximum load.   [0.10%]

8. A metal tube 1.75 m long carries a tensile load and the maximum stress in the tube must not exceed 50 MPa. Determine the extension of the tube when loaded if the modulus of elasticity for the material is 70 GPa.   [1.25 mm]

9. A piece of aluminium wire is 5 m long and has a cross-sectional area of 100 mm^2. It is subjected to increasing loads, the extension being recorded for each load applied. The results are:

Part Two

Load (kN)	0	1.12	2.94	4.76	7.00	9.10
Extension (mm)	0	0.8	2.1	3.4	5.0	6.5

Draw the load/extension graph and hence determine the modulus of elasticity for the material of the wire. [70 GPa]

10. In an experiment to determine the modulus of elasticity of a sample of copper, a wire is loaded and the corresponding extension noted. The results are:

Load (N)	0	20	34	72	94	120
Extension (mm)	0	0.7	1.2	2.5	3.3	4.2

Draw the load/extension graph and determine the modulus of elasticity of the sample if the mean diameter of the wire is 1.23 mm and its length is 4.0 m. [96 GPa]

2.9 Ductility, brittleness and malleability

Ductility is the ability of a material to be plastically deformed by elongation, without fracture. This is a property that enables a material to be drawn out into wires. For ductile materials such as mild steel, copper and gold, large extensions can result before fracture occurs with increasing tensile force. Ductile materials usually have a percentage elongation value of about 15% or more.

Brittleness is the property of a material manifested by fracture without appreciable prior plastic deformation. Brittleness is a lack of ductility, and brittle materials such as cast iron, glass, concrete, brick and ceramics, have virtually no plastic stage, the elastic stage being followed by immediate fracture. Little or no 'waist' occurs before fracture in a brittle material undergoing a tensile test.

Malleability is the property of a material whereby it can be shaped when cold by hammering or rolling. A malleable material is capable of undergoing plastic deformation without fracture. Examples of malleable materials include lead, gold, putty and mild steel.

Problem 17. Sketch typical load/extension curves for (a) an elastic non-metallic material, (b) a brittle material and (c) a ductile material. Give a typical example of each type of material.

(a) A typical load/extension curve for an elastic non-metallic material is shown in Figure 2.10(a), and an example of such a material is **polythene**.

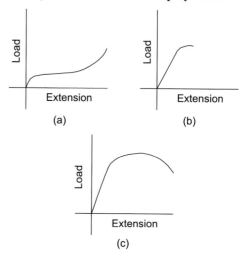

Figure 2.10

(b) A typical load/extension curve for a brittle material is shown in Figure 2.10(b), and an example of such a material is **cast iron**.

(c) A typical load/extension curve for a ductile material is shown in Figure 2.10(c), and an example of such a material is **mild steel**.

2.10 Modulus of rigidity

Experiments have shown that under pure torsion (see Chapter 11), up to the limit of proportionality, we have Hooke's law in shear, where

$$\frac{\text{shear stress}}{\text{shear strain}} = \text{rigidity (or shear) modulus}$$

or $$\frac{\tau}{\gamma} = G \qquad (2.1)$$

where τ = shear stress, γ = shear strain (see Figures 2.5 and 2.6) and G = rigidity (or shear) modulus.

2.11   Thermal strain

If a bar of length L and coefficient of linear expansion α were subjected to a temperature rise of T, its length will increase by a distance αLT, as described in Chapter 21. Thus the new length of the bar will be:

$$L + \alpha LT = L(1 + \alpha T)$$

Now, as the original length of the bar was L, then the **thermal strain** due to a temperature rise will be:

$$\varepsilon = \frac{\text{extension}}{\text{original length}} = \frac{\alpha LT}{L}$$

i.e. **thermal strain, $\varepsilon = \alpha T$**

However, if the bar were not constrained, so that it can expand freely, there will be no thermal stress.
If, however, the bar were prevented from expanding then there would be a compressive stress in the bar.

Now $\varepsilon = \dfrac{\text{original length} - \text{new length}}{\text{original length}}$

$$= \frac{L - L(1 + \alpha T)}{L}$$

$$= \frac{L - L - L\alpha T}{L}$$

i.e. strain, $\varepsilon = -\alpha T$
and, since stress = strain $\times E$,
then $\boldsymbol{\sigma = -\alpha TE}$

Problem 18. A steel prop is used to stabilise a building, as shown in Figure 2.11. (a) If the compressive stress in the bar at 20°C is 30 MPa, what will be the stress in the prop if the temperature is raised to 35°C? (b) At what temperature will the prop cease to be effective? Take $E = 2 \times 10^{11}\,\text{N/m}^2$ and $\alpha = 14 \times 10^{-6}/°\text{C}$.

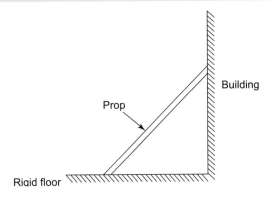

Building

Prop

Rigid floor

Figure 2.11

(a) Additional thermal strain, $\varepsilon_T = -\alpha T$
$$= -(14 \times 10^{-6}/°\text{C}) \times (35 - 20)°\text{C}$$
i.e. $\varepsilon_T = -14 \times 10^{-6} \times 15 = \boldsymbol{-210 \times 10^{-6}}$
Additional thermal stress, $\sigma_T = E\varepsilon_T$
$$= 2 \times 10^{11}\,\text{N/m}^2 \times (-210 \times 10^{-6})$$
i.e. $\sigma_T = \boldsymbol{-42\ \text{MPa}}$
Hence, the stress at 35°C = initial stress + σ_T
$$= (-30 - 42)\ \text{MPa}$$
i.e. $\boldsymbol{\sigma = -72\ \text{MPa}}$

(b) For the prop to be ineffective, it will be necessary for the temperature to fall so that there is no stress in the prop, that is, from 20°C the temperature must fall so that the initial stress of 30 MPa is nullified. Hence, drop in stress = −30 MPa

Therefore, drop in thermal strain
$$= \frac{\sigma}{E} = \frac{-30 \times 10^6\,\text{Pa}}{2 \times 10^{11}\,\text{Pa}} = -1.5 \times 10^{-4}$$

Thermal strain, $\varepsilon = \alpha T$

from which, temperature
$$T = \frac{\varepsilon}{\alpha} = \frac{-1.5 \times 10^{-4}}{14 \times 10^{-6}} = -10.7°\text{C}$$

Hence, the drop in temperature T from 20°C is −10.7°C

Therefore, **the temperature for the prop to be ineffective** = 20°C − 10.7°C = **9.3°C**

Now try the following Practise Exercise

Practise Exercise 14   Further problem on thermal strain

1. A steel rail may assumed to be stress free at 5°C. If the stress required to cause buckling of the rail is −50 MPa, at what temperature will the rail buckle? It may be assumed that the rail is rigidly fixed at its 'ends'. Take $E = 2 \times 10^{11}\,\text{N/m}^2$ and $\alpha = 14 \times 10^{-16}/°\text{C}$.
[22.86°C]

2.12   Compound bars

Compound bars are of much importance in a number of different branches of engineering, including reinforced concrete pillars, composites, bimetallic bars, and so on. In this section, solution of such problems usually involve two important considerations, namely

(a) **compatibility** (or considerations of displacements)
(b) **equilibrium**

N.B. It is necessary to introduce compatibility in this section as compound bars are, in general, statically indeterminate (see Chapter 6). The following worked problems demonstrate the method of solution.

> **Problem 19.** A solid bar of cross-sectional area A_1, Young's modulus E_1 and coefficient of linear expansion α_1 is surrounded co-axially by a hollow tube of cross-sectional area A_2, Young's modulus E_2 and coefficient of linear expansion α_2, as shown in Figure 2.12. If the two bars are secured firmly to each other, so that no slipping takes place with temperature change, determine the thermal stresses due to a temperature rise T. Both bars have an initial length L and $\alpha_1 > \alpha_2$.

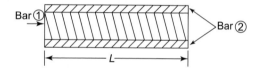

Figure 2.12 Compound bar

There are two unknown forces in these bars, namely F_1 and F_2; therefore, two simultaneous equations will be required to determine these unknown forces. The first equation can be obtained by considering the compatibility (i.e. 'deflections') of the bars, with the aid of Figure 2.13.

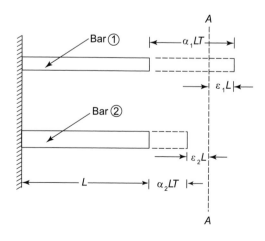

Figure 2.13 'Deflections' of compound bar

Free expansion of bar (1) $= \alpha_1 LT$

Free expansion of bar (2) $= \alpha_2 LT$

In practice, however, the final resting position of the compound bar will be somewhere in between these two positions (i.e. at the position A–A in Figure 2.13). To achieve this, it will be necessary for bar (2) to be pulled out by a distance $\varepsilon_2 L$ and for bar (2) to be pushed in by a distance $\varepsilon_1 L$, where

$$\varepsilon_1 = \text{compressive strain in (1)}$$

and $\varepsilon_2 = \text{tensile strain in (2)}$

From considerations of compatibility ('deflection') in Figure 2.13,

$$\alpha_1 LT - \varepsilon_1 L = \alpha_2 LT + \varepsilon_2 L$$

Dividing throughout by L gives:

$$\alpha_1 T - \varepsilon_1 = \alpha_2 T + \varepsilon_2$$

and $\varepsilon_1 = (\alpha_1 - \alpha_2)T - \varepsilon_2$

Now, $\sigma_1 = E_1 \varepsilon_1$

and $\sigma_2 = E_2 \varepsilon_2$ (or $\varepsilon_2 = \dfrac{\sigma_2}{E_2}$)

Hence, $\sigma_1 = (\alpha_1 - \alpha_2) E_1 T - E_1 \varepsilon_2$

i.e. $\qquad \sigma_1 = (\alpha_1 - \alpha_2) E_1 T - \sigma_2 \dfrac{E_1}{E_2}$ $\qquad$ (2.2)

To obtain the second simultaneous equation, it will be necessary to consider equilibrium of the compound bar.

Let $F_1 = $ unknown compressive force in bar (1)

and $F_2 = $ unknown tensile force in bar (2)

Now, from equilibrium considerations,

$$F_1 = F_2$$

but $\sigma_1 = \dfrac{F_1}{A_1}$ and $\sigma_2 = \dfrac{F_2}{A_2}$

Therefore, $\qquad \sigma_1 A_1 = \sigma_2 A_2$

or $\qquad\qquad \boldsymbol{\sigma_1 = \dfrac{\sigma_2 A_2}{A_1}}$ $\qquad$ (2.3)

Equating equations (2.2) and (2.3) gives

$$\frac{\sigma_2 A_2}{A_1} = (\alpha_1 - \alpha_2)E_1 T - \sigma_2 \frac{E_1}{E_2}$$

i.e. $\sigma_2 \dfrac{E_1}{E_2} + \dfrac{\sigma_2 A_2}{A_1} = (\alpha_1 - \alpha_2)E_1 T$

and $\sigma_2 \left(\dfrac{E_1}{E_2} + \dfrac{A_2}{A_1} \right) = (\alpha_1 - \alpha_2)E_1 T$

from which, $\sigma_2 = \dfrac{(\alpha_1 - \alpha_2)E_1 T}{\left(\dfrac{E_1}{E_2} + \dfrac{A_2}{A_1}\right)}$

$= \dfrac{(\alpha_1 - \alpha_2)E_1 T}{\left(\dfrac{A_1 E_1 + A_2 E_2}{E_2 A_1}\right)}$

i.e. $\sigma_2 = \dfrac{(\alpha_1 - \alpha_2)E_1 E_2 A_1 T}{(A_1 E_1 + A_2 E_2)}$ (tensile)   (2.4)

and $\sigma_1 = \dfrac{\sigma_2 A_2}{A_1}$

i.e. $\sigma_1 = \dfrac{(\alpha_1 - \alpha_2)E_1 E_2 A_2 T}{(A_1 E_1 + A_2 E_2)}$ (compressive) (2.5)

Problem 20. If the solid bar of Problem 19 did not suffer temperature change, but instead was subjected to a tensile axial force P, as shown in Figure 2.14, determine σ_1 and σ_2.

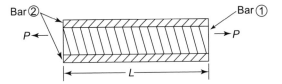

Figure 2.14 Compound bar under axial tension

There are two unknown forces in this bar, namely, F_1 and F_2; therefore, two simultaneous equations will be required.

The first of these simultaneous equations can be obtained by considering compatibility, i.e.

deflection of bar (1) = deflection of bar (2)

or $\delta_1 = \delta_2$

But $\delta_1 = \varepsilon_1 L$ and $\delta_2 = \varepsilon_2 L$

Therefore, $\varepsilon_1 L = \varepsilon_2 L$

or $\varepsilon_1 = \varepsilon_2$

Now, $\varepsilon_1 = \dfrac{\sigma_1}{E_1}$ and $\varepsilon_2 = \dfrac{\sigma_2}{E_2}$

Hence, $\dfrac{\sigma_1}{E_1} = \dfrac{\sigma_2}{E_2}$

or $\sigma_1 = \dfrac{\sigma_2 E_1}{E_2}$ (2.6)

The second simultaneous equation can be obtained by considering the equilibrium of the compound bar.

Let F_1 = tensile force in bar (1)

and F_2 = tensile force in bar (2)

Now, from equilibrium conditions

$P = F_1 + F_2$

i.e. $P = \sigma_1 A_1 + \sigma_2 A_2$ (2.7)

Substituting equation (2.6) into equation (2.7) gives:

$P = \dfrac{\sigma_2 E_1}{E_2} A_1 + \sigma_2 A_2 = \sigma_2\left(\dfrac{E_1 A_1}{E_2} + A_2\right)$

$= \sigma_2\left(\dfrac{A_1 E_1 + A_2 E_2}{E_2}\right)$

Rearranging gives: $\sigma_2 = \dfrac{P E_2}{(A_1 E_1 + A_2 E_2)}$ (2.8)

Since $\sigma_1 = \dfrac{\sigma_2 E_1}{E_2}$

then $\sigma_1 = \dfrac{P E_1}{(A_1 E_1 + A_2 E_2)}$ (2.9)

N.B. If P is a compressive force, then both σ_1 and σ_2 will be compressive stresses (i.e. negative), and vice-versa if P were tensile.

Problem 21. A concrete pillar, which is reinforced with steel rods, supports a compressive axial load of 2 MN.
(a) Determine stresses σ_1 and σ_2 given the following:
For the steel, $A_1 = 4 \times 10^{-3}$ m$_2$ and $E_1 = 2 \times 10^{11}$ N/m^2
For the concrete, $A_2 = 0.2$ m^2 and $E_2 = 2 \times 10^{10}$ N/m^2
(b) What percentage of the total load does the steel reinforcement take?

(a)   From equation (2.9),

$\sigma_1 = -\dfrac{P E_1}{(A_1 E_1 + A_2 E_2)}$

$= -\dfrac{2 \times 10^6 \times 2 \times 10^{11}}{\left(4 \times 10^{-3} \times 2 \times 10^{11} + 0.2 \times 2 \times 10^{10}\right)}$

$= -\dfrac{4 \times 10^{17}}{\left(8 \times 10^8 + 40 \times 10^8\right)} = \dfrac{4 \times 10^{17}}{48 \times 10^8} = \dfrac{10^9}{12}$

$= -83.3 \times 10^6$

i.e. **the stress in the steel,**

$\sigma_1 = -83.3$ **MPa**   (2.10)

From equation (2.8),

$$\sigma_2 = -\frac{P E_2}{(A_1 E_1 + A_2 E_2)}$$

$$= -\frac{2 \times 10^6 \times 2 \times 10^{10}}{\left(4 \times 10^{-3} \times 2 \times 10^{11} + 0.2 \times 2 \times 10^{10}\right)}$$

$$= -\frac{4 \times 10^{16}}{\left(8 \times 10^8 + 40 \times 10^8\right)} = \frac{4 \times 10^{16}}{48 \times 10^8} = \frac{10^8}{12}$$

$$= -8.3 \times 10^6$$

i.e. **the stress in the concrete,**

$$\sigma_2 = -8.3 \text{ MPa} \tag{2.11}$$

(b) Force in the steel,

$$F_1 = \sigma_1 A_1$$
$$= -83.3 \times 10^6 \times 4 \times 10^{-3} = \mathbf{3.33 \times 10^5 \, N}$$

Therefore, **the percentage total load taken by the steel reinforcement**

$$= \frac{F_1}{\text{total axial load}} \times 100\%$$

$$= \frac{3.33 \times 10^5}{2 \times 10^6} \times 100\% = \mathbf{16.65\%}$$

Problem 22. If the pillar of problem 21 were subjected to a temperature rise of 25°C, what would be the values of stresses σ_1 and σ_2? Assume the coefficients of linear expansion are, for steel, $\alpha_1 = 14 \times 10^{-6}/°C$, and for concrete, $\alpha_2 = 12 \times 10^{-6}/°C$.

As α_1 is larger than α_2, the effect of a temperature rise will cause the 'thermal stresses' in the steel to be compressive and those in the concrete to be tensile.

From equation (2.5), **the thermal stress in the steel,**

$$\sigma_1 = -\frac{(\alpha_1 - \alpha_2) E_1 E_2 A_2 T}{(A_1 E_1 + A_2 E_2)}$$

$$= -\frac{\left[\begin{array}{c}(14 \times 10^{-6} - 12 \times 10^{-6}) \\ \times 2 \times 10^{11} \times 2 \times 10^{10} \times 0.2 \times 25\end{array}\right]}{48 \times 10^8}$$

$$= -\frac{40 \times 10^{15}}{48 \times 10^8} = 0.833 \times 10^7$$

$$= -8.33 \text{ MPa} \tag{2.12}$$

From equation (2.3), **the thermal stress in the concrete,**

$$\sigma_2 = \frac{\sigma_1 A_1}{A_2} = -\frac{(-8.33 \times 10^6) \times 4 \times 10^{-3}}{0.2}$$

$$= \mathbf{0.167 \text{ MPa}} \tag{2.13}$$

From equations (2.10) to (2.13):

$$\sigma_1 = -83.3 - 8.33 = \mathbf{-91.63 \text{ MPa}}$$

and $\sigma_2 = -8.3 + 0.167 = \mathbf{-8.13 \text{ MPa}}$

Now try the following Practise Exercises

Practise Exercise 15   Further problems on compound bars

1. Two layers of carbon fibre are stuck to each other, so that their fibres lie at 90° to each other, as shown in Figure 2.15. If a tensile force of 1 kN were applied to this two-layer compound bar, determine the stresses in each. For layer 1, $E_1 = 300$ GPa and $A_1 = 10$ mm^2

   For layer 2, $E_2 = 50$ GPa and $A_2 = A_1 = 10$ mm^2

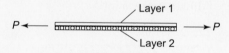

   Figure 2.15 Carbon fibre layers

   $[\sigma_1 = 85.71 \text{ MPa}, \sigma_2 = 14.29 \text{ MPa}]$

2. If the compound bar of Problem 1 were subjected to a temperature rise of 25°C, what would the resulting stresses be? Assume the coefficients of linear expansion are, for layer 1, $\alpha_1 = 5 \times 10^{-6}/°C$, and for layer 2, $\alpha_2 = 0.5 \times 10^{-6}/°C$.

   $[\sigma_1 = 80.89 \text{ MPa}, \sigma_2 = 19.11 \text{ MPa}]$

Practise Exercise 16   Short-answer questions on the effects of forces on materials

1. Name three types of mechanical force that can act on a body.

2. What is a tensile force? Name two practical examples of such a force.

3. What is a compressive force? Name two practical examples of such a force.

4. Define a shear force and name two practical examples of such a force.

5. Define elasticity and state two examples of elastic materials.

6. Define plasticity and state two examples of plastic materials.

7. Define the limit of proportionality.

8. State Hooke's law.

9. What is the difference between a ductile and a brittle material?

10. Define stress. What is the symbol used for (a) a tensile stress (b) a shear stress?

11. Strain is the ratio $\dfrac{.....}{.....}$.

12. The ratio $\dfrac{\text{stress}}{\text{strain}}$ is called ....

13. State the units of (a) stress (b) strain (c) Young's modulus of elasticity.

14. Stiffness is the ratio $\dfrac{.....}{.....}$.

15. Sketch on the same axes a typical load/extension graph for a ductile and a brittle material.

16. Define (a) ductility (b) brittleness (c) malleability.

17. Define rigidity modulus.

18. The new length L_2 of a bar of length L_1, of coefficient of linear expansion α, when subjected to a temperature rise T is: $L_2 =$ ......

19. The thermal strain ε due to a temperature rise T in material of coefficient of linear expansion α is given by: $\varepsilon =$ ......

Practise Exercise 17 **Multiple-choice questions on the effects of forces on materials**

(Answers on page 297)

1. The unit of strain is:
   (a) pascals   (b) metres
   (c) dimension-less   (d) newtons

2. The unit of stiffness is:
   (a) newtons   (b) pascals
   (c) newtons per metre   (d) dimension-less

3. The unit of Young's modulus of elasticity is:
   (a) Pascals
   (b) metres
   (c) dimension-less
   (d) newtons

4. A wire is stretched 3 mm by a force of 150 N. Assuming the elastic limit is not exceeded, the force that will stretch the wire 5 mm is:
   (a) 150 N   (b) 250 N
   (c) 90 N   (d) 450 N

5. For the wire in question 4, the extension when the applied force is 450 N is:
   (a) 1 mm   (b) 3 mm
   (c) 9 mm   (d) 12 mm

6. Due to the forces acting, a horizontal beam is in:
   (a) tension   (b) compression
   (c) shear

7. Due to forces acting, a pillar supporting a bridge is in:
   (a) tension   (b) compression
   (c) shear

8. Which of the following statements is false?
   (a) Elasticity is the ability of a material to return to its original dimensions after deformation by a load.
   (b) Plasticity is the ability of a material to retain any deformation produced in it by a load.
   (c) Ductility is the ability to be permanently stretched without fracturing.
   (d) Brittleness is the lack of ductility and a brittle material has a long plastic stage.

9. A circular rod of cross-sectional area 100 mm^2 has a tensile force of 100 kN applied to it. The stress in the rod is:
   (a) 1 MPa   (b) 1 GPa
   (c) 1 kPa   (d) 100 MPa

10. A metal bar 5.0 m long extends by 0.05 mm when a tensile load is applied to it. The percentage strain is:
    (a) 0.1   (b) 0.01
    (c) 0.001   (d) 0.0001

An aluminium rod of length 1.0 m and cross-sectional area 500 mm^2 is used to support a load of 5 kN which causes the rod to contract by 100 μm. For questions 11 to 13, select the correct answer from the following list:

(a) 100 MPa (b) 0.001
(c) 10 kPa (d) 100 GPa
(e) 0.01 (f) 10 MPa
(g) 10 GPa (h) 0.0001
(i) 10 Pa

11. The stress in the rod.

12. The strain in the rod.

13. Young's modulus of elasticity.

14. A compound bar of length L is subjected to a temperature rise of T. If $\alpha_1 > \alpha_2$, the strain in bar 1 will be:

(a) tensile (b) compressive
(c) zero (d) αT

15. For Problem 14, the stress in bar 2 will be:

(a) tensile (b) compressive
(c) zero (d) αT

Chapter 3

Tensile testing

In this chapter, additional material constants are introduced to help define the properties of materials. In addition to this, a description is given of the standard tensile test used to obtain the strength of various materials, especially the strength and material stiffness of metals. This is aided by a standard tensile test, where the relationship of axial load on a specimen and its axial deflection are described. The engineer can tell a lot from the results of this experiment. For example, the relationship between load and deflection is very different for steel and aluminium alloy and most other materials, including copper, zinc, brass, titanium, and so on. By carrying out this test, the engineer can determine the required properties of different materials, and this method, together with its interpretation, is discussed in this chapter.

At the end of this chapter you should be able to:

- describe a tensile test
- recognise from a tensile test the limit of proportionality, the elastic limit and the yield point
- plot a load/extension graph from given data
- calculate from a load/extension graph, the modulus of elasticity, the yield stress, the ultimate tensile strength, percentage elongation and the percentage reduction in area
- appreciate the meaning of proof stress

3.1 The tensile test

A **tensile test** is one in which a force is applied to a specimen of a material in increments and the corresponding extension of the specimen noted. The process may be continued until the specimen breaks into two parts and this is called **testing to destruction**. The testing is usually carried out using a universal testing machine that can apply either tensile or compressive forces to a specimen in small, accurately measured steps. **British Standard 18** gives the standard procedure for such a test. Test specimens of a material are made to standard shapes and sizes and two typical test pieces are shown in Figure 3.1. The results of a tensile test may be plotted on a load/extension graph and a typical graph for a mild steel specimen is shown in Figure 3.2.

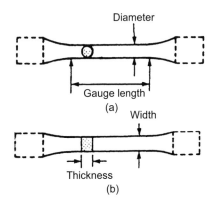

Figure 3.1

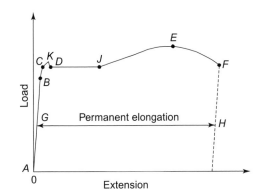

Figure 3.2

Mechanical Engineering Principles, Bird and Ross, ISBN 9780415517850

(i) Between A and B is the region in which Hooke's law applies and stress is directly proportional to strain. The gradient of AB is used when determining Young's modulus of elasticity (see Chapter 2).

(ii) Point B is the **limit of proportionality** and is the point at which stress is no longer proportional to strain when a further load is applied.

(iii) Point C is the **elastic limit** and a specimen loaded to this point will effectively return to its original length when the load is removed, i.e. there is negligible permanent extension.

(iv) Point D is called the **yield point** and at this point there is a sudden extension to J, with no increase in load. The yield stress of the material is given by:

yield stress =
$$\frac{\textbf{load where yield begins to take place}}{\textbf{original cross} - \textbf{sectional area}}$$

The yield stress gives an indication of the ductility of the material (see Chapter 2).

(v) For mild steel, the extension up to the point J is some 40 times larger than the extension up to the point B.

(vi) Shortly after point J, the material strain hardens, where the slope of the load-extension curve is about 1/50th the slope of the curve from A to B, for materials such as mild steel.

(vii) Between points D and E extension takes place over the whole gauge length of the specimen.

(viii) Point E gives the maximum load which can be applied to the specimen and is used to determine the ultimate tensile strength (UTS) of the specimen (often just called the tensile strength)

$$\textbf{UTS} = \frac{\textbf{maximum load}}{\textbf{original cross-sectional area}}$$

(ix) Between points E and F the cross-sectional area of the specimen decreases, usually about half way between the ends, and a **waist** or **neck** is formed before fracture.

Percentage reduction in area
$$= \frac{\begin{array}{c}\textbf{(original cross-sectional area)} -\\ \textbf{(final cross-sectional area)}\end{array}}{\textbf{original cross-sectional area}} \times \textbf{100\%}$$

The percentage reduction in area provides information about the malleability of the material (see Chapter 2). The value of stress at point F is greater than at point E since although the load on the specimen is decreasing as the extension increases, the cross-sectional area is also reducing.

(x) At point F the specimen fractures.

(xi) Distance GH is called the **permanent elongation** and

permanent elongation
$$= \frac{\textbf{increase in length during test to destruction}}{\textbf{original cross-sectional area}}$$
$$\times \textbf{100\%}$$

(xii) The point K is known as the **upper yield point.** It occurs for constant load experiments, such as when a hydraulic tensile testing machine is used. It does not occur for constant strain experiments, such as when a Hounsfield tensometer is used.

3.2 Worked problems on tensile testing

Problem 1. A tensile test is carried out on a mild steel specimen. The results are shown in the following table of values:

Load (kN)	0	10	23	32
Extension (mm)	0	0.023	0.053	0.074

Plot a graph of load against extension, and from the graph determine (a) the load at an extension of 0.04 mm, and (b) the extension corresponding to a load of 28 kN.

The load/extension graph is shown in Figure 3.3. From the graph:

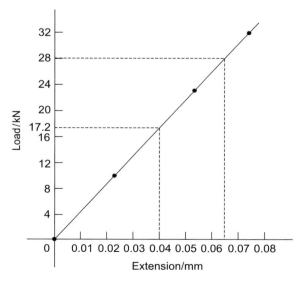

Figure 3.3

(a)   when the extension is 0.04 mm, the load is **17.2 kN**
(b)   when the load is 28 kN, the extension is **0.065 mm**

> **Problem 2.**   A tensile test is carried out on a mild steel specimen of gauge length 40 mm and cross-sectional area 100 mm². The results obtained for the specimen up to its yield point are given below:
>
> | Load (kN) | 0 | 8 | 19 | 29 | 36 |
> |---|---|---|---|---|---|
> | Extension (mm) | 0 | 0.015 | 0.038 | 0.060 | 0.072 |
>
> The maximum load carried by the specimen is 50 kN and its length after fracture is 52 mm. Determine (a) the modulus of elasticity, (b) the ultimate tensile strength, (c) the percentage elongation of the mild steel.

The load/extension graph is shown in Figure 3.4.

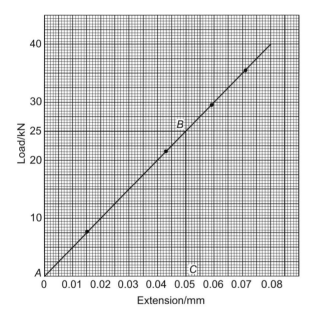

Figure 3.4

(a)   Gradient of straight line is given by:

$$\frac{BC}{AC} = \frac{25000\,\text{N}}{0.05 \times 10^{-3}\,\text{m}} = 500 \times 10^6\,\text{N/m}$$

Young's modulus of elasticity

$$= (\text{gradient of graph})\left(\frac{L}{A}\right),\ \text{where}$$

$L = 40$ mm (gauge length) $= 0.040$ m and area, $A = 100$ mm² $= 100 \times 10^{-6}$ m².

Young's modulus of elasticity

$$= (500 \times 10^6)\left(\frac{0.040}{100 \times 10^{-6}}\right) = 200 \times 10^9\,\text{Pa}$$

$$= \textbf{200 GPa}$$

(b)   Ultimate tensile strength

$$= \frac{\text{maximum load}}{\text{original cross-sectional area}}$$

$$= \frac{50000\,\text{N}}{100 \times 10^{-6}\,\text{m}^2} = 500 \times 10^6\,\text{Pa} = \textbf{500 MPa}$$

(c)   Percentage elongation

$$= \frac{\text{increase in length}}{\text{original length}} \times 100$$

$$= \frac{52 - 40}{40} \times 100$$

$$= \frac{12}{40} \times 100 = \textbf{30\%}$$

> **Problem 3.**   The results of a tensile test are: Diameter of specimen 15 mm; gauge length 40 mm; load at limit of proportionality 85 kN; extension at limit of proportionality 0.075 mm; maximum load 120 kN; final length at point of fracture 55 mm. Determine (a) Young's modulus of elasticity (b) the ultimate tensile strength (c) the stress at the limit of proportionality (d) the percentage elongation.

(a)   Young's modulus of elasticity is given by:

$$E = \frac{\text{stress}}{\text{strain}} = \frac{\dfrac{F}{A}}{\dfrac{x}{L}} = \frac{FL}{Ax}$$

where the load at the limit of proportionality, $F = 85$ kN $= 85000$ N,

$L =$ gauge length $= 40$ mm $= 0.040$ m,

$A =$ cross-sectional area

$$= \frac{\pi d^2}{4} = \frac{\pi (0.015)^2}{4} = 0.0001767\ \text{m}^2,\ \text{and}$$

$x =$ extension $= 0.075$ mm $= 0.000075$ m.

Hence, Young's modulus of elasticity

$$E = \frac{FL}{Ax} = \frac{(85000)(0.040)}{(0.0001767)(0.000075)}$$

$$= 256.6 \times 10^9\,\text{Pa} = \textbf{256.6 GPa}$$

(b)   Ultimate tensile strength

$$= \frac{\text{maximum load}}{\text{original cross-sectional area}} = \frac{120000}{0.0001767}$$

$$= 679 \times 10^6\,\text{Pa} = \textbf{679 MPa}$$

(c) Stress at limit of proportionality

$$= \frac{\text{load at limit of proportionality}}{\text{cross-sectional area}}$$

$$= \frac{85000}{0.0001767} = 481 \times 10^6 \, \text{Pa} = \textbf{481 MPa}$$

(d) Percentage elongation

$$= \frac{\text{increase in length}}{\text{original length}} \times 100$$

$$= \frac{(55 - 40)\,\text{mm}}{40\,\text{mm}} \times 100 = \textbf{37.5\%}$$

Now try the following Practise Exercise

Practise Exercise 18   Further problems on tensile testing

1. What is a tensile test? Make a sketch of a typical load/extension graph for a mild steel specimen to the point of fracture and mark on the sketch the following: (a) the limit of proportionality (b) the elastic limit (c) the yield point.

2. In a tensile test on a zinc specimen of gauge length 100 mm and diameter 15 mm, a load of 100 kN produced an extension of 0.666 mm. Determine (a) the stress induced (b) the strain (c) Young's modulus of elasticity.
    [(a) 566 MPa   (b) 0.00666   (c) 85 GPa]

3. The results of a tensile test are: Diameter of specimen 20 mm, gauge length 50 mm, load at limit of proportionality 80 kN, extension at limit of proportionality 0.075 mm, maximum load 100 kN, and final length at point of fracture 60 mm. Determine (a) Young's modulus of elasticity (b) the ultimate tensile strength (c) the stress at the limit of proportionality (d) the percentage elongation.
    $$\begin{bmatrix} \text{(a) 169.8 GPa} & \text{(b) 318.3 MPa} \\ \text{(c) 254.6 MPa} & \text{(d) 20\%} \end{bmatrix}$$

3.3 Further worked problems on tensile testing

Problem 4.   A rectangular zinc specimen is subjected to a tensile test and the data from the test

is shown below. Width of specimen 40 mm; breadth of specimen 2.5 mm; gauge length 120 mm.

Load (kN)									
10	17	25	30	35	37.5	38.5	37	34	32
Extension (mm)									
0.15	0.25	0.35	0.55	1.00	1.50	2.50	3.50	4.50	5.00

Fracture occurs when the extension is 5.0 mm and the maximum load recorded is 38.5 kN. Plot the load/extension graph and hence determine (a) the stress at the limit of proportionality (b) Young's modulus of elasticity (c) the ultimate tensile strength (d) the percentage elongation (e) the stress at a strain of 0.01 (f) the extension at a stress of 200 MPa.

A load/extension graph is shown in Figure 3.5.

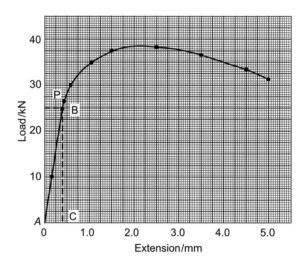

Figure 3.5

(a) The limit of proportionality occurs at point P on the graph, where the initial gradient of the graph starts to change. This point has a load value of 26.5 kN.

Cross-sectional area of specimen
$= 40 \, \text{mm} \times 2.5 \, \text{mm} = 100 \, \text{mm}^2 = 100 \times 10^{-6} \, \text{m}^2$

Stress at the limit of proportionality is given by:

$$\sigma = \frac{\text{force}}{\text{area}} = \frac{26.5 \times 10^3 \, \text{N}}{100 \times 10^{-6} \, \text{m}^2}$$

$$= 265 \times 10^6 \text{Pa} = \textbf{265 MPa}$$

(b) Gradient of straight line portion of graph is given by:

$$\frac{BC}{AC} = \frac{25000 \, \text{N}}{0.35 \times 10^{-3} \, \text{m}} = 71.43 \times 10^6 \, \text{N/m}$$

Young's modulus of elasticity

$$= (\text{gradient of graph})\left(\frac{L}{A}\right)$$

$$= \left(71.43 \times 10^6\right)\left(\frac{120 \times 10^{-3}}{100 \times 10^{-6}}\right)$$

$$= 85.72 \times 10^9 \, \text{Pa} = \textbf{85.72 GPa}$$

(c) Ultimate tensile strength

$$= \frac{\text{maximum load}}{\text{original cross-sectional area}}$$

$$= \frac{38.5 \times 10^3 \, \text{N}}{100 \times 10^{-6} \, \text{m}^2}$$

$$= 385 \times 10^6 \, \text{Pa} = \textbf{385 MPa}$$

(d) Percentage elongation

$$= \frac{\text{extension at fracture point}}{\text{original length}} \times 100$$

$$= \frac{5.0 \, \text{mm}}{120 \, \text{mm}} \times 100 = \textbf{4.17\%}$$

(e) Strain $\varepsilon = \dfrac{\text{extension} \, x}{\text{original length} \, l}$, from which,

extension $x = \varepsilon l = 0.01 \times 120$

$$= 1.20 \, \text{mm}.$$

From the graph, the load corresponding to an extension of 1.20 mm is 36 kN.

Stress at a strain of 0.01 is given by:

$$\sigma = \frac{\text{force}}{\text{area}} = \frac{36000 \, \text{N}}{100 \times 10^{-6} \, \text{m}^2}$$

$$= 360 \times 10^6 \, \text{Pa} = \textbf{360 MPa}$$

(f) When the stress is 200 MPa, then

force = area × stress

$$= (100 \times 10^{-6})(200 \times 10^6) = 20 \, \text{kN}$$

From the graph, the corresponding extension is **0.30 mm**

Problem 5. A mild steel specimen of cross-sectional area 250 mm² and gauge length 100 mm is subjected to a tensile test and the following data is obtained: within the limit of proportionality, a load of 75 kN produced an extension of 0.143 mm, load at yield point = 80 kN, maximum load on specimen = 120 kN, final cross-sectional area of waist at fracture = 90 mm², and the gauge length had increased to 135 mm at fracture.

Determine for the specimen: (a) Young's modulus of elasticity (b) the yield stress (c) the tensile strength (d) the percentage elongation and (e) the percentage reduction in area.

(a) Force $F = 75$ kN $= 75000$ N, gauge length $L = 100$ mm $= 0.1$ m, cross-sectional area $A = 250$ mm² $= 250 \times 10^{-6}$ m², and extension $x = 0.143$ mm $= 0.143 \times 10^{-3}$ m.

Young's modulus of elasticity,

$$E = \frac{\text{stress}}{\text{strain}} = \frac{F/A}{x/L} = \frac{FL}{Ax}$$

$$= \frac{(75000)(0.1)}{(250 \times 10^{-6})(0.143 \times 10^{-3})}$$

$$= 210 \times 10^9 \, \text{Pa} = \textbf{210 GPa}$$

(b) Yield stress $= \dfrac{\text{load when yield begins to take place}}{\text{original cross-sectional area}}$

$$= \frac{80000 \, \text{N}}{250 \times 10^{-6} \, \text{m}^2}$$

$$= 320 \times 10^6 \, \text{Pa} = \textbf{320 MPa}$$

(c) Tensile strength $= \dfrac{\text{maximum load}}{\text{original cross-sectional area}}$

$$= \frac{120000 \, \text{N}}{250 \times 10^{-6} \, \text{m}^2}$$

$$= 480 \times 10^6 \, \text{Pa} = \textbf{480 MPa}$$

(d) Percentage elongation

$$= \frac{\text{increase in length during test to destruction}}{\text{original length}}$$

$$= \left(\frac{135 - 100}{100}\right) \times 100 = \textbf{35\%}$$

(e) Percentage reduction in area

$$= \frac{\begin{array}{c}(\text{original cross-sectional area})\\ -(\text{final cross-sectional area})\end{array}}{\text{original cross-sectional area}} \times 100$$

$$= \left(\frac{250 - 90}{250}\right) \times 100 = \left(\frac{160}{250}\right) \times 100 = \textbf{64\%}$$

Now try the following Practise Exercise

Practise Exercise 19 Further problem on tensile testing

1. A tensile test is carried out on a specimen of mild steel of gauge length 40 mm and diameter 7.42 mm. The results are:

Load (kN)								
0	10	17	25	30	34	37.5	38.5	36

Extension (mm)								
0	0.05	0.08	0.11	0.14	0.20	0.40	0.60	0.90

At fracture the final length of the specimen is 40.90 mm. Plot the load/ extension graph and determine (a) the modulus of elasticity for mild steel (b) the stress at the limit of proportionality (c) the ultimate tensile strength (d) the percentage elongation.

$$\begin{bmatrix}\text{(a) 210 GPa} & \text{(b) 650 MPa}\\ \text{(c) 890 MPa} & \text{(d) 2.25\%}\end{bmatrix}$$

2. An aluminium alloy specimen of gauge length 75 mm and of diameter 11.28 mm was subjected to a tensile test, with these results:

Load (kN)									
0	2.0	6.5	11.5	13.6	16.0	18.0	19.0	20.5	19.0

Extension (mm)									
0	0.012	0.039	0.069	0.080	0.107	0.133	0.158	0.225	0.310

The specimen fractured at a load of 19.0 kN.

Determine (a) the modulus of elasticity of the alloy (b) the percentage elongation.

[(a) 125 GPa (b) 0.413%]

3. An aluminium test piece 10 mm in diameter and gauge length 50 mm gave the following results when tested to destruction:

Load at yield point 4.0 kN, maximum load 6.3 kN, extension at yield point 0.036 mm, diameter at fracture 7.7 mm.

Determine (a) the yield stress (b) Young's modulus of elasticity (c) the ultimate tensile strength (d) the percentage reduction in area.

$$\begin{bmatrix}\text{(a) 50.93 MPa} & \text{(b) 70.7 GPa}\\ \text{(c) 80.2 MPa} & \text{(d) 40.7\%}\end{bmatrix}$$

3.4 Proof stress

Certain materials, such as high-tensile steel, aluminium alloy, titanium, and so on, do not exhibit a definite yield point in their stress-strain curves, unlike mild steel, as shown in Figure 3.6.

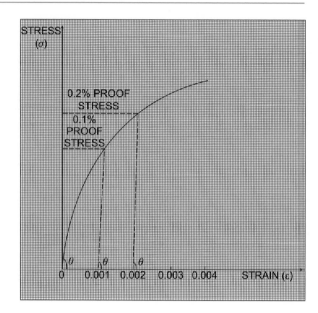

Figure 3.6 Stress-Strain curves for high tension steels, aluminium alloys, etc.

In this case, the equivalent yield stresses are taken in the form of either 0.1% or a 0.2% proof stress.

For 0.1% **proof stress**, a strain of $\varepsilon = 0.1\% = \frac{0.1}{100} = 0.001$, is set off on the horizontal strain axis of Figure 3.6 and a straight line is drawn parallel to the straight part of the stress-strain curve from this point, near to the bottom of the stress-strain curve where the angle is θ. The intersection of this broken line with the stress-strain curve gives the 0.1% proof stress, as shown in Figure 3.6.

For the 0.2% proof stress, a strain of $\varepsilon = 0.2\% = \frac{0.2}{100} = 0.002$, is set off on the horizontal strain axis of Figure 3.6 and a straight line is drawn from this point at an angle θ to the horizontal and parallel to the linear part of the bottom of the stress-strain curve as shown in Figure 3.6. The 0.2% proof stress is **not** twice the value of the 0.1% proof stress! It may be about 10% larger.

Now try the following Practise Exercises

Practise Exercise 20 Short-answer questions on tensile testing

1. What is a tensile test?

2. Which British Standard gives the standard procedure for a tensile test?

3. With reference to a load/extension graph for mild steel, state the meaning of (a) the limit of proportionality (b) the elastic limit (c) the yield point (d) the percentage elongation.

4. Define ultimate tensile strength.

5. Yield stress is the ratio $\dfrac{\ldots\ldots}{\ldots\ldots}$

6. Define 'percentage reduction in area'.

7. Briefly explain, with a diagram, what is meant by '0.1% proof stress'.

Practise Exercise 21   Multiple-choice questions on tensile testing

(Answers on page 297)

A brass specimen having a cross-sectional area of 100 mm^2 and gauge length 100 mm is subjected to a tensile test from which the following information is obtained:

Load at yield point = 45 kN, maximum load = 52.5 kN, final cross-sectional area of waist at fracture = 75 mm^2, and gauge length at fracture = 110 mm.

For questions 1 to 4, select the correct answer from the following list:
(a) 600 MPa   (b) 525 MPa
(c) 33.33%   (d) 10%
(e) 9.09%   (f) 450 MPa
(g) 25%   (h) 700 MPa

1. The yield stress.

2. The percentage elongation.

3. The percentage reduction in area.

4. The ultimate tensile strength.

References

Videos of the Tensile Test and Poisson's Ratio experiment can be obtained by visiting the website below:
www.routledge.com/cw/bird

Chapter 4

Forces acting at a point

In this chapter the fundamental quantities, scalars and vectors, which form the very basis of all branches of engineering, are introduced. A force is a vector quantity and in this chapter, the resolution of forces is introduced. Resolving forces is very important in structures, where the principle is used to determine the strength of roof trusses, bridges, cranes, etc. Great lengths are gone to, to explain this very fundamental skill, where step-by-step methods are adopted to help the reader understand this very important procedure. The resolution of forces is also used in studying the motion of vehicles and other particles in dynamics and in the case of the navigation of ships, aircraft, etc., the vectors take the form of displacements, velocities and accelerations. This chapter gives a sound introduction to the manipulation and use of scalars and vectors, by both graphical and analytical methods.

At the end of this chapter you should be able to:

- distinguish between scalar and vector quantities
- define 'centre of gravity' of an object
- define 'equilibrium' of an object
- understand the terms 'coplanar' and 'concurrent'
- determine the resultant of two coplanar forces using (a) the triangle of forces method (b) the parallelogram of forces method
- calculate the resultant of two coplanar forces using (a) the cosine and sine rules (b) resolution of forces
- determine the resultant of more than two coplanar forces using (a) the polygon of forces method (b) calculation by resolution of forces
- determine unknown forces when three or more coplanar forces are in equilibrium

4.1 Scalar and vector quantities

Quantities used in engineering and science can be divided into two groups:

(a) **Scalar quantities** have a size (or magnitude) only and need no other information to specify them. Thus, 10 centimetres, 50 seconds, 7 litres and 3 kilograms are all examples of scalar quantities.

(b) **Vector quantities** have both a size or magnitude and a direction, called the line of action of the quantity. Thus, a velocity of 50 kilometres per hour due east, an acceleration of 9.81 metres per second squared vertically downwards and a force of 15 Newtons at an angle of 30 degrees are all examples of vector quantities.

4.2 Centre of gravity and equilibrium

The **centre of gravity** of an object is a point where the resultant gravitational force acting on the body may be taken to act. For objects of uniform thickness lying in a horizontal plane, the centre of gravity is vertically in line with the point of balance of the object. For a thin uniform rod the point of balance and hence the centre of gravity is halfway along the rod as shown in Figure 4.1(a).

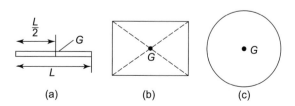

(a)   (b)   (c)

Figure 4.1

Mechanical Engineering Principles, Bird and Ross, ISBN 9780415517850

A thin flat sheet of a material of uniform thickness is called a **lamina** and the centre of gravity of a rectangular lamina lies at the point of intersection of its diagonals, as shown in Figure 4.1(b). The centre of gravity of a circular lamina is at the centre of the circle, as shown in Figure 4.1(c).

An object is in **equilibrium** when the forces acting on the object are such that there is no tendency for the object to move. The state of equilibrium of an object can be divided into three groups.

(i) If an object is in **stable equilibrium** and it is slightly disturbed by pushing or pulling (i.e. a disturbing force is applied), the centre of gravity is raised and when the disturbing force is removed, the object returns to its original position. Thus a ball bearing in a hemispherical cup is in stable equilibrium, as shown in Figure 4.2(a).

(a) Stable
equilibrium

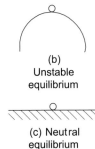

(b)
Unstable
equilibrium

(c) Neutral
equilibrium

Figure 4.2

(ii) An object is in **unstable equilibrium** if, when a disturbing force is applied, the centre of gravity is lowered and the object moves away from its original position. Thus, a ball bearing balanced on top of a hemispherical cup is in unstable equilibrium, as shown in Figure 4.2(b).

(iii) When an object in **neutral equilibrium** has a disturbing force applied, the centre of gravity remains at the same height and the object does not move when the disturbing force is removed.

Thus, a ball bearing on a flat horizontal surface is in neutral equilibrium, as shown in Figure 4.2(c).

4.3 Forces

When forces are all acting in the same plane, they are called **coplanar**. When forces act at the same time and at the same point, they are called **concurrent forces.**

Force is a **vector quantity** and thus has both a magnitude and a direction. A vector can be represented graphically by a line drawn to scale in the direction of the line of action of the force.

To distinguish between vector and scalar quantities, various ways are used.

These include:

(i) **bold print**,

(ii) two capital letters with an arrow above them to denote the sense of direction, for example, $\overrightarrow{AB}$ where A is the starting point and B the end point of the vector,

(iii) a line over the top of letters, for example, $\overline{AB}$ or $\bar{a}$

(iv) letters with an arrow above, for example, $\vec{a}$, $\vec{A}$

(v) underlined letters, for example, $\underline{a}$

(vi) $xi + jy$, where i and j are axes at right-angles to each other; for example, $3i + 4j$ means 3 units in the i direction and 4 units in the j direction, as shown in Figure 4.3.

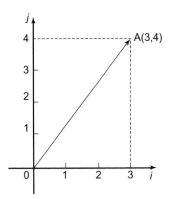

Figure 4.3

(vii) a column matrix $\begin{pmatrix} a \\ b \end{pmatrix}$; for example, the vector **OA** shown in Figure 4.3 could be represented by $\begin{pmatrix} 3 \\ 4 \end{pmatrix}$

Thus, in Figure 4.3,

$$\textbf{OA} \equiv \overrightarrow{OA} \equiv \overline{OA} \equiv 3i + 4j \equiv \begin{pmatrix} 3 \\ 4 \end{pmatrix}$$

The method adopted in this text is to denote vector quantities in **bold print**.

Part Two

Thus, **ab** in Figure 4.4 represents a force of 5 Newton's acting in a direction due east.

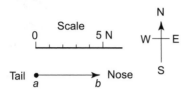

Figure 4.4

4.4　The resultant of two coplanar forces

For two forces acting at a point, there are three possibilities.

(a)　For forces acting in the same direction and having the same line of action, the single force having the same effect as both of the forces, called the **resultant force** or just the **resultant**, is the arithmetic sum of the separate forces. Forces of F_1 and F_2 acting at point P, as shown in Figure 4.5(a), have exactly the same effect on point P as force F shown in Figure 4.5(b), where $F = F_1 + F_2$ and acts in the same direction as F_1 and F_2. Thus F is the resultant of F_1 and F_2

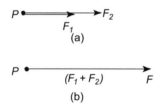

Figure 4.5

(b)　For forces acting in opposite directions along the same line of action, the resultant force is the arithmetic difference between the two forces. Forces of F_1 and F_2 acting at point P as shown in Figure 4.6(a) have exactly the same effect on point P as

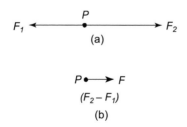

Figure 4.6

force F shown in Figure 4.6(b), where $F = F_2 - F_1$ and acts in the direction of F_2, since F_2 is greater than F_1. Thus F is the resultant of F_1 and F_2

(c)　When two forces do not have the same line of action, the magnitude and direction of the resultant force may be found by a procedure called vector addition of forces. There are two graphical methods of performing **vector addition**, known as the **triangle of forces** method (see Section 4.5) and the **parallelogram of forces** method (see Section 4.6).

Problem 1.　Determine the resultant force of two forces of 5 kN and 8 kN, (a) acting in the same direction and having the same line of action (b) acting in opposite directions but having the same line of action.

(a)　The vector diagram of the two forces acting in the same direction is shown in Figure 4.7(a), which assumes that the line of action is horizontal, although since it is not specified, could be in any direction. From above, the resultant force F is given by: $F = F_1 + F_2$, i.e. $\boldsymbol{F} = (5 + 8)$ kN = **13 kN** in the direction of the original forces.

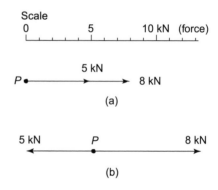

Figure 4.7

(b)　The vector diagram of the two forces acting in opposite directions is shown in Figure 4.7(b), again assuming that the line of action is in a horizontal direction. From above, the resultant force F is given by: $F = F_2 - F_1$, i.e. $\boldsymbol{F} = (8 - 5)$ kN = **3 kN** in the direction of the 8 kN force.

4.5　Triangle of forces method

A simple procedure for the triangle of forces method of vector addition is as follows:

(i) Draw a vector representing one of the forces, using an appropriate scale and in the direction of its line of action.

(ii) From the **nose** of this vector and using the same scale, draw a vector representing the second force in the direction of its line of action.

(iii) The resultant vector is represented in both magnitude and direction by the vector drawn from the tail of the first vector to the nose of the second vector.

> **Problem 2.** Determine the magnitude and direction of the resultant of a force of 15 N acting horizontally to the right and a force of 20 N, inclined at an angle of 60° to the 15 N force. Use the triangle of forces method.

Using the procedure given above and with reference to Figure 4.8:

(i) **ab** is drawn 15 units long horizontally

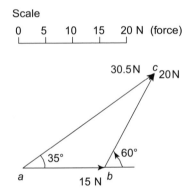

Figure 4.8

(ii) From *b*, **bc** is drawn 20 units long, inclined at an angle of 60° to *ab*. (Note, in angular measure, an angle of 60° from *ab* means 60° in an anticlockwise direction.)

(iii) By measurement, the resultant *ac* is 30.5 units long inclined at an angle of 35° to *ab*. That is, the resultant force is **30.5 N**, inclined at an angle of **35°** to the 15 N force.

> **Problem 3.** Find the magnitude and direction of the two forces given, using the triangle of forces method.
> First force: 1.5 kN acting at an angle of 30°
> Second force: 3.7 kN acting at an angle of –45°

From the above procedure and with reference to Figure 4.9:

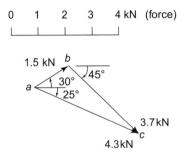

Figure 4.9

(i) **ab** is drawn at an angle of 30° and 1.5 units in length.

(ii) From *b*, **bc** is drawn at an angle of –45° and 3.7 units in length. (Note, an angle of –45° means a clockwise rotation of 45° from a line drawn horizontally to the right)

(iii) By measurement, the resultant *ac* is 4.3 units long at an angle of –25°. That is, the resultant force is **4.3 kN** at an angle of **–25°**

Now try the following Practise Exercise

> **Practise Exercise 22   Further problems on the triangle of forces method**
>
> 1. Determine the magnitude and direction of the resultant of the forces 1.3 kN and 2.7 kN, having the same line of action and acting in the same direction.
>   [4.0 kN in the direction of the forces]
>
> 2. Determine the magnitude and direction of the resultant of the forces 470 N and 538 N having the same line of action but acting in opposite directions.
>   [68 N in the direction of the 538 N force]
>
> In questions 3 to 5, use the triangle of forces method to determine the magnitude and direction of the resultant of the forces given.
>
> 3. 13 N at 0° and 25 N at 30°   [36.8 N at 20°]
>
> 4. 5 N at 60° and 8 N at 90°   [12.6 N at 79°]
>
> 5. 1.3 kN at 45° and 2.8 kN at –30°
>   [3.4 kN at –8°]

4.6 The parallelogram of forces method

A simple procedure for the parallelogram of forces method of vector addition is as follows:

(i) Draw a vector representing one of the forces, using an appropriate scale and in the direction of its line of action.

(ii) From the **tail** of this vector and using the same scale draw a vector representing the second force in the direction of its line of action.

(iii) Complete the parallelogram using the two vectors drawn in (i) and (ii) as two sides of the parallelogram.

(iv) The resultant force is represented in both magnitude and direction by the vector corresponding to the diagonal of the parallelogram drawn from the tail of the vectors in (i) and (ii).

> **Problem 4.** Use the parallelogram of forces method to find the magnitude and direction of the resultant of a force of 250 N acting at an angle of 135° and a force of 400 N acting at an angle of –120°.

From the procedure given above and with reference to Figure 4.10:

(i) **ab** is drawn at an angle of 135° and 250 units in length

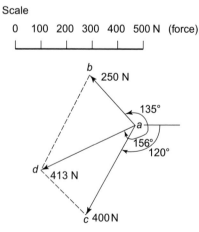

Figure 4.10

(ii) **ac** is drawn at an angle of –120° and 400 units in length

(iii) **bd** and **cd** are drawn to complete the parallelogram

(iv) **ad** is drawn. By measurement, **ad** is 413 units long at an angle of –156°.

That is, the resultant force is **413 N** at an angle of **–156°**

Now try the following Practise Exercise

> **Practise Exercise 23   Further problems on the parallelogram of forces method**
>
> In questions 1 to 5, use the parallelogram of forces method to determine the magnitude and direction of the resultant of the forces given.
>
> 1. 1.7 N at 45° and 2.4 N at –60°
>
> [2.6 N at –20°]
>
> 2. 9 N at 126° and 14 N at 223°
>
> [15.7 N at –172°]
>
> 3. 23.8 N at –50° and 14.4 N at 215°
>
> [26.7 N at –82°]
>
> 4. 0.7 kN at 147° and 1.3 kN at –71°
>
> [0.86 kN at –101°]
>
> 5. 47 N at 79° and 58 N at 247°
>
> [15.5 N at –152°]

4.7 Resultant of coplanar forces by calculation

An alternative to the graphical methods of determining the resultant of two coplanar forces is by **calculation**. This can be achieved by **trigonometry** using the **cosine rule** and the **sine rule**, as shown in Problem 5 following, or by **resolution of forces** (see Section 4.10).

> **Problem 5.** Use the cosine and sine rules to determine the magnitude and direction of the of a force of 8 kN acting at an angle of 50° to the horizontal and a force of 5 kN acting at an angle of –30° to the horizontal.

The space diagram is shown in Figure 4.11(a). A sketch is made of the vector diagram, **0a** representing the 8 kN force in magnitude and direction and **ab** representing the 5 kN force in magnitude and direction. The resultant is given by length **0b**. By the cosine rule,

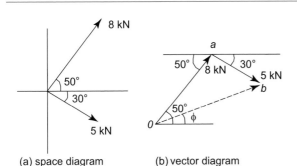

(a) space diagram   (b) vector diagram

Figure 4.11

$$0b^2 = oa^2 + ab^2 - 2(oa)(ab)\cos\angle 0ab$$
$$= 8^2 + 5^2 - 2(8)(5)\cos 100°$$
$$(\text{since } \angle 0ab = 180° - 50° - 30° = 100°)$$
$$= 64 + 25 - (-13.892) = 102.892$$

Hence, $0b = \sqrt{102.892} = 10.14$ kN

By the sine rule, $\dfrac{5}{\sin\angle a0b} = \dfrac{10.14}{\sin 100°}$

from which, $\sin\angle a0b = \dfrac{5\sin 100°}{10.14} = 0.4856$

Hence $\angle a0b = \sin^{-1}(0.4856) = 29.05°$. Thus angle ϕ in Figure 4.11(b) is $50° - 29.05° = 20.95°$

Hence the resultant of the two forces is 10.14 kN acting at an angle of 20.95° to the horizontal

---

Now try the following Practise Exercise

Practise Exercise 24   Further problems on the resultant of coplanar forces by calculation

1. Forces of 7.6 kN at 32° and 11.8 kN at 143° act at a point. Use the cosine and sine rules to calculate the magnitude and direction of their resultant.   [11.52 kN at 105°]

In questions 2 to 5, calculate the resultant of the given forces by using the cosine and sine rules.

2. 13 N at 0° and 25 N at 30°
                                   [36.84 N at 19.83°]

3. 1.3 kN at 45° and 2.8 kN at –30°
                                   [3.38 kN at – 8.19°]

4. 9 N at 126° and 14 N at 223°
                                   [15.69 N at – 171.67°]

5. 0.7 kN at 147° and 1.3 kN at – 71°
                                   [0.86 kN at – 100.94°]

4.8 Resultant of more than two coplanar forces

For the three coplanar forces F_1, F_2 and F_3 acting at a point as shown in Figure 4.12, the vector diagram is drawn using the nose-to-tail method of Section 4.5. The procedure is:

(i)   Draw **0a** to scale to represent force F_1 in both magnitude and direction (see Figure 4.13)

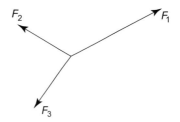

Figure 4.12

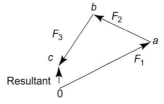

Figure 4.13

(ii)   From the nose of **0a**, draw **ab** to represent force F_2
(iii) From the nose of **ab**, draw **bc** to represent force F_3
(iv)   The resultant vector is given by length **0c** in Figure 4.13. The direction of resultant **oc** is from where we started, i.e. point 0, to where we finished, i.e. point c. When acting by itself, the resultant force, given by **0c**, has the same effect on the point as forces F_1, F_2 and F_3 have when acting together. The resulting vector diagram of Figure 4.13 is called the **polygon of forces**.

Problem 6. Determine graphically the magnitude and direction of the resultant of the following three coplanar forces, which may be considered as acting at a point:

Force A, 12 N acting horizontally to the right; force B, 7 N inclined at 60° to force A; force C, 15 N inclined at 150° to force A.

The space diagram is shown in Figure 4.14. The vector diagram shown in Figure 4.15, is produced as follows:

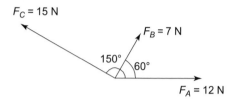

Figure 4.14

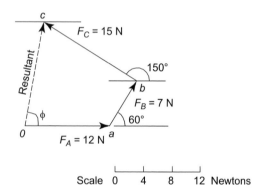

Figure 4.15

(i) **0a** represents the 12 N force in magnitude and direction

(ii) From the nose of **0a**, **ab** is drawn inclined at 60° to **0a** and 7 units long

(iii) From the nose of **ab**, **bc** is drawn 15 units long inclined at 150° to **oa** (i.e. 150° to the horizontal)

(iv) **0c** represents the resultant; by measurement, the resultant is 13.8 N inclined at $\phi = 80°$ to the horizontal.

Thus the resultant of the three forces, F_A, F_B and F_C is a force of 13.8 N at 80° to the horizontal.

Problem 7. The following coplanar forces are acting at a point, the given angles being measured from the horizontal: 100 N at 30°, 200 N at 80°, 40 N at –150°, 120 N at –100° and 70 N at –60°.

Determine graphically the magnitude and direction of the resultant of the five forces.

The five forces are shown in the space diagram of Figure 4.16. Since the 200 N and 120 N forces have the same line of action but are in opposite sense, they can be represented by a single force of 200 – 120, i.e. 80 N acting at 80° to the horizontal. Similarly, the 100 N and 40 N forces can be represented by a force of 100 – 40, i.e. 60 N acting at 30° to the horizontal. Hence the space diagram of Figure 4.16 may be represented by the space diagram of Figure 4.17. Such a simplification of the vectors is not essential but it is easier to construct the vector diagram from a space diagram having three forces, than one with five.

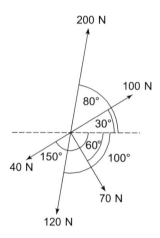

Figure 4.16

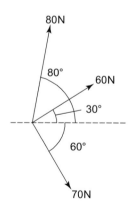

Figure 4.17

The vector diagram is shown in Figure 4.18, **0a** representing the 60 N force, **ab** representing the 80N force and **bc** the 70 N force. The resultant, **0c**, is found by measurement to represent a force of 112 N and angle ϕ is 25°.

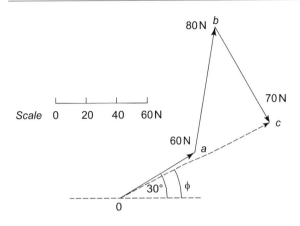

Figure 4.18

Thus, the five forces shown in Figure 4.16 may be represented by a single force of 112 N at 25° to the horizontal.

Now try the following Practise Exercise

> **Practise Exercise 25   Further problems on the resultant of more than two coplanar forces**
>
> In questions 1 to 3, determine graphically the magnitude and direction of the resultant of the coplanar forces given which are acting at a point.
>
> 1.   Force A, 12 N acting horizontally to the right, force B, 20 N acting at 140° to force A, force C, 16 N acting 290° to force A.
>   [3.1 N at −45° to force A]
>
> 2.   Force 1, 23 kN acting at 80° to the horizontal, force 2, 30 kN acting at 37° to force 1, force 3, 15 kN acting at 70° to force 2.
> $\begin{bmatrix} 53.5 \text{ kN at } 37° \text{ to force } 1 \\ (\text{i.e. } 117° \text{ to the horizontal}) \end{bmatrix}$
>
> 3.   Force P, 50 kN acting horizontally to the right, force Q, 20 kN at 70° to force P, force R, 40 kN at 170° to force P, force S, 80 kN at 300° to force P.   [72 kN at −37° to force P]
>
> 4.   Four horizontal wires are attached to a telephone pole and exert tensions of 30 N to the south, 20 N to the east, 50 N to the north-east and 40 N to the north-west. Determine the resultant force on the pole and its direction.
>   [43.2 N at 39° east of north]

4.9   Coplanar forces in equilibrium

When three or more coplanar forces are acting at a point and the vector diagram closes, there is no resultant. The forces acting at the point are in **equilibrium**.

> **Problem 8.**   A load of 200 N is lifted by two ropes connected to the same point on the load, making angles of 40° and 35° with the vertical. Determine graphically the tensions in each rope when the system is in equilibrium.

The space diagram is shown in Figure 4.19. Since the system is in equilibrium, the vector diagram must close. The vector diagram, shown in Figure 4.20, is drawn as follows:

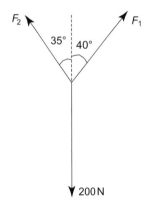

Figure 4.19

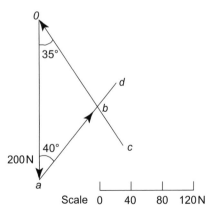

Figure 4.20

(i)   The load of 200 N is drawn vertically as shown by **0a**

(ii)   The direction only of force F_1 is known, so from point a, **ad** is drawn at 40° to the vertical

(iii) The direction only of force F_2 is known, so from point 0, **0c** is drawn at 35° to the vertical

(iv) Lines **ad** and **0c** cross at point *b*; hence the vector diagram is given by triangle *oab*. By measurement, **ab** is 119 N and **0b** is 133 N.

Thus the tensions in the ropes are F_1 = 119 N and F_2 = 133 N

> **Problem 9.** Five coplanar forces are acting on a body and the body is in equilibrium. The forces are: 12 kN acting horizontally to the right, 18 kN acting at an angle of 75°, 7 kN acting at an angle of 165°, 16 kN acting from the nose of the 7 kN force, and 15 kN acting from the nose of the 16 kN force. Determine the directions of the 16 kN and 15 kN forces relative to the 12 kN force.

With reference to Figure 4.21, **0a** is drawn 12 units long horizontally to the right. From point *a*, **ab** is drawn 18 units long at an angle of 75°. From *b*, **bc** is drawn 7 units long at an angle of 165°. The direction of the 16 kN force is not known, thus arc *pq* is drawn with a compass, with centre at *c*, radius 16 units. Since the forces are at equilibrium, the polygon of forces must close. Using a compass with centre at 0, *arc rs* is drawn having a radius 15 units. The point where the arcs intersect is at *d*.

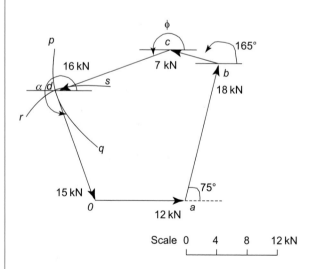

Scale 0   4   8   12 kN

Figure 4.21

By measurement, angle ϕ = 198° and α = 291°

Thus the 16 kN force acts at an angle of 198° (or –162°) to the 12 kN force, and the 15 kN force acts at an angle of 291° (or –69°) to the 12 kN force.

Now try the following Practise Exercise

> **Practise Exercise 26   Further problems on coplanar forces in equilibrium**
>
> 1. A load of 12.5 N is lifted by two strings connected to the same point on the load, making angles of 22° and 31° on opposite sides of the vertical. Determine the tensions in the strings.   [5.86 N, 8.06 N]
>
> 2. A two-legged sling and hoist chain used for lifting machine parts is shown in Figure 4.22. Determine the forces in each leg of the sling if parts exerting a downward force of 15 kN are lifted.   [9.96 kN, 7.77 kN]

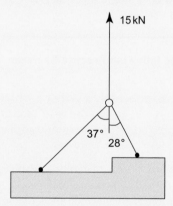

Figure 4.22

> 3. Four coplanar forces acting on a body are such that it is in equilibrium. The vector diagram for the forces is such that the 40 N force acts vertically upwards, the 40 N force acts at 65° to the 60 N force, the 100 N force acts from the nose of the 40 N force and the 90 N force acts from the nose of the 100 N force. Determine the direction of the 100 N and 90 N forces relative to the 60 N force.
>
> $$\begin{bmatrix} \text{100 N force at 148° to the 60 N force,} \\ \text{90 N force at 277° to the 60 N force} \end{bmatrix}$$

4.10   Resolution of forces

A vector quantity may be expressed in terms of its **horizontal** and **vertical components**. For example, a vector representing a force of 10 N at an angle of

60° to the horizontal is shown in Figure 4.23. If the horizontal line **0a** and the vertical line **ab** are constructed as shown, then **0a** is called the horizontal component of the 10 N force, and **ab** the vertical component of the 10 N force.

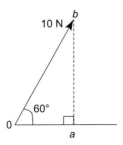

Figure 4.23

By trigonometry, $\cos 60° = \dfrac{0a}{0b}$, hence the horizontal component, $0a = 10 \cos 60°$

Also, $\sin 60° = \dfrac{ab}{0b}$, hence the vertical component, $ab = 10 \sin 60°$

This process is called **finding the horizontal and vertical components of a vector** or **the resolution of a vector**, and can be used as an alternative to graphical methods for calculating the resultant of two or more coplanar forces acting at a point.

For example, to calculate the resultant of a 10 N force acting at 60° to the horizontal and a 20 N force acting at −30° to the horizontal (see Figure 4.24) the procedure is as follows:

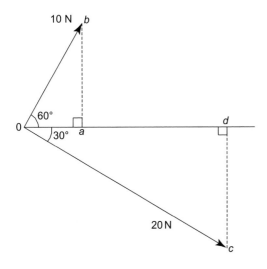

Figure 4.24

(i)   Determine the horizontal and vertical components of the 10 N force, i.e. horizontal component,

$0a = 10 \cos 60° = 5.0$ N, and vertical component, $ab = 10 \sin 60° = 8.66$ N

(ii)   Determine the horizontal and vertical components of the 20 N force, i.e. horizontal component, $0d = 20 \cos(-30°) = 17.32$ N, and vertical component, $cd = 20 \sin(-30°) = -10.0$ N

(iii)   Determine the total horizontal component, i.e.

$$0a + 0d = 5.0 + 17.32 = 22.32 \text{ N}$$

(iv)   Determine the total vertical component, i.e.

$$ab + cd = 8.66 + (-10.0) = -1.34 \text{ N}$$

(v)   Sketch the total horizontal and vertical components as shown in Figure 4.25. The resultant of the two components is given by length **0r** and, by Pythagoras' theorem,

$$0r = \sqrt{22.32^2 + 1.34^2} = 22.36 \text{ N}$$

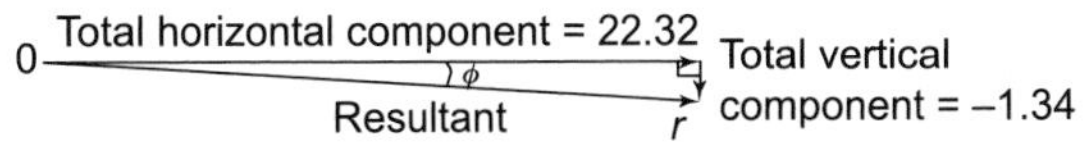

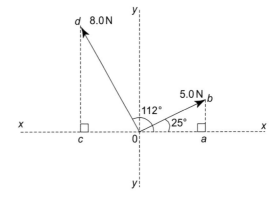

Figure 4.25

and using trigonometry,

$$\text{angle } \phi = \tan^{-1}\frac{1.34}{22.32} = 3.44°$$

Hence the resultant of the 10 N and 20 N forces shown in Figure 4.24 is **22.36 N at an angle of −3.44° to the horizontal**.

Problem 10. Forces of 5.0 N at 25° and 8.0 N at 112° act at a point. By resolving these forces into horizontal and vertical components, determine their resultant.

The space diagram is shown in Figure 4.26.

Figure 4.26

(i)   The horizontal component of the 5.0 N force, $0a = 5.0 \cos 25° = 4.532$, and the vertical component of the 5.0 N force, $ab = 5.0 \sin 25° = 2.113$

(ii)   The horizontal component of the 8.0 N force, $0c = 8.0 \cos 112° = -2.997$
The vertical component of the 8.0 N force, $cd = 8.0 \sin 112° = 7.417$

(iii)   Total horizontal component
$= 0a + 0c = 4.532 + (-2.997) = +1.535$

(iv)   Total vertical component
$= ab + cd = 2.113 + 7.417 = +9.530$

(v)   The components are shown sketched in Figure 4.27.

Figure 4.27

By Pythagoras' theorem, $r = \sqrt{1.535^2 + 9.530^2}$
$= 9.653,$

and by trigonometry, angle $\phi = \tan^{-1}\dfrac{9.530}{1.535}$
$= 80.85°$

Hence the resultant of the two forces shown in Figure 4.26 is a force of 9.653 N acting at 80.85° to the horizontal.

Problems 9 and 10 demonstrate the use of resolution of forces for calculating the resultant of two co-planar forces acting at a point. However the method may be used for more than two forces acting at a point, as shown in Problem 11.

Problem 11. Determine by resolution of forces the resultant of the following three coplanar forces acting at a point: 200 N acting at 20° to the horizontal; 400 N acting at 165° to the horizontal; 500 N acting at 250° to the horizontal.

A tabular approach using a calculator may be made as shown below:

Horizontal component

Force 1	$200 \cos 20° = 187.94$
Force 2	$400 \cos 165° = -386.37$
Force 3	$500 \cos 250° = -171.01$
Total horizontal component	$= -369.44$

Vertical component

Force 1	$200 \sin 20° = 68.40$
Force 2	$400 \sin 165° = 103.53$
Force 3	$500 \sin 250° = -469.85$
Total vertical component	$= -297.92$

The total horizontal and vertical components are shown in Figure 4.28.

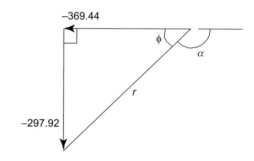

Figure 4.28

Resultant $r = \sqrt{369.44^2 + 297.92^2}$
$= 474.60,$ and

angle $\phi = \tan^{-1}\dfrac{297.92}{369.44} = 38.88°,$

from which, $\alpha = 180° - 38.88° = 141.12°$

Thus the resultant of the three forces given is 474.6 N acting at an angle of -141.12° (or +218.88°) to the horizontal.

Problem 12. Determine by resolution the magnitude and direction of the following six concurrent coplanar forces: 150 N acting at 25° to the horizontal, 250 N acting at 90° to the horizontal, 200 N acting at 120° to the horizontal, 300 N acting at 200° to the horizontal, 60 N acting at 280° to the horizontal, and 220 N acting at 320° to the horizontal.

A tabular approach using a calculator may be made as shown below:

Horizontal components of the forces

$$150 \cos 25° = \quad 135.9$$
$$250 \cos 90° = \quad\quad 0$$
$$200 \cos 120° = -100$$
$$300 \cos 200° = -281.9$$
$$60 \cos 280° = \quad 10.42$$
$$220 \cos 320° = \quad 168.5$$

Total horizontal component, $H = \underline{\ -67.1 \text{ N}\ }$

Vertical components of the forces

$$150 \sin 25° = \quad 63.4$$
$$250 \sin 90° = \quad 250$$
$$200 \sin 120° = \quad 173.2$$
$$300 \sin 200° = -102.6$$
$$60 \sin 280° = \quad -59.1$$
$$220 \sin 320° = -141.4$$

Total vertical component, $V = \underline{\ 183.5 \text{ N}\ }$

H and V are shown plotted in Figure 4.29 and from Pythagoras' theorem:

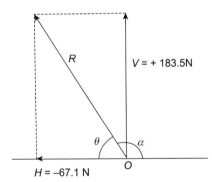

Figure 4.29

$R^2 = H^2 + V^2$ and **resultant,**

$$\boldsymbol{R} = \sqrt{H^2 + V^2} = \sqrt{(-67.1)^2 + (183.5)^2} = \textbf{195.4 N}$$

Angle $\theta = \tan^{-1}\left(\dfrac{V}{H}\right) = \tan^{-1}\left(\dfrac{183.5}{67.1}\right)$

$$= \tan^{-1}(2.7347) = 69.91°$$

Measuring from the positive horizontal axis, **angle of the resultant R,** $\alpha = 180° - 69.91° = \textbf{110.09°}$

Thus the resultant of the six forces given is 195.4 N acting at an angle of 110.09° to the horizontal.

Now try the following Practise Exercise

Practise Exercise 27   Further problems on resolution of forces

1. Resolve a force of 23.0 N at an angle of 64° into its horizontal and vertical components.
    [10.08 N, 20.67 N]

2. Forces of 5 N at 21° and 9 N at 126° act at a point. By resolving these forces into horizontal and vertical components, determine their resultant.   [9.09 N at 93.92°]

In questions 3 and 4, determine the magnitude and direction of the resultant of the coplanar forces given which are acting at a point, by resolution of forces.

3. Force A, 12 N acting horizontally to the right, force B, 20 N acting at 140° to force A, force C, 16 N acting 290° to force A.
    [3.06 N at −45.40° to force A]

4. Force 1, 23 kN acting at 80° to the horizontal, force 2, 30 kN acting at 37° to force 1, force 3, 15 kN acting at 70° to force 2.
    $\left[\begin{array}{l}53.50 \text{ kN at } 117.27° \\ \text{to the horizontal}\end{array}\right]$

5. Determine, by resolution of forces, the resultant of the following three coplanar forces acting at a point: 10 kN acting at 32° to the horizontal, 15 kN acting at 170° to the horizontal; 20 kN acting at 240° to the horizontal.
    [18.82 kN at 210.03° to the horizontal]

6. The following coplanar forces act at a point: force A, 15 N acting horizontally to the right, force B, 23 N at 81° to the horizontal, force C, 7 N at 210° to the horizontal, force D, 9 N at 265° to the horizontal, and force E, 28 N at 324° to the horizontal. Determine the resultant of the five forces by resolution of the forces.
    [34.96 N at −10.23° to the horizontal]

7. At a certain point, 12 different values of coplanar and concurrent, radially outward tensile forces are applied. The first force is applied horizontally to the right and the

remaining 11 forces are applied at equally spaced intervals of 30° anti-clockwise. Starting from the first force and then 30° anti-clockwise to the second force, and so on, so that the 12 forces encompass a complete circle of 360°. The magnitude of the 12 forces, in order of sequence, are: 30 kN, 250 kN, 200 kN, 180 kN, 160 kN, 140 kN, 120 kN, 100 kN, 80 kN, 60 kN, 40 kN and 20 kN. Determine (a) the sum of the horizontal components of the forces, H, (b) the sum of the vertical components of the forces, V, (c) the magnitude, R, and the direction of the resultant force, θ

$$\begin{bmatrix} H = -64.02 \text{ kN} \quad \text{(b)} \ V = 462.85 \text{ kN} \\ \text{(c)} \ R = 467.26 \text{ kN at } 97.88° \\ \text{anticlockwise from the horizontal} \end{bmatrix}$$

4.11   Summary

(a)   To determine the **resultant of two coplanar forces** acting at a point, four methods are commonly used. They are:

by drawing:
(1)   triangle of forces method, and
(2)   parallelogram of forces method, and

by calculation:
(3)   use of cosine and sine rules, and
(4)   resolution of forces.

(b)   To determine the **resultant of more than two co-planar forces** acting at a point, two methods are commonly used. They are:

by drawing:
(1)   polygon of forces method, and

by calculation:
(2)   resolution of forces.

Now try the following Practise Exercises

Practise Exercise 28   Short-answer questions on forces acting at a point

1.   Give one example of a scalar quantity and one example of a vector quantity.

2.   Explain the difference between a scalar and a vector quantity.

3.   What is meant by the centre of gravity of an object?

4.   Where is the centre of gravity of a rectangular lamina?

5.   What is meant by neutral equilibrium?

6.   State the meaning of the term 'coplanar'.

7.   What is a concurrent force?

8.   State what is meant by a triangle of forces.

9.   State what is meant by a parallelogram of forces.

10.   State what is meant by a polygon of forces.

11.   When a vector diagram is drawn representing coplanar forces acting at a point, and there is no resultant, the forces are in ....

12.   Two forces of 6 N and 9 N act horizontally to the right. The resultant is .... N acting .....

13.   A force of 10 N acts at an angle of 50° and another force of 20 N acts at an angle of 230°. The resultant is a force ...... N acting at an angle of .... °.

14.   What is meant by 'resolution of forces'?

15.   A coplanar force system comprises a 20 kN force acting horizontally to the right, 30 kN at 45°, 20 kN at 180° and 25 kN at 225°. The resultant is a force of ...... N acting at an angle of .... ° to the horizontal.

Practise Exercise 29   Multiple-choice questions on forces acting at a point

(Answers on page 297)

1.   A physical quantity which has direction as well as magnitude is known as a:
(a) force   (b) vector
(c) scalar   (d) weight.

2.   Which of the following is not a scalar quantity?
(a) velocity   (b) potential energy
(c) work   (d) kinetic energy.

3. Which of the following is not a vector quantity?

   (a) displacement
   (b) density
   (c) velocity
   (d) acceleration.

4. Which of the following statements is false?

   (a) Scalar quantities have size or magnitude only
   (b) Vector quantities have both magnitude and direction
   (c) Mass, length and time are all scalar quantities
   (d) Distance, velocity and acceleration are all vector quantities.

5. If the centre of gravity of an object which is slightly disturbed is raised and the object returns to its original position when the disturbing force is removed, the object is said to be in

   (a) neutral equilibrium
   (b) stable equilibrium
   (c) static equilibrium
   (d) unstable equilibrium.

6. Which of the following statements is false?

   (a) The centre of gravity of a lamina is at its point of balance.
   (b) The centre of gravity of a circular lamina is at its centre.
   (c) The centre of gravity of a rectangular lamina is at the point of intersection of its two sides.
   (d) The centre of gravity of a thin uniform rod is halfway along the rod.

7. The magnitude of the resultant of the vectors shown in Figure 4.30 is:

   (a) 2 N   (b) 12 N
   (c) 35 N   (d) –2 N

Figure 4.30

8. The magnitude of the resultant of the vectors shown in Figure 4.31 is:

   (a) 7 N   (b) 5 N
   (c) 1 N   (d) 12 N

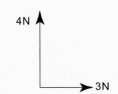

Figure 4.31

9. Which of the following statements is false?

   (a) There is always a resultant vector required to close a vector diagram representing a system of coplanar forces acting at a point, which are not in equilibrium.
   (b) A vector quantity has both magnitude and direction.
   (c) A vector diagram representing a system of coplanar forces acting at a point when in equilibrium does not close.
   (d) Concurrent forces are those which act at the same time at the same point.

10. Which of the following statements is false?

   (a) The resultant of coplanar forces of 1 N, 2 N and 3 N acting at a point can be 4 N.
   (b) The resultant of forces of 6 N and 3 N acting in the same line of action but opposite in sense is 3 N.
   (c) The resultant of forces of 6 N and 3 N acting in the same sense and having the same line of action is 9 N.
   (d) The resultant of coplanar forces of 4 N at 0°, 3 N at 90° and 8 N at 180° is 15 N.

11. A space diagram of a force system is shown in Figure 4.32. Which of the vector diagrams in Figure 4.33 does not represent this force system?

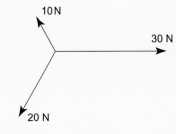

Figure 4.32

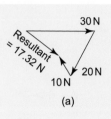

(a)

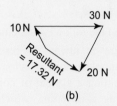

(b)

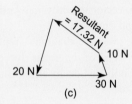

(c)

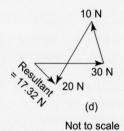

(d)

Not to scale

Figure 4.33

12. With reference to Figure 4.34, which of the following statements is false?

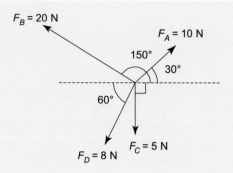

Figure 4.34

(a) The horizontal component of F_A is 8.66 N
(b) The vertical component of F_B is 10 N
(c) The horizontal component of F_C is 0
(d) The vertical component of F_D is 4 N

13. The resultant of two forces of 3 N and 4 N can never be equal to:

(a) 2.5 N   (b) 4.5 N
(c) 6.5 N   (d) 7.5 N

14. The magnitude of the resultant of the vectors shown in Figure 4.35 is:

(a) 5 N   (b) 13 N
(c) 1 N   (d) 63 N

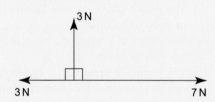

Figure 4.35

Chapter 5

Simply supported beams

This chapter is very important for the design of beams which carry loads, acting transversely to them. These structures are of importance in the design of buildings, bridges, cranes, ships, aircraft, automobiles, and so on. The chapter commences with defining the moment of a force, and then using equilibrium considerations, demonstrates the principle of moments. This is a very important skill widely used in designing structures. Several examples are given, with increasing complexity to help the reader acquire these valuable skills. This chapter describes skills which are fundamental and extremely important in many branches of engineering.

At the end of this chapter you should be able to:

- define a 'moment' of a force and state its unit
- calculate the moment of a force from $M = F \times d$
- understand the conditions for equilibrium of a beam
- state the principle of moments
- perform calculations involving the principle of moments
- recognise typical practical applications of simply supported beams with point loadings
- perform calculations on simply supported beams having point loads
- perform calculations on simply supported beams with couples

5.1 The moment of a force

When using a spanner to tighten a nut, a force tends to turn the nut in a clockwise direction. This turning effect of a force is called the **moment of a force** or more briefly, just a **moment**. The size of the moment acting on the nut depends on two factors:

(a) the size of the force acting at right angles to the shank of the spanner, and
(b) the perpendicular distance between the point of application of the force and the centre of the nut.

In general, with reference to Figure 5.1, the moment M of a force acting about a point P is: force $\times$ perpendicular distance between the line of action of the force and P,

i.e. $$M = F \times d$$

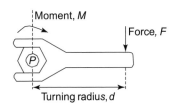

Figure 5.1

The unit of a moment is the **newton metre (N m)**. Thus, if force F in Figure 5.1 is 7 N and distance d is 3 m, then the moment M is 7 N $\times$ 3 m, i.e. 21 N m.

> **Problem 1.** A force of 15 N is applied to a spanner at an effective length of 140 mm from the centre of a nut. Calculate (a) the moment of the force applied to the nut (b) the magnitude of the force required to produce the same moment if the effective length is reduced to 100 mm.

From above, $M = F \times d$, where M is the turning moment, F is the force applied at right angles to the spanner and d is the effective length between the force and the centre of the nut. Thus, with reference to Figure 5.2(a):

Mechanical Engineering Principles, Bird and Ross, ISBN 9780415517850

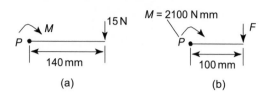

(a)   (b)

Figure 5.2

(a)   Turning moment,
$$M = 15\text{ N} \times 140\text{ mm} = 2100\text{ N mm}$$
$$= 2100\text{ N mm} \times \frac{1\text{ m}}{1000\text{ mm}} = \textbf{2.1 N m}$$

(b)   Turning moment, M is 2100 N mm and the effective length d becomes 100 mm (see Figure 5.2(b)). Applying $M = F \times d$ gives:
2100 N mm = $F \times$ 100 mm from which,

force, $F = \dfrac{2100\text{ N mm}}{100\text{ mm}} = \textbf{21 N}$

Problem 2.   A moment of 25 N m is required to operate a lifting jack. Determine the effective length of the handle of the jack if the force applied to it is: (a) 125 N (b) 0.4 kN

From above, moment $M = F \times d$, where F is the force applied at right angles to the handle and d is the effective length of the handle. Thus:

(a)   25 N m = 125 N $\times d$, from which **effective length,**

$d = \dfrac{25\text{ N m}}{125\text{ N}} = \dfrac{1}{5}\text{ m} = \dfrac{1}{5} \times 1000\text{ mm} = \textbf{200 mm}$

(b)   Turning moment M is 25 N m and the force F becomes 0.4 kN, i.e. 400 N.

Since $M = F \times d$, then 25 N m = 400 N $\times d$

Thus,

effective length, $d = \dfrac{25\text{ N m}}{400\text{ N}} = \dfrac{1}{16}\text{ m}$

$= \dfrac{1}{16} \times 1000\text{ mm} = \textbf{62.5 mm}$

Now try the following Practise Exercise

Practise Exercise 30   Further problems on the moment of a force

1.   Determine the moment of a force of 25 N applied to a spanner at an effective length of 180 mm from the centre of a nut.
[4.5 N m]

2.   A moment of 7.5 N m is required to turn a wheel. If a force of 37.5 N applied to the rim of the wheel can just turn the wheel, calculate the effective distance from the rim to the hub of the wheel.   [200 mm]

3.   Calculate the force required to produce a moment of 27 N m on a shaft, when the effective distance from the centre of the shaft to the point of application of the force is 180 mm.   [150 N]

5.2   Equilibrium and the principle of moments

If more than one force is acting on an object and the forces do not act at a single point, then the turning effect of the forces, that is, the moment of each of the forces, must be considered.

Figure 5.3 shows a beam with its support (known as its **pivot** or **fulcrum**) at P, acting vertically upwards, and forces F_1 and F_2 acting vertically downwards at distances a and b, respectively, from the fulcrum.

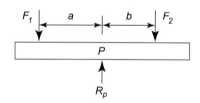

Figure 5.3

A beam is said to be in **equilibrium** when there is no tendency for it to move. There are two conditions for equilibrium:

(i)   The sum of the forces acting vertically downwards must be equal to the sum of the forces acting vertically upwards, i.e. for Figure 5.3, $R_P = F_1 + F_2$

(ii)   The total moment of the forces acting on a beam must be zero; for the total moment to be zero: *the sum of the clockwise moments about any point must be equal to the sum of the anticlockwise, or counter-clockwise, moments about that point.*

This statement is known as the **principle of moments**. Hence, taking moments about P in Figure 5.3,

$F_2 \times b$ = the clockwise moment, and

$F_1 \times a$ = the anticlockwise, or counter-clockwise, moment

Thus for equilibrium: $F_1 a = F_2 b$

Problem 3. A system of forces is as shown in Figure 5.4.

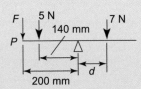

Figure 5.4

(a) If the system is in equilibrium find the distance d.

(b) If the point of application of the 5 N force is moved to point P, distance 200 mm from the support, and the 5 N force is replaced by an unknown force F, find the value of F for the system to be in equilibrium.

(a) From above, the clockwise moment M_1 is due to a force of 7 N acting at a distance d from the support; the support is called the **fulcrum**,

i.e. $M_1 = 7\,\text{N} \times d$

The anticlockwise moment M_2 is due to a force of 5 N acting at a distance of 140 mm from the fulcrum,

i.e. $M_2 = 5\,\text{N} \times 140\,\text{mm}$

Applying the principle of moments, for the system to be in equilibrium about the fulcrum:

clockwise moment = anticlockwise moment

i.e. $7\,\text{N} \times d = 5 \times 140\,\text{N mm}$

Hence, **distance, $d = \dfrac{5 \times 140\,\text{N mm}}{7\,\text{N}} = \textbf{100 mm}$**

(b) When the 5 N force is replaced by force F at a distance of 200 mm from the fulcrum, the new value of the anticlockwise moment is $F \times 200$. For the system to be in equilibrium:

clockwise moment = anticlockwise moment

i.e. $(7 \times 100)\,\text{N mm} = F \times 200\,\text{mm}$

Hence,

new value of force, $F = \dfrac{700\,\text{N mm}}{200\,\text{mm}} = \textbf{3.5 N}$

Problem 4. A beam is supported on its fulcrum at the point A, which is at mid-span, and forces act as shown in Figure 5.5. Calculate (a) force F for the beam to be in equilibrium (b) the new position of the 23 N force when F is decreased to 21 N, if equilibrium is to be maintained.

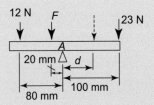

Figure 5.5

(a) The clockwise moment, M_1, is due to the 23 N force acting at a distance of 100 mm from the fulcrum, i.e. $M_1 = 23 \times 100 = 2300\,\text{N mm}$

There are two forces giving the anticlockwise moment M_2. One is the force F acting at a distance of 20 mm from the fulcrum and the other a force of 12 N acting at a distance of 80 mm.

Thus, $M_2 = (F \times 20) + (12 \times 80)\,\text{N mm}$

Applying the principle of moments about the fulcrum:

clockwise moment = anticlockwise moments

i.e. $2300 = (F \times 20) + (12 \times 80)$

Hence $F \times 20 = 2300 - 960$

i.e.   **force, $F = \dfrac{1340}{20} = \textbf{67 N}$**

(b) The clockwise moment is now due to a force of 23 N acting at a distance of, say, d from the fulcrum. Since the value of F is decreased to 21 N, the anticlockwise moment is $(21 \times 20) + (12 \times 80)\,\text{N mm}$.

Applying the principle of moments,

$23 \times d = (21 \times 20) + (12 \times 80)$

i.e. **distance, $d = \dfrac{420 + 960}{23}$**

$= \dfrac{1380}{23} = \textbf{60 mm}$

Problem 5. For the centrally supported uniform beam shown in Figure 5.6, determine the values of forces F_1 and F_2 when the beam is in equilibrium.

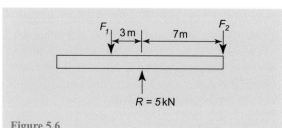

Figure 5.6

At equilibrium: (i) $R = F_1 + F_2$ i.e. $5 = F_1 + F_2$ (1)

and (ii) $F_1 \times 3 = F_2 \times 7$ (2)

From equation (1), $F_2 = 5 - F_1$

Substituting for F_2 in equation (2) gives:

$$F_1 \times 3 = (5 - F_1) \times 7$$

i.e. $3F_1 = 35 - 7F_1$

$$10F_1 = 35$$

from which, $F_1 = 3.5$ kN

Since $F_2 = 5 - F_1, \quad F_2 = 1.5$ kN

Thus at equilibrium,
force F_1 = 3.5 kN and force F_2 = 1.5 kN

Now try the following Practise Exercise

Practise Exercise 31 Further problems on equilibrium and the principle of moments

1. Determine distance d and the force acting at the support A for the force system shown in Figure 5.7, when the system is in equilibrium.

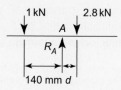

Figure 5.7

[50 mm, 3.8 kN]

2. If the 1 kN force shown in Figure 5.7 is replaced by a force F at a distance of 250 mm to the left of R_A, find the value of F for the system to be in equilibrium. [560 N]

3. Determine the values of the forces acting at A and B for the force system shown in Figure 5.8.

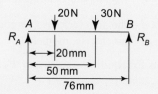

Figure 5.8

$[R_A = R_B = 25 \text{ N}]$

4. The forces acting on a beam are as shown in Figure 5.9. Neglecting the mass of the beam, find the value of R_A and distance d when the beam is in equilibrium.

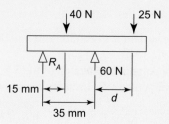

Figure 5.9

[5 N, 25 mm]

5.3 Simply supported beams having point loads

A **simply supported beam** is said to be one that rests on two knife-edge supports and is free to move horizontally.

Two typical simply supported beams having loads acting at given points on the beam, called **point loading**, are shown in Figure 5.10.

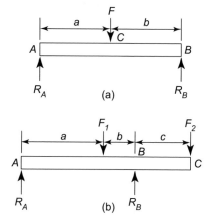

Figure 5.10

A man whose mass exerts a force F vertically downwards, standing on a wooden plank which is simply supported at its ends, may, for example, be represented by the beam diagram of Figure 5.10(a) if the mass of the plank is neglected. The forces exerted by the supports on the plank, R_A and R_B, act vertically upwards, and are called **reactions**.

When the forces acting are all in one plane, the algebraic sum of the moments can be taken about **any** point.

For the beam in Figure 5.10(a) at equilibrium:

(i) $R_A + R_B = F$

and   (ii)   taking moments about A, $F \times a = R_B (a + b)$
(Alternatively, taking moments about C,
$R_A a = R_B b$)

For the beam in Figure 5.10(b), at equilibrium

(i) $R_A + R_B = F_1 + F_2$

and   (ii)   taking moments about B,
$R_A (a + b) + F_2 c = F_1 b$

Typical **practical applications** of simply supported beams with point loadings include bridges, beams in buildings, and beds of machine tools.

Problem 6.   A beam is loaded as shown in Figure 5.11.

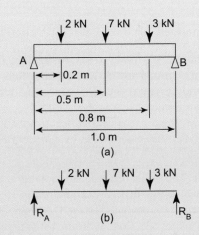

Figure 5.11

Determine (a) the force acting on the beam support at B (b) the force acting on the beam support at A, neglecting the mass of the beam.

A beam supported as shown in Figure 5.11 is called a simply supported beam.

(a)   Taking moments about point A and applying the principle of moments gives:

clockwise moments = anticlockwise moments

$(2 \times 0.2) + (7 \times 0.5) + (3 \times 0.8)$ kN m $= R_B \times 1.0$ m
where R_B is the force supporting the beam at B, as shown in Figure 5.11(b).

Thus $(0.4 + 3.5 + 2.4)$ kN m $= R_B \times 1.0$ m

i.e. $R_B = \dfrac{6.3\,\text{kN m}}{1.0\,\text{m}} = \textbf{6.3 kN}$

(b)   For the beam to be in equilibrium, the forces acting upwards must be equal to the forces acting downwards, thus

$R_A + R_B = (2 + 7 + 3)$ kN $= 12$ kN

$R_B = 6.3$ kN,

thus $R_A = 12 - R_B = 12 - 6.3$

$= \textbf{5.7 kN}$

Problem 7.   For the beam shown in Figure 5.12 calculate (a) the force acting on support A (b) distance d, neglecting any forces arising from the mass of the beam.

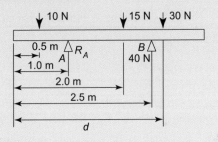

Figure 5.12

(a)   From Section 5.2,

(the forces acting in an upward direction) = (the forces acting in a downward direction)

Hence $(R_A + 40)$ N $= (10 + 15 + 30)$ N

$R_A = 10 + 15 + 30 - 40 = \textbf{15 N}$

(b)   Taking moments about the left-hand end of the beam and applying the principle of moments gives:

clockwise moments = anticlockwise moments

$(10 \times 0.5) + (15 \times 2.0)$ N m $+ 30$ N $\times d$

$= (15 \times 1.0) + (40 \times 2.5)$ N m

i.e. 35 N m $+ 30$ N $\times d = 115$ N m

from which, **distance, d** $= \dfrac{(115 - 35)\,\text{N m}}{30\,\text{N}}$

$= \textbf{2.67 m}$

Problem 8. A metal bar *AB* is 4.0 m long and is simply supported at each end in a horizontal position. It carries loads of 2.5 kN and 5.5 kN at distances of 2.0 m and 3.0 m, respectively, from *A*.

Neglecting the mass of the beam, determine the reactions of the supports when the beam is in equilibrium.

The beam and its loads are shown in Figure 5.13.

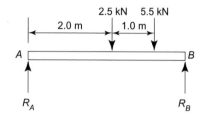

Figure 5.13

At equilibrium, $R_A + R_B = 2.5 + 5.5 = 8.0$ kN (1)

Taking moments about *A*,

clockwise moments = anticlockwise moments

i.e. $(2.5 \times 2.0) + (5.5 \times 3.0) = 4.0 R_B$

or $5.0 + 16.5 = 4.0 R_B$

from which, $R_B = \dfrac{21.5}{4.0} = 5.375$ kN

From equation (1), $R_A = 8.0 - 5.375 = 2.625$ kN

Thus the reactions at the supports at equilibrium are 2.625 kN at *A* and 5.375 kN at *B*

Problem 9. A beam *PQ* is 5.0 m long and is simply supported at its ends in a horizontal position as shown in Figure 5.14. Its mass is equivalent to a force of 400 N acting at its centre as shown. Point loads of 12 kN and 20 kN act on the beam in the positions shown. When the beam is in equilibrium, determine (a) the reactions of the supports, R_P and R_Q, and (b) the position to which the 12 kN load must be moved for the force on the supports to be equal.

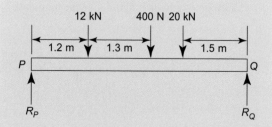

Figure 5.14

(a) At equilibrium,

$R_P + R_Q = 12 + 0.4 + 20 = 32.4$ kN (1)

Taking moments about *P*:

clockwise moments = anticlockwise moments

i.e. $(12 \times 1.2) + (0.4 \times 2.5)$
$+ (20 \times 3.5) = (R_Q \times 5.0)$

$14.4 + 1.0 + 70.0 = 5.0\ R_Q$

from which, $R_Q = \dfrac{85.4}{5.0} = \mathbf{17.08}$ **kN**

From equation (1), $\boldsymbol{R_P} = 32.4 - R_Q$

$= 32.4 - 17.08$

$= \mathbf{15.32}$ **kN**

(b) For the reactions of the supports to be equal,

$$R_P = R_Q = \frac{32.4}{2} = 16.2 \text{ kN}$$

Let the 12 kN load be at a distance *d* metres from *P* (instead of at 1.2 m from *P*). Taking moments about point *P* gives:

$(12 \times d) + (0.4 \times 2.5) + (20 \times 3.5) = 5.0\ R_Q$

i.e. $12d + 1.0 + 70.0 = 5.0 \times 16.2$

and $12d = 81.0 - 71.0$

from which, $d = \dfrac{10.0}{12} = 0.833$ m

Hence the 12 kN load needs to be moved to a position 833 mm from *P* for the reactions of the supports to be equal (i.e. 367 mm to the left of its original position).

Problem 10. A uniform steel girder *AB* is 6.0 m long and has a mass equivalent to 4.0 kN acting at its centre. The girder rests on two supports at *C* and *B* as shown in Figure 5.15. A point load of 20.0 kN is attached to the beam as shown. Determine the value of the force *F* that causes the beam to just lift off the support *B*.

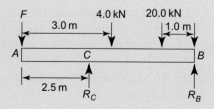

Figure 5.15

At equilibrium, $R_C + R_B = F + 4.0 + 20.0$.

When the beam is just lifting off of the support B, then $R_B = 0$, hence $R_C = (F + 24.0)$kN.

Taking moments about A:

Clockwise moments = anticlockwise moments

i.e. $(4.0 \times 3.0) + (5.0 \times 20.0) = (R_C \times 2.5) + (R_B \times 6.0)$

i.e. $12.0 + 100.0 = (F + 24.0) \times 2.5 + (0)$

i.e. $\dfrac{112.0}{2.5} = (F + 24.0)$

from which, $F = 44.8 - 24.0 = 20.8$ kN

i.e. **the value of force F which causes the beam to just lift off the support B is 20.8 kN**

Now try the following Practise Exercise

Practise Exercise 32 Further problems on simply supported beams having point loads

1. Calculate the force R_A and distance d for the beam shown in Figure 5.16. The mass of the beam should be neglected and equilibrium conditions assumed.

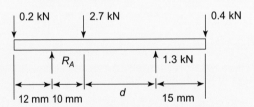

Figure 5.16

[2.0 kN, 24 mm]

2. For the force system shown in Figure 5.17, find the values of F and d for the system to be in equilibrium.

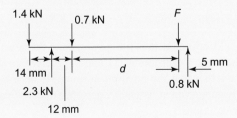

Figure 5.17

[1.0 kN, 64 mm]

3. For the force system shown in Figure 5.18, determine distance d for the forces R_A and R_B to be equal, assuming equilibrium conditions. [80 m]

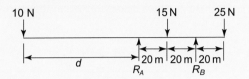

Figure 5.18

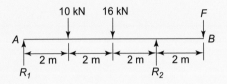

Figure 5.19

4. A simply supported beam AB is loaded as shown in Figure 5.19. Determine the load F in order that the reaction at A is zero. [36 kN]

5. A uniform wooden beam, 4.8 m long, is supported at its left-hand end and also at 3.2 m from the left-hand end. The mass of the beam is equivalent to 200 N acting vertically downwards at its centre. Determine the reactions at the supports. [50 N, 150 N]

6. For the simply supported beam PQ shown in Figure 5.20, determine (a) the reaction at each support (b) the maximum force which can be applied at Q without losing equilibrium.

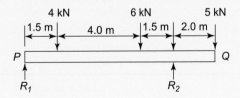

Figure 5.20

[(a) R_1 = 3 kN, R_2 = 12 kN (b) 15.5 kN]

7. A uniform beam AB is 12.0 m long and is supported at distances of 2.0 m and 9.0 m from A. Loads of 60 kN, 104 kN, 50 kN and 40 kN act vertically downwards at A, 5.0 m from A, 7.0 m from A and at B. Neglecting the mass

of the beam, determine the reactions at the supports. [133.7 kN, 120.3 kN]

8. A uniform girder carrying point loads is shown in Figure 5.21. Determine the value of load F which causes the beam to just lift off the support B.

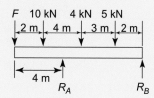

Figure 5.21

[3.25 kN]

5.4 Simply supported beams with couples

The procedure adopted here is a simple extension to Section 5.3, but it must be remembered that the units of a couple are in: N m, N mm, kN m, etc, unlike that of a force. The method of calculating reactions on beams due to couples will now be explained with the aid of worked problems.

Problem 11. Determine the end reactions for the simply supported beam of Figure 5.22, which is subjected to an anti-clockwise couple of 5 kN m applied at mid-span.

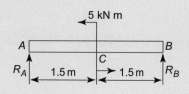

Figure 5.22

Taking moments about B:

Now the reaction R_A exerts a clockwise moment about B given by: $R_A \times 3$ m

Additionally, the couple of 5 kN m is anti-clockwise and its moment is 5 kN m regardless of where it is placed.

Clockwise moments about B =
anti-clockwise moments about B

i.e. $R_A \times 3$ m $= 5$ kN m (5.1)

from which, $R_A = \dfrac{5}{3}$ kN

or $\boldsymbol{R_A = 1.667}$ **kN** (5.2)

Resolving forces vertically gives:

Upward forces = downward forces

i.e. $R_A + R_B = 0$ (5.3)

It should be noted that in equation (5.3) the 5 kN m couple does not appear; this is because it is a couple and not a force.

From equations (5.2) and (5.3), $\boldsymbol{R_B = -R_A = -1.667}$ **kN**

i.e. R_B acts in the opposite direction to R_A, so that R_B and R_A also form a couple that resists the 5 kN m couple.

Problem 12. Determine the end reactions for the simply supported beam of Figure 5.23, which is subjected to an anti-clockwise couple of 5 kN m at the point C.

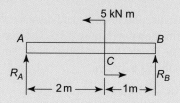

Figure 5.23

Taking moments about B gives:

$$R_A \times 3 \text{ m} = 5 \text{ kN m} \quad (5.4)$$

from which, $R_A = \dfrac{5}{3}$ kN

or $\boldsymbol{R_A = 1.667}$ **kN**

Resolving forces vertically gives:

i.e. $R_A + R_B = 0$

from which, $\boldsymbol{R_B = -R_A = -1.667}$ **kN**

It should be noted that the answers for the reactions are the same for Problems 11 and 12, thereby proving by induction that the position of a couple on a beam, simply supported at its ends, does not affect the values of the reactions.

Problem 13. Determine the reactions for the simply supported beam of Figure 5.24.

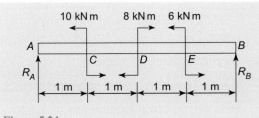

Figure 5.24

Taking moments about B gives:

$$R_A \times 4 \text{ m} + 8 \text{ kN m} = 10 \text{ kN m} + 6 \text{ kN m}$$

i.e. $4 R_A = 10 + 6 - 8 = 8$

from which, $R_A = \dfrac{8}{4} = \textbf{2 kN}$

Resolving forces vertically gives:

$$R_A + R_B = 0$$

from which, $\boldsymbol{R_B = -R_A = -2 \text{ kN}}$

Problem 14.   Determine the reactions for the simply supported beam of Figure 5.25.

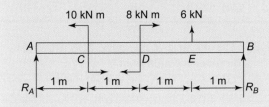

Figure 5.25

Taking moments about B gives:

$$R_A \times 4 \text{ m} + 8 \text{ kN m} + 6 \text{ kN} \times 1 \text{ m} = 10 \text{ kN m}$$

i.e. $4 R_A = 10 - 8 - 6 = -4$

from which, $R_A = -\dfrac{4}{4} = \textbf{-1 kN}$ (acting downwards)

Resolving forces vertically gives:

$$R_A + R_B + 6 = 0$$

from which, $R_B = -R_A - 6 = -(-1) - 6$

i.e. $\boldsymbol{R_B = 1 - 6 = -5 \text{ kN}}$ (acting downwards)

Now try the following Practise Exercises

Practise Exercise 33   Further problems on simply supported beams with couples

For each of the following problems, determine the reactions acting on the simply supported beams:

1.

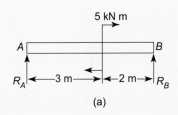

Figure 5.26

$$[R_A = -1 \text{ kN}, R_B = 1 \text{ kN}]$$

2.

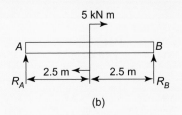

Figure 5.27

$$[R_A = -1 \text{ kN}, R_B = 1 \text{ kN}]$$

3.

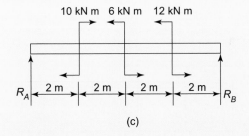

Figure 5.28

$$[R_A = 1 \text{ kN}, R_B = -1 \text{ kN}]$$

4.

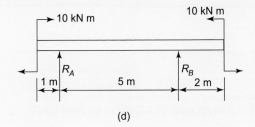

Figure 5.29

$$[R_A = 0, R_B = 0]$$

Part Two

5.

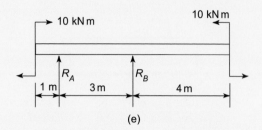

(e)

Figure 5.30

$$[R_A = 0, R_B = 0]$$

6.

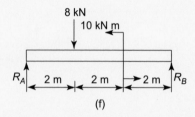

(f)

Figure 5.31

$$[R_A = 7 \text{ kN}, R_B = 1 \text{ kN}]$$

7.

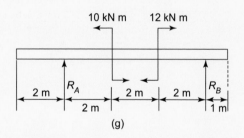

(g)

Figure 5.32

$$[R_A = -333 \text{ N}, R_B = 333 \text{ N}]$$

Practise Exercise 34   Short-answer questions on simply supported beams

1. The moment of a force is the product of .... and ......

2. When a beam has no tendency to move it is in ....

3. State the two conditions for equilibrium of a beam.

4. State the principle of moments.

5. What is meant by a simply supported beam?

6. State two practical applications of simply supported beams.

7. Why does a couple not have a vertical component of force?

Practise Exercise 35   Multiple-choice questions on simply supported beams

(Answers on page 297)

1. A force of 10 N is applied at right angles to the handle of a spanner, 0.5 m from the centre of a nut. The moment on the nut is:
    (a) 5 N m   (b) 2 N/m
    (c) 0.5 m/N   (d) 15 N m

2. The distance d in Figure 5.33 when the beam is in equilibrium is:
    (a) 0.5 m   (b) 1.0 m
    (c) 4.0 m   (d) 15 m

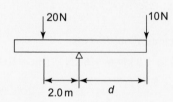

Figure 5.33

3. With reference to Figure 5.34, the clockwise moment about A is:
    (a) 70 N m   (b) 10 N m
    (c) 60 N m   (d) $5 \times R_B$ N m

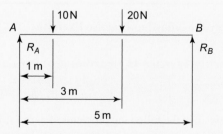

Figure 5.34

4. The force acting at B (i.e. R_B) in Figure 5.34 is:
   (a) 16 N   (b) 20 N
   (c) 5 N   (d) 14 N

5. The force acting at A (i.e. R_A) in Figure 5.34 is:
   (a) 16 N   (b) 10 N
   (c) 15 N   (d) 14 N

6. Which of the following statements is false for the beam shown in Figure 5.35 if the beam is in equilibrium?
   (a) The anticlockwise moment is 27 N
   (b) The force F is 9 N
   (c) The reaction at the support R is 18 N
   (d) The beam cannot be in equilibrium for the given conditions

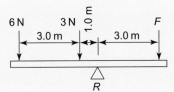

Figure 5.35

7. With reference to Figure 5.36, the reaction R_A is:
   (a) 10 N   (b) 30 N
   (c) 20 N   (d) 40 N

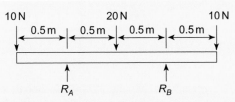

Figure 5.36

8. With reference to Figure 5.36, when moments are taken about R_A, the sum of the anticlockwise moments is:
   (a) 25 N m   (b) 20 N m
   (c) 35 N m   (d) 30 N m

9. With reference to Figure 5.36, when moments are taken about the right-hand end, the sum of the clockwise moments is:
   (a) 10 N m   (b) 20 N m
   (c) 30 N m   (d) 40 N m

10. With reference to Figure 5.36, which of the following statements is false?
    (a) $(5 + R_B) = 25$ N m
    (b) $R_A = R_B$
    (c) $(10 \times 0.5) = (10 \times 1) + (10 \times 1.5) + R_A$
    (d) $R_A + R_B = 40$ N

11. A beam simply supported at its ends is subjected to two intermediate couples of 4 kN m clockwise and 4 kN m anticlockwise. The values of the end reactions are:
    (a) 4 kN   (b) 8 kN
    (c) zero   (d) unknown

Revision Test 2   Forces, tensile testing and beams

This Revision Test covers the material contained in Chapters 2 to 5. *The marks for each question are shown in brackets at the end of each question.*

1. A metal bar having a cross-sectional area of 80 mm^2 has a tensile force of 20 kN applied to it. Determine the stress in the bar. (4)

2. (a) A rectangular metal bar has a width of 16 mm and can support a maximum compressive stress of 15 MPa; determine the minimum breadth of the bar when loaded with a force of 6 kN.

   (b) If the bar in (a) is 1.5 m long and decreases in length by 0.18 mm when the force is applied, determine the strain and the percentage strain. (7)

3. A wire is stretched 2.50 mm by a force of 400 N. Determine the force that would stretch the wire 3.50 mm, assuming that the elastic limit is not exceeded. (5)

4. A copper tube has an internal diameter of 140 mm and an outside diameter of 180 mm and is used to support a load of 4 kN. The tube is 600 mm long before the load is applied. Determine, in micrometres, by how much the tube contracts when loaded, taking the modulus of elasticity for copper as 96 GPa. (8)

5. The results of a tensile test are: diameter of specimen 21.7 mm; gauge length 60 mm; load at limit of proportionality 50 kN; extension at limit of proportionality 0.090 mm; maximum load 100 kN; final length at point of fracture 75 mm.
   Determine (a) Young's modulus of elasticity (b) the ultimate tensile strength (c) the stress at the limit of proportionality (d) the percentage elongation. (10)

6. A force of 25 N acts horizontally to the right and a force of 15 N is inclined at an angle of 30° to the 25 N force. Determine the magnitude and direction of the resultant of the two forces using (a) the triangle of forces method (b) the parallelogram of forces method and (c) by calculation (12)

7. Determine graphically the magnitude and direction of the resultant of the following three coplanar forces, which may be considered as acting at a point.

Force P, 15 N acting horizontally to the right

Force Q, 8 N inclined at 45° to force P

Force R, 20 N inclined at 120° to force P. (7)

8. Determine by resolution of forces the resultant of the following three coplanar forces acting at a point: 120 N acting at 40° to the horizontal; 250 N acting at 145° to the horizontal; 300 N acting at 260° to the horizontal. (8)

9. A moment of 18 N m is required to operate a lifting jack. Determine the effective length of the handle of the jack (in millimetres) if the force applied to it is (a) 90 N (b) 0.36 kN (6)

10. For the centrally supported uniform beam shown in Figure RT2.1, determine the values of forces F_1 and F_2 when the beam is in equilibrium. (7)

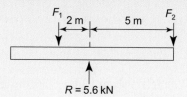

Figure RT2.1

11. For the beam shown in Figure RT2.2 calculate (a) the force acting on support Q, (b) distance d, neglecting any forces arising from the mass of the beam. (7)

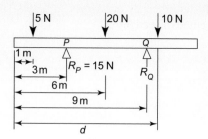

Figure RT2.2

12. A beam of length 3 m is simply supported at its ends. If a clockwise couple of 4 kN m is placed at a distance of 1 m from the left hand support, determine the end reactions. (4)

Chapter 6

Forces in structures

In this chapter it will be shown how the principles described in Chapter 4 on forces acting at a point can be used to determine the internal forces in the members of a truss, due to externally applied loads. In countries where there is a lot of rain, such structures are used to support the sloping roofs of the building; the externally applied loads acting on the pin-jointed trusses are usually due to snow and self-weight, and also due to wind. Tie bars and struts are defined, Bow's notation is explained and internal forces in a truss are calculated by different methods.

At the end of this chapter you should be able to:

- recognise a pin-jointed truss
- recognise a mechanism
- define a tie bar
- define a strut
- understand Bow's notation
- calculate the internal forces in a truss by a graphical method
- calculate the internal forces in a truss by the 'method of joints'
- calculate the internal forces in a truss by the 'method of sections'

6.1 Introduction

As stated above, in this chapter it will be shown how the principles described in Chapter 4 can be used to determine the internal forces in the members of a truss, due to externally applied loads. The definition of a truss is that it is a frame where the joints are assumed to be frictionless and pin-jointed, and that all external loads are applied to the pin joints. In countries where there

is a lot of rain, such structures are used to support the sloping roofs of the building, as shown in Figure 6.1.

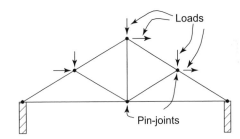

Figure 6.1 Pin-jointed truss

The externally applied loads acting on the pin-jointed trusses are usually due to snow and self-weight, and also due to wind, as shown in Figure 6.1. In Figure 6.1, the snow and self-weight loads act vertically downwards and the wind loads are usually assumed to act horizontally. Thus, for structures such as that shown in Figure 6.1, where the externally applied loads are assumed to act at the pin-joints, the internal members of the framework resist the externally applied loads in tension or compression.

Members of the framework that resist the externally applied loads in tension are called **ties** and members of the framework which resist the externally applied loads in compression are called **struts**, as shown in Figure 6.2.

Figure 6.2 Ties and struts with external loads

The internal resisting forces in the ties and struts will act in the opposite direction to the externally applied loads, as shown in Figure 6.3.

Mechanical Engineering Principles, Bird and Ross, ISBN 9780415517850

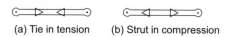

Figure 6.3 Internal resisting forces in ties and struts

The methods of analysis used in this chapter breakdown if the joints are rigid (welded), or if the loads are applied between the joints. In these cases, flexure occurs in the members of the framework, and other methods of analysis have to be used, as described in Chapters 5 and 7. It must be remembered, however, that even if the joints of the framework are smooth and pin-jointed and also if externally applied loads are placed at the pin-joints, members of the truss in compression can fail through structural failure (see references [1] and [2] on page 86).

It must also be remembered that the methods used here cannot be used to determine forces in **statically indeterminate** pin-jointed trusses, nor can they be used to determine forces in **mechanisms.** Statically indeterminate structures are so called because they cannot be analysed by the principles of statics alone. Typical mechanisms are shown in Figure 6.4; these are not classified as structures because they are not firm and can be moved easily under external loads.

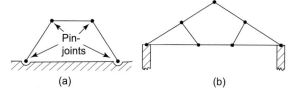

Figure 6.4 Mechanisms

To make the mechanism of Figure 6.4(a) into a simple statically determinate structure, it is necessary to add one diagonal member joined to a top joint and an 'opposite' bottom joint. To make the mechanism of Figure 6.4(b) into a statically determinate structure, it is necessary to add two members from the top joint to each of the two bottom joints near the mid-length of the bottom horizontal.

Three methods of analysis will be used in this chapter – one graphical and two analytical methods.

6.2 Worked problems on mechanisms and pin-jointed trusses

Problem 1. Show how the mechanism of Figure 6.4(a) can be made into a statically determinate structure.

The two solutions are shown by the broken lines of Figures 6.5(a) and (b), which represent the placement of additional members.

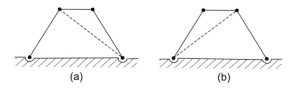

Figure 6.5

Problem 2. Show how the mechanism of Figure 6.4(b) can be made into a statically determinate truss.

The solution is shown in Figures 6.6, where the broken lines represent the placement of two additional members.

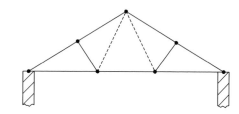

Figure 6.6

Problem 3. Show how the mechanism of Figure 6.4(a) can be made into a statically indeterminate truss.

The solution is shown in Figure 6.7, where the broken lines represent the addition of two members, which are not joined where they cross.

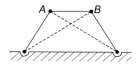

Figure 6.7

Problem 4. Why is the structure of Figure 6.7 said to be statically indeterminate?

As you can only resolve vertically and horizontally at the joints A and B, you can only obtain four simultaneous equations. However, as there are five members, each with an unknown force, you have one unknown force too many. Thus, using the principles of statics alone, the structure cannot be satisfactorily analysed; such structures are therefore said to be **statically indeterminate.**

6.3   Graphical method

In this case, the method described in Chapter 4 will be used to analyse statically determinate plane pin-jointed trusses. The method will be described with the aid of worked examples.

> **Problem 5.**   Determine the internal forces that occur in the plane pin-jointed truss of Figure 6.8, due to the externally applied vertical load of 3 kN.

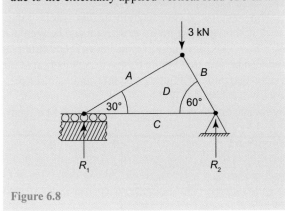

Figure 6.8

Firstly, we will fill the spaces between the forces with upper case letters of the alphabet, as shown in Figure 6.8. It should be noted that the only reactions are the vertical reactions R_1 and R_2; this is because the only externally applied load is the vertical load of 3 kN, and there is no external horizontal load. The capital letters A, B, C and D can be used to represent the forces between them, providing they are taken in a clockwise direction about each joint. Thus the letters AB represent the vertical load of 3 kN. Now as this load acts vertically downwards, it can be represented by a vector $\boldsymbol{ab}$, where the magnitude of ab is 3 kN and it points in the direction from a to b. As $\boldsymbol{ab}$ is a vector, it will have a direction as well as a magnitude. Thus $\boldsymbol{ab}$ will point downwards from a to b as the 3 kN load acts downwards.

This method of representing forces is known as **Bow's notation.**

To analyse the truss, we must first consider the joint ABD; this is because this joint has only two unknown forces, namely the internal forces in the two members that meet at the joint ABD. Neither joints BCD and CAD can be considered first, because each of these joints has more than two unknown forces.

Now the 3 kN load is between the spaces A and B, so that it can be represented by the lower case letters ab, point from a to b and of magnitude 3 kN, as shown in Figure 6.9.

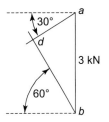

Figure 6.9

Similarly, the force in the truss between the spaces B and D, namely the vector $\boldsymbol{bd}$, lies at 60° to the horizontal and the force in the truss between the spaces D and A, namely the vector $\boldsymbol{da}$, lies at 30° to the horizontal. Thus, in Figure 6.9, if the vectors $\boldsymbol{bd}$ and $\boldsymbol{ad}$ are drawn, they will cross at the point d, where the point d will obviously lie to the left of the vector ab, as shown. Hence, if the vector $\boldsymbol{ab}$ is drawn to scale, the magnitudes of the vectors $\boldsymbol{bd}$ and $\boldsymbol{da}$ can be measured from the scaled drawing. The direction of the force in the member between the spaces B and D at the joint ABD point upwards because the vector from b to d points upwards. Similarly, the direction of the force in the member between the spaces D and A at the joint ABD is also upwards because the vector from d to a points upwards. These directions at the joint ABD are shown in Figure 6.10. Now as the framework is in equilibrium, the internal forces in the members BD and DA at the joints (2) and (1) respectively, will be equal and opposite to the internal forces at the joint ABD; these are shown in Figure 6.10.

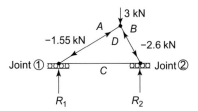

Figure 6.10

Comparing the directions of the arrows in Figure 6.10 with those of Figure 6.3, it can be seen that the members BD and DA are in compression and are defined as struts. It should also be noted from Figure 6.10, that when a member of the framework, say, BD, is so defined, we are referring to the top joint, because we must **always work around a joint in a clockwise manner**; thus the arrow at the top of *BD* points upwards, because in Figure 6.9, the vector **bd** points upwards from *b* to *d*. Similarly, if the same member is referred to as *DB*, then we are referring to the bottom of this member at the joint (2), because we must always work clockwise around a joint. Hence, at joint (2), the arrow points downwards, because the vector **db** points downwards from *d* to *b* in Figure 6.9.

To determine the unknown forces in the horizontal member between joints (1) and (2), either of these joints can be considered, as both joints now only have two unknown forces. Let us consider joint (1), i.e. joint *ADC*. Now the vector **ad** can be measured from Figure 6.9 and drawn to scale in Figure 6.11.

Figure 6.11

Now the unknown force between the spaces *D* and *C*, namely the vector **dc** is horizontal and the unknown force between the spaces *C* and *A*, namely the vector **ca** is vertical, hence, by drawing to scale and direction, the point *c* can be found. This is because the point *c* in Figure 6.11 lies below the point *a* and to the right of *d*.

In Figure 6.11, the vector **ca** represents the magnitude and direction of the unknown reaction R_1 and the vector **dc** represents the magnitude and direction of the force in the horizontal member at joint (1); these forces are shown in Figure 6.12, where $R_1 = 0.82$ kN and $dc = 1.25$ kN.

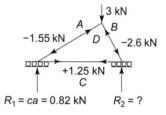

Figure 6.12

Comparing the directions of the internal forces in the bottom of the horizontal member with Figure 6.3, it can be seen that this member is in tension and therefore, it is a tie.

The reaction R_2 can be determined by considering joint (2), i.e. joint *BCD*, as shown in Figure 6.13, where the vector **bc** represents the unknown reaction R_2 which is measured as 2.18 kN.

Figure 6.13

The complete vector diagram for the whole framework is shown in Figure 6.14, where it can be seen that $R_1 + R_2 = 3$ kN, as required by the laws of equilibrium. It can also be seen that Figure 6.14 is a combination of the vector diagrams of Figures 6.9, 6.11 and 6.13. Experience will enable this problem to be solved more quickly by producing the vector diagram of Figure 6.14 directly.

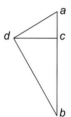

Figure 6.14

The table below contains a summary of all the measured forces.

Member	Force (kN)
bd	−2.6
da	−1.55
cd	1.25
R_1	0.82
R_2	2.18

Problem 6.   Determine the internal forces in the members of the truss of Figure 6.15 due to the externally applied horizontal force of 4 kN at the joint *ABE*.

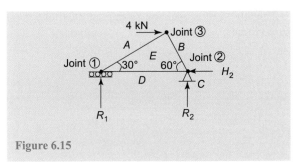

Figure 6.15

In this case, the spaces between the unknown forces are A, B, C, D and E. It should be noted that the reaction at joint (1) is vertical because the joint is on rollers, and that there are two reactions at joint (2) because it is firmly anchored to the ground and there is also a horizontal force of 4 kN which must be balanced by the unknown horizontal reaction H_2. If this unknown horizontal reaction did not exist, the structure would 'float' into space due to the 4 kN load.

Consider joint ABE, as there are only two unknown forces here, namely the forces in the members BE and EA. Working clockwise around this joint, the vector diagram for this joint is shown in Figure 6.16.

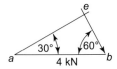

Figure 6.16

Joint (2) cannot be considered next, as it has three unknown forces, namely H_2, R_2 and the unknown member force DE. Hence, joint (1) must be considered next; it has two unknown forces, namely R_1 and the force in member ED. As the member AE and its direction can be obtained from Figure 6.16, it can be drawn to scale in Figure 6.17. By measurement, de = 3 kN.

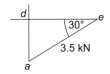

Figure 6.17

As R_1 is vertical, then the vector da is vertical, hence, the position d can be found in the vector diagram of Figure 6.17, where R_1 = da (pointing downwards).

To determine R_2 and H_2, joint (2) can now be considered, as shown by the vector diagram for the joint in Figure 6.18.

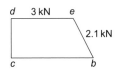

Figure 6.18

The complete diagram for the whole framework is shown in Figure 6.19, where it can be seen that this diagram is the sum of the vector diagrams of Figures 6.16 to 6.18.

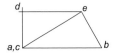

Figure 6.19

The table below contains a summary of all the measured forces

Member	Force (kN)
be	−2.1
ae	3.5
de	−3.0
R_1	−1.8
R_2	1.8
H_2	4.0

Couple and moment

Prior to solving Problem 7, it will be necessary for the reader to understand the nature of a **couple**; this is described in Chapter 10, page 126.

The magnitude of a couple is called its **moment**; this is described in Chapter 5, page 61.

Problem 7. Determine the internal forces in the pin-jointed truss of Figure 6.20.

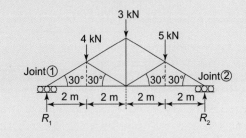

Figure 6.20

In this case, there are more than two unknowns at every joint; hence it will first be necessary to calculate the unknown reactions R_1 and R_2.

To determine R_1, take moments about joint (2):

Clockwise moments about joint (2) = counter-clockwise (or anti-clockwise) moments about joint (2)

i.e. $R_1 \times 8\,\text{m} = 4\,\text{kN} \times 6\,\text{m} + 3\,\text{kN} \times 4\,\text{m} + 5\,\text{kN} \times 2\,\text{m}$

$= 24 + 12 + 10 = 46\,\text{kN m}$

Therefore, $R_1 = \dfrac{46\,\text{kN m}}{8\,\text{m}} =$ **5.75 kN**

Resolving forces vertically:

Upward forces = downward forces

i.e. $R_1 + R_2 = 4 + 3 + 5 = 12\,\text{kN}$

However, $R_1 = 5.75\,\text{kN}$, from above,

hence, $5.75\,\text{kN} + R_2 = 12\,\text{kN}$

from which, $R_2 = 12 - 5.75 =$ **6.25 kN**

Placing these reactions on Figure 6.21, together with the spaces between the lines of action of the forces, we can now begin to analyse the structure.

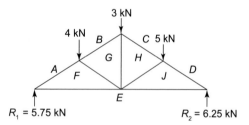

Figure 6.21

Starting at either joint *AFE* or joint *DEJ*, where there are two or less unknowns, the drawing to scale of the vector diagram can commence. It must be remembered to work around each joint in turn, in a clockwise manner, and only to tackle a joint when it has two or less unknowns. The complete vector diagram for the entire structure is shown in Figure 6.22.

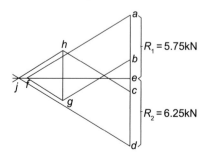

Figure 6.22

The table below contains a summary of all the measured forces.

Member	Force (kN)
af	−11.5
fe	10.0
jd	−12.5
ej	10.9
bg	−7.5
gf	−4.0
ch	−7.6
hg	4.6
jh	−5.0
R_1	5.75
R_2	6.25

Now try the following Practise Exercise

Determine the internal forces in the following pin-jointed trusses using a graphical method:

1.

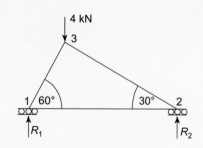

Figure 6.23

$$\begin{bmatrix} R_1 = 3.0\,\text{kN},\ R_2 = 1.0\,\text{kN}, \\ 1-2,\ 1.7\,\text{kN},\ 1-3,\ -3.5\,\text{kN}, \\ 2-3,\ -2.0\,\text{kN} \end{bmatrix}$$

2.

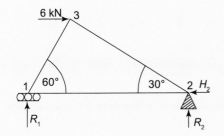

Figure 6.24

$$\begin{bmatrix} R_1 = -2.6 \text{ kN}, R_2 = 2.6 \text{ kN}, \\ H_2 = 6.0 \text{ kN}, 1-2, -1.5 \text{ kN}, \\ 1-3, 3.0 \text{ kN}, 2-3, -5.2 \text{ kN} \end{bmatrix}$$

3.

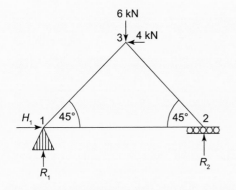

Figure 6.25

$$\begin{bmatrix} R_1 = 5.0 \text{ kN}, R_2 = 1.0 \text{ kN}, \\ H_1 = 4.0 \text{ kN} \; 1-2, 1.0 \text{ kN}, \\ 1-3, -7.1 \text{ kN}, 2-3, -1.4 \text{ kN} \end{bmatrix}$$

4.

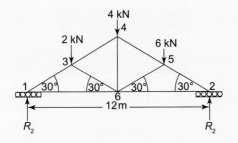

Figure 6.26

$$\begin{bmatrix} R_1 = 5.0 \text{ kN}, R_2 = 7.0 \text{ kN}, 1-3, -10.0 \text{ kN}, \\ 1-6, 8.7 \text{ kN}, 3-4, -8.0 \text{ kN}, 3-6, \\ -2.0 \text{ kN}, 4-6, 4.0 \text{ kN}, 4-5, -8.0 \text{ kN}, 5-6, \\ -6.0 \text{ kN}, 5-2, -14.0 \text{ kN}, 6-2, 12.1 \text{ kN} \end{bmatrix}$$

6.4 Method of joints (a mathematical method)

In this method, all unknown internal member forces are initially assumed to be in tension. Next, an imaginary cut is made around a joint that has two or less unknown forces, so that a free body diagram is obtained for this joint. Next, by resolving forces in respective vertical and horizontal directions at this joint, the unknown forces can be calculated. To continue the analysis, another joint is selected with two or less unknowns and the process repeated, remembering that this may only be possible because some of the unknown member forces have been previously calculated. By selecting, in turn, other joints where there are two or less unknown forces, the entire framework can be analysed.

It must be remembered that if the calculated force in a member is **negative**, then that member is in **compression**. Vice-versa is true for a member in tension.

To demonstrate the method, some pin-jointed trusses will now be analysed in Problems 8 to 10.

> **Problem 8.** Solve Problem 5, Figure 6.8 on page 75, by the method of joints.

Firstly, assume all unknowns are in tension, as shown in Figure 6.27.

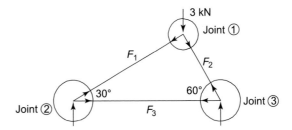

Figure 6.27

Next, make imaginary cuts around the joints, as shown by the circles in Figure 6.27. This action will give us three free body diagrams. The first we consider is around joint (1), because this joint has only two unknown forces; see Figure 6.28.

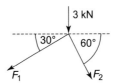

Figure 6.28

Resolving forces horizontally at joint (1):

Forces to the left = forces to the right

i.e. $F_1 \cos 30° = F_2 \cos 60°$

i.e. $0.866\, F_1 = 0.5\, F_2$

from which, $F_1 = \dfrac{0.5\, F_2}{0.866}$

i.e. $\boldsymbol{F_1 = 0.577\, F_2}$ (6.1)

Resolving forces vertically at joint (1):

Upward forces = downward forces

i.e. $0 = 3\text{ kN} + F_1 \sin 30° + F_2 \sin 60°$

i.e. $0 = 3 + 0.5\, F_1 + 0.866\, F_2$ (6.2)

Substituting equation (6.1) into equation (6.2) gives:

$$0 = 3 + 0.5 \times 0.577\, F_2 + 0.866\, F_2$$

i.e. $-3 = 1.1545\, F_2$

from which, $F_2 = -\dfrac{3}{1.1545}$

i.e. $\boldsymbol{F_2 = -2.6\text{ kN (compressive)}}$ (6.3)

Substituting equation (6.3) into equation (6.1) gives:

$$F_1 = 0.577 \times (-2.6)$$

i.e. $\boldsymbol{F_1 = -1.5\text{ kN (compressive)}}$

Considering joint (2), as it now has two or less unknown forces; see Figure 6.29.

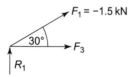

Figure 6.29

Resolving horizontally:

Forces to the left = forces to the right

i.e. $0 = F_1 \cos 30° + F_3$

However, $F_1 = -1.5$ kN,

hence, $0 = -1.5 \times 0.866 + F_3$

from which, $\boldsymbol{F_3 = 1.30\text{ kN (tensile)}}$

These results are similar to those obtained by the graphical method used in Problem 5; see Figure 6.12 on page 76, and the table below.

Member	Force (kN)
F_1	−1.5
F_2	−2.6
F_3	1.3

Problem 9. Solve Problem 6, Figure 6.15 on page 77, by the method of joints.

Firstly, we will assume that all unknown internal forces are in tension, as shown by Figure 6.30.

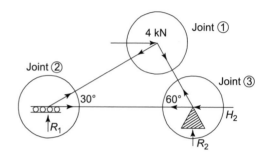

Figure 6.30

Next, we will isolate each joint by making imaginary cuts around each joint, as shown by the circles in Figure 6.30; this will result in three free body diagrams. The first free body diagram will be for joint (1), as this joint has two or less unknown forces; see Figure 6.31.

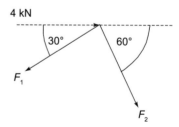

Figure 6.31

Resolving forces horizontally:

$$F_1 \cos 30° = 4\text{ kN} + F_2 \cos 60°$$

i.e. $0.866\, F_1 = 4 + 0.5\, F_2$

from which, $F_1 = \dfrac{4 + 0.5\, F_2}{0.866}$

or $F_1 = 4.619\text{ kN} + 0.577\, F_2$ (6.4)

Resolving forces vertically:

$$0 = F_1 \sin 30° + F_2 \sin 60°$$

i.e.　　　　$0 = 0.5\,F_1 + 0.866\,F_2$

or　　　　$-F_1 = \dfrac{0.866\,F_2}{0.5}$

or　　　　$F_1 = -1.732\,F_2$　　　　(6.5)

Equating equations (6.4) and (6.5) gives:

$$4.619 \text{ kN} + 0.577\,F_2 = -1.732\,F_2$$

i.e.　　　　$4.619 = -1.732\,F_2 - 0.577\,F_2$

　　　　　　　$= -2.309\,F_2$

Hence　　　$F_2 = -\dfrac{4.619}{2.309}$

i.e.　　　　$\mathbf{F_2 = -2 \text{ kN (compressive)}}$　　(6.6)

Substituting equation (6.6) into equation (6.5) gives:

$$F_1 = -1.732 \times (-2)$$

i.e.　　　　$\mathbf{F_1 = 3.465 \text{ kN}}$　　　　(6.7)

Consider next joint (2), as this joint now has two or less unknown forces; see Figure 6.32.

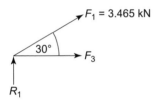

Figure 6.32

Resolving forces vertically:

$$R_1 + F_1 \sin 30° = 0$$

or　　　　　$R_1 = -F_1 \sin 30°$　　　　(6.8)

Substituting equation (6.7) into equation (6.8) gives:

$$R_1 = -3.465 \times 0.5$$

i.e.　　　$\mathbf{R_1 = -1.733 \text{ kN (acting downwards)}}$

Resolving forces horizontally:

$$0 = F_1 \cos 30° + F_3$$

or　　　　$F_3 = -F_1 \cos 30°$　　　　(6.9)

Substituting equation (6.7) into equation (6.9) gives:

$$F_3 = -3.465 \times 0.866$$

i.e.　　　$\mathbf{F_3 = -3 \text{ kN (compressive)}}$　　(6.10)

Consider next joint 3; see Figure 6.33.

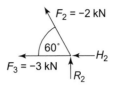

Figure 6.33

Resolving forces vertically:

$$F_2 \sin 60° + R_2 = 0$$

i.e.　　　　$R_2 = -F_2 \sin 60°$　　　　(6.11)

Substituting equation (6.6) into equation (6.11) gives:

$$R_2 = -(-2) \times 0.866$$

i.e.　　$\mathbf{R_2 = 1.732 \text{ kN (acting upwards)}}$

Resolving forces horizontally:

$$F_3 + F_2 \cos 60° + H_2 = 0$$

i.e.　　　　$H_2 = -F_3 - F_2 \times 0.5$　　(6.12)

Substituting equations (6.6) and (6.10) into equation (6.12) gives:

$$H_2 = -(-3) - (-2) \times 0.5$$

i.e.　　　　$\mathbf{H_2 = 4 \text{ kN}}$

These calculated forces are of similar value to those obtained by the graphical solution for Problem 6, as shown in the table below.

Member	Force (kN)
F_1	3.47
F_2	−2.0
F_3	−3.0
R_1	−1.73
R_2	1.73
H_2	4.0

Problem 10.　Solve Problem 7, Figure 6.20 on page 77, by the method of joints.

Firstly, assume all unknown member forces are in tension, as shown in Figure 6.34.

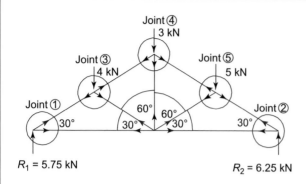

Figure 6.34

Next, we will isolate the forces acting at each joint by making imaginary cuts around each of the five joints as shown in Figure 6.34.

As there are no joints with two or less unknown forces, it will be necessary to calculate the unknown reactions R_1 and R_2 prior to using the method of joints.

Using the same method as that described for the solution of Problem 7, we have

$$R_1 = 5.75 \text{ kN and } R_2 = 6.25 \text{ kN}$$

Now either joint (1) or joint (2) can be considered, as each of these joints has two or less unknown forces.

Consider joint (1); see Figure 6.35.

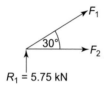

Figure 6.35

Resolving forces vertically:

$$5.75 + F_1 \sin 30° = 0$$

i.e. $F_1 \sin 30° = -5.75$

or $0.5 F_1 = -5.75$

i.e. $F_1 = -\dfrac{5.75}{0.5}$

i.e. $\boldsymbol{F_1 = -11.5 \text{ kN (compressive)}}$ (6.13)

Resolving forces horizontally:

$$0 = F_2 + F_1 \cos 30°$$

i.e. $F_2 = -F_1 \cos 30°$ (6.14)

Substituting equation (6.13) into equation (6.14) gives:

$$F_2 = -F_1 \cos 30° = -(-11.5) \times 0.866$$

i.e. $\boldsymbol{F_2 = 9.96 \text{ kN (tensile)}}$

Consider joint (2); see Figure 6.36.

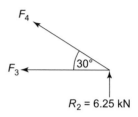

Figure 6.36

Resolving forces vertically:

$$R_2 + F_4 \sin 30° = 0$$

i.e. $R_2 + 0.5 F_4 = 0$

or $F_4 = -\dfrac{R_2}{0.5}$ (6.15)

Since $R_2 = 6.25$, $F_4 = -\dfrac{6.25}{0.5}$

i.e. $\boldsymbol{F_4 = -12.5 \text{ kN (compressive)}}$ (6.16)

Resolving forces horizontally:

$$F_3 + F_4 \cos 30° = 0$$

i.e. $F_3 = -F_4 \cos 30°$ (6.17)

Substituting equation (6.16) into equation (6.17) gives:

$$F_3 = -(-12.5) \times 0.866$$

i.e. $\boldsymbol{F_3 = 10.83 \text{ kN (tensile)}}$

Consider joint (3); see Figure 6.37.

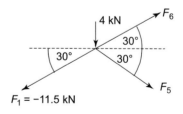

Figure 6.37

Resolving forces vertically:

$$F_6 \sin 30° = F_1 \sin 30° + F_5 \sin 30° + 4$$

i.e. $F_6 = F_1 + F_5 + \dfrac{4}{\sin 30°}$ (6.18)

Substituting equation (6.13) into equation (6.18) gives:

$$F_6 = -11.5 + F_5 + 8 \qquad (6.19)$$

Resolving forces horizontally:

$$F_1 \cos 30° = F_5 \cos 30° + F_6 \cos 30°$$

i.e. $\qquad F_1 = F_5 + F_6 \qquad (6.20)$

Substituting equation (6.13) into equation (6.20) gives:

$$-11.5 = F_5 + F_6$$

or $\qquad F_6 = -11.5 - F_5 \qquad (6.21)$

Equating equations (6.19) and (6.21) gives:

$$-11.5 + F_5 + 8 = -11.5 - F_5$$

or $\qquad F_5 + F_5 = -11.5 + 11.5 - 8$

i.e. $\qquad 2F_5 = -8$

from which, $\qquad \boldsymbol{F_5 = -4kN \text{ (compressive)}} \quad (6.22)$

Substituting equation (6.22) into equation (6.21) gives:

$$F_6 = -11.5 - (-4)$$

i.e. $\qquad \boldsymbol{F_6 = -7.5 \text{ kN (compressive)}} \qquad (6.23)$

Consider joint (4); see Figure 6.38.

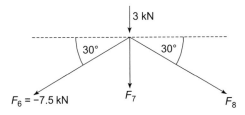

Figure 6.38

Resolving forces horizontally:

$$F_6 \cos 30° = F_8 \cos 30°$$

i.e. $\qquad F_6 = F_8$

but from equation (6.23), $\quad F_6 = -7.5 \text{ kN}$

Hence, $\quad \boldsymbol{F_8 = -7.5 \text{ kN (compressive)}} \qquad (6.24)$

Resolving forces vertically:

$$0 = 3 + F_6 \sin 30° + F_7 + F_8 \sin 30°$$

i.e. $\quad 0 = 3 + 0.5 F_6 + F_7 + 0.5 F_8 \qquad (6.25)$

Substituting equations (6.23) and (6.24) into equation (6.25) gives:

$$0 = 3 + 0.5 \times -7.5 + F_7 + 0.5 \times -7.5$$

or $\qquad F_7 = -3 + 0.5 \times 7.5 + 0.5 \times 7.5$

$$= -3 + 7.5$$

from which, $\quad \boldsymbol{F_7 = 4.5 \text{ kN (tensile)}} \qquad (6.26)$

Consider joint (5); see Figure 6.39.

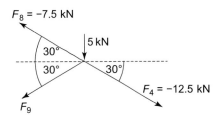

Figure 6.39

Resolving forces horizontally:

$$F_8 \cos 30° + F_9 \cos 30° = F_4 \cos 30°$$

i.e. $\qquad F_8 + F_9 = F_4 \qquad (6.27)$

Substituting equations (6.24) and (6.16) into equation (6.27) gives:

$$-7.5 + F_9 = -12.5$$

i.e. $\qquad F_9 = -12.5 + 7.5$

i.e. $\qquad \boldsymbol{F_9 = -5 \text{ kN (compressive)}}$

The results compare favourably with those obtained by the graphical method used in Problem 7; see the table below.

Member	Force (kN)
F_1	−11.5
F_2	9.96
F_3	10.83
F_4	−12.5
F_5	−4.0
F_6	−7.5
F_7	4.5
F_8	−7.5
F_9	−5.0

Now try the following Practise Exercise

6.5 The method of sections (a mathematical method)

In this method, an imaginary cut is made through the framework and the equilibrium of this part of the structure is considered through a free body diagram. No more than three unknown forces can be determined through any cut section, as only three equilibrium considerations can be made, namely

(a) resolve forces horizontally
(b) resolve forces vertically
(c) take moments about a convenient point.

Worked problem 11 demonstrates the method of sections.

Problem 11. Determine the unknown member forces F_2, F_5 and F_6 of the truss of Figure 6.34, Problem 10, page 81, by the method of sections.

Firstly, all members will be assumed to be in tension and an imaginary cut will be made through the framework, as shown by Figure 6.40.

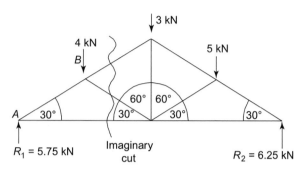

Figure 6.40

Taking moments about B; see Figure 6.41.

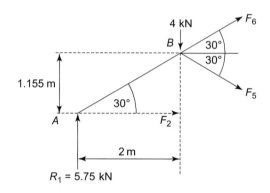

Figure 6.41

Clockwise moments = anti-clockwise moments

Hence, $\quad 5.75 \text{ kN} \times 2 \text{ m} = F_2 \times 1.155 \text{ m}$

(where $2 \tan 30° = 1.155$ m from Figure 6.41)

i.e. $\quad \boldsymbol{F_2 = \dfrac{5.75 \times 2}{1.155} = 9.96 \text{ kN (tensile)}} \quad (6.28)$

Resolving forces vertically:

$$5.75 \text{ kN} + F_6 \sin 30° = F_5 \sin 30° + 4 \text{ kN}$$

i.e. $\quad F_5 = F_6 + \dfrac{5.75}{0.5} - \dfrac{4}{0.5}$

i.e. $\quad F_5 = F_6 + 3.5 \quad (6.29)$

Resolving forces horizontally:

$$0 = F_2 + F_5 \cos 30° + F_6 \cos 30°$$

from which, $F_5 \cos 30° = -F_2 - F_6 \cos 30°$

and $\quad F_5 = -\dfrac{F_2}{\cos 30°} - F_6 \quad (6.30)$

Substituting equation (6.28) into equation (6.30) gives:

$$F_5 = -\frac{9.96}{0.866} - F_6$$

or $\qquad F_5 = -11.5 - F_6 \qquad$ (6.31)

Equating equation (6.29) to equation (6.31) gives:

$$F_6 + 3.5 = -11.5 - F_6$$

from which, $\qquad 2F_6 = -11.5 - 3.5 = -15$

and $\qquad F_6 = -\dfrac{15}{2} = \textbf{-7.5 kN (compressive)}$ (6.32)

Substituting equation (6.32) into equation (6.31) gives:

$$F_5 = -11.5 - (-7.5)$$
$$= \textbf{-4 kN (compressive)} \qquad (6.33)$$

i.e. $F_2 = \textbf{9.96 kN}$, $F_5 = \textbf{-4 kN}$ and $F_6 = \textbf{-7.5 kN}$

The above answers can be seen to be the same as those obtained in Problem 10.

Now try the following Practise Exercises

Practise Exercise 38   Further problems on the method of sections

Determine the internal member forces of the following trusses, by the method of sections:

1. Figure 6.23 (page 78)

$$\begin{bmatrix} R_1 = 3.0 \text{ kN}, R_2 = 1.0 \text{ kN}, \\ 1-2,\ 1.73 \text{ kN},\ 1-3, \\ -3.46 \text{ kN},\ 2-3,\ -2.0 \text{ kN} \end{bmatrix}$$

2. Figure 6.24 (page 79)

$$\begin{bmatrix} R_1 = -2.61 \text{ kN}, R_2 = 2.61 \text{ kN}, \\ H_2 = 6 \text{ kN},\ 1-2,\ -1.5 \text{ kN}, \\ 1-3,\ 3.0 \text{ kN},\ 2-3,\ -5.2 \text{ kN} \end{bmatrix}$$

3. Figure 6.25 (page 79)

$$\begin{bmatrix} R_1 = 5.0 \text{ kN}, R_2 = 1.0 \text{ kN}, \\ H_1 = 4.0 \text{ kN},\ 1-2,\ 1.0 \text{ kN}, \\ 1-3,\ -7.07 \text{ kN},\ 2-3,\ -1.41 \text{ kN} \end{bmatrix}$$

4. Figure 6.26 (page 79)

$$\begin{bmatrix} R_1 = 5.0 \text{ kN}, R_2 = 7.0 \text{ kN},\ 1-3, \\ -10.0 \text{ kN},\ 1-6,\ 8.7 \text{ kN},\ 3-4, \\ -8.0 \text{ kN},\ 3-6,\ -2.0 \text{ kN},\ 4-6, \\ 4.0 \text{ kN},\ 4-5,\ -8.0 \text{ kN},\ 5-6, \\ -6.0 \text{ kN},\ 5-2,\ -14.0 \text{ kN},\ 6-2,\ 12.1 \text{ kN} \end{bmatrix}$$

Practise Exercise 39   Short-answer questions on forces in structures

1. Where must the loads be applied on a pin-jointed truss?

2. If there are three unknown forces in a truss, how many simultaneous equations are required to determine these unknown forces?

3. When is a truss said to be statically indeterminate?

4. For a plane pin-jointed truss, what are the maximum number of unknowns that can exist at a joint to analyse that joint without analysing another joint before it?

Practise Exercise 40   Multiple-choice questions on forces in frameworks

(Answers on page 297)

1. The truss of Figure 6.42 is a:
    (a) mechanism
    (b) statically determinate
    (c) statically indeterminate

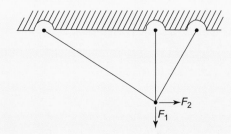

Figure 6.42

2. The value of F_1 in Figure 6.43 is:
    (a) 1 kN $\qquad\qquad$ (b) 0.5 kN
    (c) 0.707 kN

Figure 6.43

3. The value of F_2 in Figure 6.43 is:
   (a) 0.707 kN   (b) 0.5 kN
   (c) 0

4. If the Young's modulus is doubled in the members of a pin-jointed truss, and the external loads remain the same, the internal forces in the truss will:
   (a) double
   (b) halve
   (c) stay the same

5. If the Young's modulus is doubled in the members of a pin-jointed truss, and the loads remain the same, the deflection of the truss will:

   (a) double   (b) halve
   (c) stay the same

6. If the external loads in a certain truss are doubled, the internal forces will:
   (a) double   (b) halve
   (c) stay the same

References

[1] CASE, J, LORD CHILVER and ROSS, C.T.F. *Strength of Materials and Structures* Butterworth/Heinemann, 1999.
[2] ROSS, C.T.F., *Mechanics of Solids*, Woodhead Publishing Cambridge, U.K. 1999.

Chapter 7

Bending moment and shear force diagrams

The members of the structures in the previous chapter withstood the externally applied loads in either tension or compression; this was because they were not subjected to bending. In practise many structures are subjected to bending action; such structures include beams and rigid-jointed frameworks (- a rigid-jointed framework is one which has its joints welded or riveted or bolted together; such structures are beyond the scope of this text - see reference [1], on page 86). In this chapter, bending moments and shearing forces are defined and calculated, before the plotting of bending moment and shearing force diagrams are explained.

At the end of this chapter you should be able to:

- define a rigid-jointed framework
- define bending moment
- define sagging and hogging
- define shearing force
- calculate bending moments
- calculate shearing forces
- plot bending moment diagrams
- plot shearing force diagrams
- define the point of contraflexure

7.1 Bending moment (*M*)

The units of bending moment are N mm, N m, kN m, etc. When a beam is subjected to the couples shown in Figure 7.1, the beam will suffer flexure due to the bending moment of magnitude *M*.

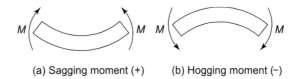

(a) Sagging moment (+)   (b) Hogging moment (−)

Figure 7.1 Bending moments

If the beam is in equilibrium and it is subjected to a clockwise couple of magnitude M on the left of the section, then from equilibrium considerations, the couple on the right of the section will be of exactly equal magnitude and of opposite direction to the couple on the left of the section. Thus, when calculating the bending moment at a particular point on a beam in equilibrium, we need only calculate the magnitude of the resultant of all the couples on one side of the beam under consideration. This is because as the beam is in equilibrium, the magnitude of the resultant of all the couples on the other side of the beam is exactly equal and opposite. The beam in Figure 7.1(a) is said to be **sagging** and the beam in Figure 7.1(b) is said to be **hogging**.

The **sign convention** adopted in this text is:

(a) **sagging moments** are said to be **positive**
(b) **hogging moments** are said to be **negative**

7.2 Shearing force (*F*)

Whereas a beam can fail due to its bending moments being excessive, it can also fail due to other forces being too large, namely the shearing forces; these

Mechanical Engineering Principles, Bird and Ross, ISBN 9780415517850

are shown in Figure 7.2. The units of shearing force are N, kN, MN, etc.

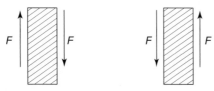

(a) Positive shearing force   (b) Negative shearing force

Figure 7.2 Shearing forces

It can be seen from Figures 7.2(a) and (b) that the shearing forces F act in a manner similar to that exerted by a pair of garden shears when they are used to cut a branch of a shrub or a plant through shearing action. This mode of failure is different to that caused by bending action.

In the case of the garden shears, it is necessary for the blades to be close together and sharp, so that they do not bend the branch at this point. If the garden shears are old and worn the branch can bend and may lie between the blades. Additionally, if the garden shears are not sharp, it may be more difficult to cut the branch because the shearing stress exerted by the blades will be smaller as the contact area between the blades and the branch will be larger.

The shearing action is illustrated by the sketch of Figure 7.3.

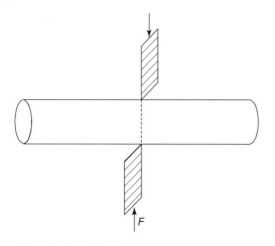

Figure 7.3 Shearing action

Once again, if the beam is in equilibrium, then the shearing forces either side of the point being considered will be exactly equal and opposite, as shown in Figures 7.2(a) and (b). The sign convention for shearing force is that it is said to be **positive if the right hand is going down**; see Figure 7.2(a).

Thus, when calculating the shearing force at a particular point on a horizontal beam, we need to calculate the **resultant** of all the vertical forces on **one side of the beam**, as the resultant of all the vertical forces on the other side of the beam will be exactly equal and opposite. The calculation of bending moments and shearing forces and the plotting of their respective diagrams are demonstrated in the following worked problems.

7.3 Worked problems on bending moment and shearing force diagrams

Problem 1. Calculate and sketch the bending moment and shearing force diagrams for the horizontal beam shown in Figure 7.4, which is simply supported at its ends.

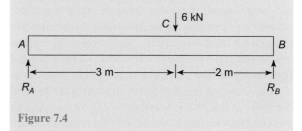

Figure 7.4

Firstly, it will be necessary to calculate the magnitude of reactions R_A and R_B

Taking moments about B gives:

$$\text{Clockwise moments about } B = \text{anti-clockwise moments about } B$$

i.e. $\quad R_A \times 5 \text{ m} = 6 \text{ kN} \times 2 \text{ m} = 12 \text{ kN m}$

from which, $\quad \boldsymbol{R_A = \dfrac{12}{5} = 2.4 \text{ kN}}$

Resolving forces vertically gives:

$$\text{Upward forces} = \text{downward forces}$$

i.e. $\quad R_A + R_B = 6 \text{ kN}$

i.e. $\quad 2.4 + R_B = 6$

from which, $\quad \boldsymbol{R_B = 6 - 2.4 = 3.6 \text{ kN}}$

As there is a discontinuity at point C in Figure 7.4, due to the concentrated load of 6 kN, it will be necessary to consider the length of the beam AC separately from the length of the beam CB. The reason for this is that the equations for bending moment and shearing force for span AC are different to the equation for the span *CB*; this is caused by the concentrated load of 6 kN.

For the present problem, to demonstrate the nature of bending moment and shearing force, these values will be calculated on both sides of the point of the beam under consideration. It should be noted that normally, the bending moment and shearing force at any point on the beam, are calculated only due to the resultant couples or forces, respectively, on one side of the beam.

Consider span *AC*:

Bending moment

Consider a section of the beam at a distance *x* from the left end *A*, where the value of *x* lies between *A* and *C*, as shown in Figure 7.5.

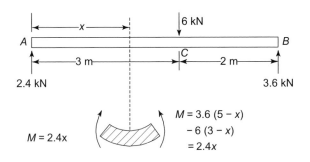

Figure 7.5

From Figure 7.5, it can be seen that the reaction R_A causes a clockwise moment of magnitude $R_A \times x = 2.4x$ on the left of this section and as shown in the lower diagram of Figure 7.5. It can also be seen from the upper diagram of Figure 7.5, that the forces on the right of this section on the beam causes an anti-clockwise moment equal to $R_B \times (5-x)$ or $3.6(5-x)$ and a clockwise moment of $6 \times (3-x)$, resulting in an anti-clockwise moment of:

$$3.6(5-x) - 6(3-x) = 3.6 \times 5 - 3.6x - 6 \times 3 + 6x$$
$$= 18 - 3.6x - 18 + 6x$$
$$= 2.4x$$

Thus, the left side of the beam at this section is subjected to a clockwise moment of magnitude 2.4*x* and the right side of this section is subjected to an anti-clockwise moment of 2.4*x*, as shown by the lower diagram of Figure 7.5. As the two moments are of equal magnitude but of opposite direction, they cause the beam to be subjected to a bending moment $M = 2.4x$. As this bending moment causes the beam to sag

between *A* and *C*, the bending moment is assumed to be positive, or at any distance *x* between *A* and *C*:

$$\text{Bending moment} = M = +2.4x \qquad (7.1)$$

Shearing force

Here again, because there is a discontinuity at *C*, due to the concentrated load of 6 kN at *C*, we must consider a section of the beam at a distance *x* from the left end *A*, where *x* varies between *A* and *C*, as shown in Figure 7.6.

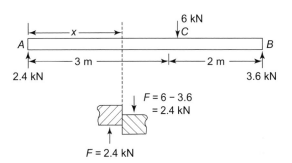

Figure 7.6

From Figure 7.6, it can be seen that the resultant of the vertical forces on the left of the section at *x* are 2.4 kN acting upwards. This force causes the left of the section at *x* to slide upwards, as shown in the lower diagram of Figure 7.6. Similarly, if the vertical forces on the right of the section at *x* are considered, it can be seen that the 6 kN acts downwards and that $R_B = 3.6$ kN acts upwards, giving a resultant of 2.4 kN acting downwards. The effect of the two shearing forces acting on the left and the right of the section at *x*, causes the shearing action shown in the lower diagram of Figure 7.6. As this shearing action causes the right side of the section to glide downwards, it is said to be a positive shearing force.

Summarising, at any distance *x* between *A* and *C*:

$$F = \text{shearing force} = +2.4 \text{ kN} \qquad (7.2)$$

Consider span *CB*:

Bending moment

At any distance *x* between *C* and *B*, the resultant moment caused by the forces on the left of *x* is given by:

$$M = R_A \times x - 6(x-3) = 2.4x - 6(x-3)$$
$$= 2.4x - 6x + 18$$

i.e. $\qquad M = 18 - 3.6x \text{ (clockwise)} \qquad (7.3)$

The effect of this resultant moment on the left of *x* is shown in the lower diagram of Figure 7.7.

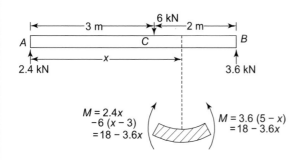

Figure 7.7

Now from Figure 7.7, it can be seen that on the right side of x, there is an anti-clockwise moment of:

$$M = R_B \times (5 - x) = 3.6(5 - x) = 18 - 3.6x$$

i.e. $M = 18 - 3.6x$ (anti-clockwise)   (7.4)

The effect of the moment of equation (7.3) and that of the moment of equation (7.4), is to cause the beam to sag at this point as shown by the lower diagram of Figure 7.7, i.e. M is positive between C and B, and

$$M = +18 - 3.6x \qquad (7.5)$$

Shearing force

Consider a distance x between C and B, as shown in Figure 7.8.

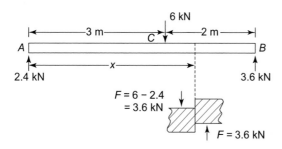

Figure 7.8

From Figure 7.8, it can be seen that at x, there are two vertical forces to the left of this section, namely the 6 kN load acting downwards and the 2.4 kN load acting upwards, resulting in a net value of 3.6 kN acting downwards, as shown by the lower diagram of Figure 7.8. Similarly, by considering the vertical forces acting on the beam to the right of x, it can be seen that there is one vertical force, namely the 3.6 kN load acting upwards, as shown by the lower diagram of Figure 7.8. Thus, as the right hand of the section is tending to slide upwards, the shearing force is said to be negative, i.e. between C and B,

$$F = -3.6 \text{ kN} \qquad (7.6)$$

It should be noted that at C, there is a discontinuity in the value of the shearing force, where over an infinitesimal length the shearing force changes from +2.4 kN to – 3.6 kN, from left to right.

Bending moment and shearing force diagrams

The bending moment and shearing force diagrams are simply diagrams representing the variation of bending moment and shearing force, respectively, along the length of the beam. In the bending moment and shearing force diagrams, the values of the bending moments and shearing forces are plotted vertically and the value of x is plotted horizontally, where $x = 0$ at the left end of the beam and $x = $ the whole length of the beam at its right end.

In the case of the beam of Figure 7.4, bending moment distribution between A and C is given by equation (7.1), i.e. $M = 2.4x$, where the value of x varies between A and C.

At A, $x = 0$, therefore $M_A = 2.4 \times 0 = 0$

and at C, $x = 3$ m, therefore $M_C = 2.4 \times 3 = 7.2$ kN m

Additionally, as the equation $M = 2.4x$ is a straight line, the bending moment distribution between A and C will be as shown by the left side of Figure 7.9(a).

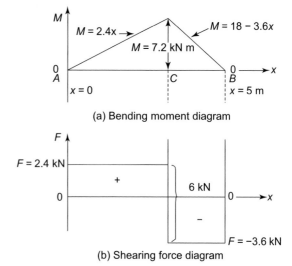

(a) Bending moment diagram

(b) Shearing force diagram

Figure 7.9 Bending moment and shearing force diagrams

Similarly, the expression for the variation of bending moment between C and B is given by equation (7.3), i.e. $M = 18 - 3.6x$, where the value of x varies between C and B. The equation can be seen to be a straight line between C and B.

At C, $x = 3$ m,

therefore $M_C = 18 - 3.6 \times 3 = 18 - 10.8 = 7.2$ kN m

At B, $x = 5$ m,

therefore $M_B = 18 - 3.6 \times 5 = 18 - 18 = 0$

Therefore, plotting of the equation $M = 18 - 3.6x$ between C and B results in the straight line on the right of Figure 7.9(a), i.e. the bending moment diagram for this beam has now been drawn.

In the case of the beam of Figure 7.4, the shearing force distribution along its length from A to C is given by equation (7.2), i.e. $F = 2.4$ kN, i.e. F is constant between A and C. Thus the shearing force diagram between A and C is given by the horizontal line shown on the left of C in Figure 7.9(b).

Similarly, the shearing force distribution to the right of C is given by equation (7.6), i.e. $F = -3.6$ kN, i.e. F is a constant between C and B, as shown by the horizontal line to the right of C in Figure 7.9(b). At the point C, the shearing force is indeterminate and changes from +2.4 kN to –3.6 kN over an infinitesimal length.

> **Problem 2.** Determine expressions for the distributions of bending moment and shearing force for the horizontal beam of Figure 7.10. Hence, sketch the bending and shearing force diagrams.
>
>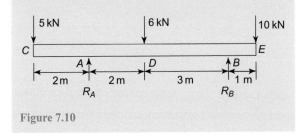
>
> **Figure 7.10**

Firstly, it will be necessary to calculate the unknown reactions R_A and R_B

Taking moments about B gives:

$$R_A \times 5 \text{ m} + 10 \text{ kN} \times 1 \text{ m} = 5 \text{ kN} \times 7 \text{ m} + 6 \text{ kN} \times 3 \text{ m}$$

i.e. $5 R_A + 10 = 35 + 18$

$5 R_A = 35 + 18 - 10 = 43$

from which, $R_A = \dfrac{43}{5} = \textbf{8.6 kN}$

Resolving forces vertically gives:

$$R_A + R_B = 5 \text{ kN} + 6 \text{ kN} + 10 \text{ kN}$$

i.e. $8.6 + R_B = 21$

from which, $R_B = 21 - 8.6 = \textbf{12.4 kN}$

For the range C to A, see Figure 7.11.

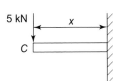

Figure 7.11

To calculate the bending moment distribution (M), only the resultant of the moments to the left of the section at x will be considered, as the resultant of the moments on the right of the section of x will be exactly equal and opposite.

Bending moment (BM)

From Figure 7.11, at any distance x,

$$M = -5 \times x \text{ (hogging)} = \textbf{-5}\boldsymbol{x} \qquad (7.7)$$

Equation (7.7) is a straight line between C and A.

At C, $x = 0$, therefore $M_C = -5 \times 0 = \textbf{0 kN m}$

At A, $x = 2$ m, therefore $M_A = -5 \times 2 = \textbf{-10 kN m}$

Shearing force (SF)

To calculate the shearing force distribution (F) at any distance x, only the resultant of the vertical forces to the left of x will be considered, as the resultant of the vertical forces to the right of x will be exactly equal and opposite.

From Figure 7.11, at any distance x,

$$F = \textbf{-5 kN} \qquad (7.8)$$

It is negative, because as the left of the section tends to slide downwards, the right of the section tends to slide upwards. (Remember, right hand down is positive).

For the range A to D, see Figure 7.12.

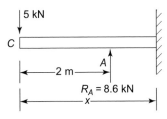

Figure 7.12

Bending moment (BM)

At any distance x between A and D

$$M = -5 \times x + R_A \times (x - 2)$$
$$= -5x + 8.6(x - 2)$$
$$= -5x + 8.6x - 17.2$$

i.e. $M = 3.6x - 17.2$

(a straight line between A and D) (7.9)

At A, $x = 2$ m, $M_A = 3.6 \times 2 - 17.2$

$= 7.2 - 17.2 = -10$ **kN m**

At D, $x = 4$ m, $M_D = 3.6 \times 4 - 17.2$

$= 14.4 - 17.2 = -2.8$ **kN m**

Shearing force (SF)

At any distance x between A and D,   (7.10)

$F = -5$ kN $+ 8.6$ kN $= 3.6$ **kN** (constant)

For the range D to B, see Figure 7.13.

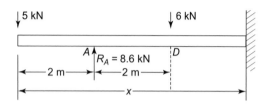

Figure 7.13

Bending moment (BM)

At x, $M = -5 \times x + 8.6 \times (x - 2) - 6 \times (x - 4)$
$= -5x + 8.6x - 17.2 - 6x + 24$

i.e. $M = -2.4x + 6.8$ (7.11)

(a straight line between D and B)

At D, $x = 4$ m, therefore $M_D = -2.4 \times 4 + 6.8$
$= -9.6 + 6.8$

i.e. $M_D = -2.8$ **kN m**

At B, $x = 7$ m, therefore $M_B = -2.4 \times 7 + 6.8$
$= -16.8 + 6.8$

i.e. $M_B = -10$ **kN m**

Shearing force (SF)

At x, $F = -5 + 8.6 - 6 = -2.4$ **kN** (constant)   (7.12)

For the range B to E, see Figure 7.14.

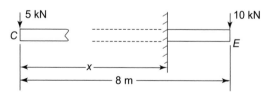

Figure 7.14

Bending moment (BM)

In this case it will be convenient to consider only the resultant of the couples to the right of x (-remember

that only one side need be considered, and in this case, there is only one load to the right of x).

At x, $M = -10 \times (8 - x) = -80 + 10x$ (7.13)

Equation (7.13) can be seen to be a straight line between B and E.

At B, $x = 7$ m,

therefore $M_B = -80 + 10 \times 7 = -10$ **kN m**

At E, $x = 8$ m,

therefore $M_E = -80 + 10 \times 8 = 0$ **kN m**

Shearing force (SF)

At x, $F = +10$ **kN** (constant)   (7.14)

Equation (7.14) is positive because the shearing force is causing the right side to slide downwards.

The bending moment and shearing force diagrams are plotted in Figure 7.15 with the aid of equations (7.7) to (7.14) and the associated calculations at C, A, D, B and E.

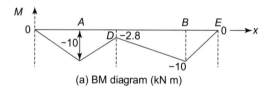

(a) BM diagram (kN m)

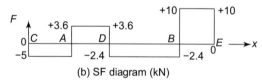

(b) SF diagram (kN)

Figure 7.15

Problem 3.   Determine expressions for the bending moment and shearing force distributions for the beam of Figure 7.16. Hence, sketch the bending moment and shearing force diagrams.

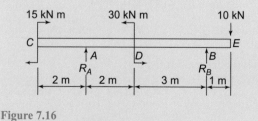

Figure 7.16

Firstly, it will be necessary to calculate the reactions R_A and R_B

Taking moments about B gives:

$$15 \text{ kN m} + R_A \times 5 \text{ m} + 10 \text{ kN} \times 1 \text{ m} = 30 \text{ kN m}$$

i.e. $5 R_A = 30 - 10 - 15 = 5$

from which, $R_A = \dfrac{5}{5} = \mathbf{1 \text{ kN}}$

Resolving forces vertically gives:

$$R_A + R_B = 10 \text{ kN}$$

i.e. $1 + R_B = 10$

from which, $R_A = 10 - 1 = \mathbf{9 \text{ kN}}$

For the span C to A, see Figure 7.17

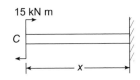

Figure 7.17

Bending moment (BM)

At x, $M = \mathbf{15 \text{ kN m}}$ (constant)   (7.15)

Shearing force (SF)

At $x = 0$, $F = \mathbf{0 \text{ kN}}$ (7.16)

For the span A to D, see Figure 7.18.

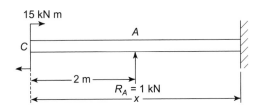

Figure 7.18

Bending moment (BM)

At x, $\begin{aligned} M &= 15 \text{ kN m} + R_A \times (x-2) \\ &= 15 + 1(x-2) \\ &= 15 + x - 2 \end{aligned}$

i.e. $M = \mathbf{13 + x}$ (a straight line)   (7.17)

At A, $x = 2$ m,

therefore $M_A = 13 + 2 = \mathbf{15 \text{ kN m}}$

At D, $x = 4$ m,

therefore $M_{D_{(-)}} = 13 + 4 = \mathbf{17 \text{ kN m}}$

Note that $M_{D_{(-)}}$ means that M is calculated to the left of D

Shearing force (SF)

At x, $F = \mathbf{1 \text{ kN}}$ (constant)   (7.18)

For the span D to B, see Figure 7.19.

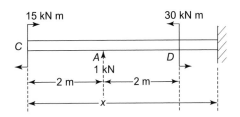

Figure 7.19

Bending moment (BM)

At x, $\begin{aligned} M &= 15 \text{ kN m} + 1 \text{ kN m} \times (x-2) - 30 \text{ kN m} \\ &= 15 + x - 2 - 30 \end{aligned}$

i.e $M = \mathbf{x - 17}$ (a straight line)   (7.19)

At D, $x = 4$ m,

therefore $M_{D_{(+)}} = 4 - 17 = \mathbf{-13 \text{ kN m}}$

Note that $M_{D_{(+)}}$ means that M is calculated just to the right of D.

At B, $x = 7$ m,

therefore $M_{B_{(-)}} = 7 - 17 = \mathbf{-10 \text{ kN m}}$

Shearing force (SF)

At x, $F = \mathbf{-1 \text{ kN}}$ (constant)   (7.20)

For the span B to E, see Figure 7.20.

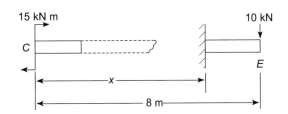

Figure 7.20

In this case we will consider the right of the beam as there is only one force to the right of the section at x.

$$M = -10 \times (8 - x) = -80 + 10x \text{ (a straight line)}$$

At x, $F = \mathbf{10 \text{ kN}}$ (positive as the right hand is going down, and constant).

Plotting the above equations, for the various spans, results in the bending moment and shearing force diagrams of Figure 7.21.

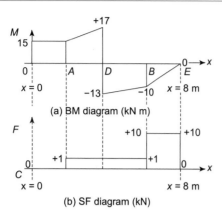

(a) BM diagram (kN m)

(b) SF diagram (kN)

Figure 7.21

Problem 4. Calculate and plot the bending moment and shearing force distributions for the cantilever of Figure 7.22.

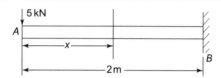

Figure 7.22

In the cantilever of Figure 7.22, the left hand end is free and the right hand end is firmly fixed; the right hand end is called the **constrained end.**

Bending moment (BM)

At x in Figure 7.22, $M = -5$ kN $\times x$

i.e. $\qquad M = -5x$ (a straight line) $\qquad$ (7.21)

Shearing force (SF)

At x in Figure 7.22, $F = -5$ kN (a constant) $\qquad$ (7.22)

For equations (7.21) and (7.22), it can be seen that the bending moment and shearing force diagrams are as shown in Figure 7.23.

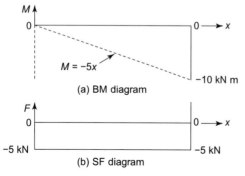

(a) BM diagram

(b) SF diagram

Figure 7.23

Problem 5. Determine the bending moment and shearing force diagram for the cantilever shown in Figure 7.24, which is rigidly constrained at the end B.

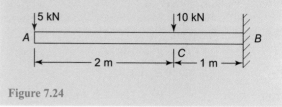

Figure 7.24

For the span A to C, see Figure 7.25.

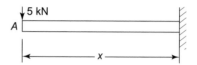

Figure 7.25

Bending moment (BM)

At x, $\qquad M = -5$ kN $\times x$

i.e. $\qquad M = -5x$ (a straight line) $\qquad$ (7.23)

Shearing force (SF)

At x, $\qquad F = -5$ kN (constant) $\qquad$ (7.24)

For the span C to B, see Figure 7.26.

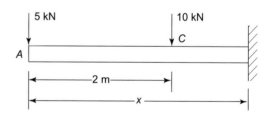

Figure 7.26

Bending moment (BM)

At x, $\qquad M = -5$ kN $\times x - 10$ kN $\times (x - 2)$

$\qquad = -5x - 10x + 20$

i.e. $\qquad M = 20 - 15x$ (a straight line) $\qquad$ (7.25)

At C, $x = 2$ m, therefore $M_C = 20 - 15 \times 2$

$\qquad = 20 - 30$

i.e. $\qquad M_C = -10$ kN m

At B, $x = 3$ m, therefore $M_B = 20 - 15 \times 3$

$\qquad = 20 - 45$

i.e. $\qquad M_B = -25$ kN m

Shearing force (SF)

At x in Figure 7.26, $F = -5$ kN $- 10$ kN

i.e. $\qquad F = -15$ **kN** (constant) $\qquad$ (7.26)

From equations (7.23) to (7.26) and the associated calculations, the bending moment and shearing force diagrams can be plotted, as shown in Figure 7.27.

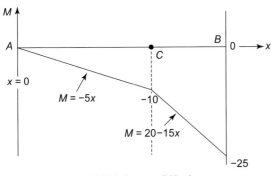

(a) BM diagram (kN m)

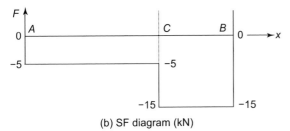

(b) SF diagram (kN)

Figure 7.27

> Problem 6. A uniform section, 6 m long, horizontal beam is simply-supported at its ends. If this beam is subjected to two vertically applied downward loads, one of magnitude 5 kN at 2 m from the left support, and the other of magnitude 10 kN, 4 m from the left support, calculate and plot the bending moment and shear force diagrams.

The loaded beam is shown in Figure 7.28.

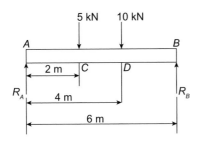

Figure 7.28

Taking moments about B:

$$R_A \times 6 = 5 \times 4 + 10 \times 2$$

i.e. $\quad 6R_A = 40$ and $R_A = \dfrac{40}{6} = \mathbf{6.67}$ **kN**

$$R_A + R_B = 5 + 10 = 15 \text{ kN}$$

hence $\qquad R_B = 15 - 6.67 = \mathbf{8.33}$ **kN**

Bending moments

At A, **BM** $= 0$; At C, **BM** $= R_A \times 2 = 6.67 \times 2$
$$= \mathbf{13.34 \text{ kN m}}$$

At D, **BM** $= R_A \times 4 - 5 \times 2 = 6.67 \times 4 - 5 \times 2$
$$= \mathbf{16.68 \text{ kN m}}$$

At B, **BM** $= 0$

Shearing force

At A, **SF** $= + R_A = \mathbf{6.67}$ **kN**

At C, **SF** $= 6.67 - 5 = \mathbf{1.67}$ **kN**

At D, **SF** $= 6.67 - 5 - 10 = \mathbf{-8.33}$ **kN**

At B, **SF** $= \mathbf{-8.33}$ **kN**

The bending moment and shearing force diagram is shown in Figure 7.29.

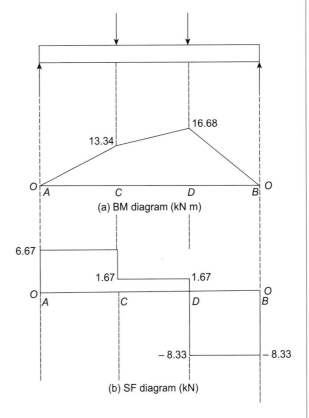

(a) BM diagram (kN m)

(b) SF diagram (kN)

Figure 7.29

Part Two

Now try the following Practise Exercise

Practise Exercise 41   Further problems on bending moment and shearing force diagrams

Determine expressions for the bending moment and shearing force distributions for each of the following simply supported beams; hence, or otherwise, plot the bending moment and shearing force diagrams.

1.

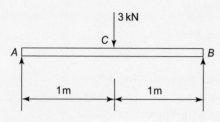

Figure 7.30

[see Figure 7.45(a) on page 100)]

2.

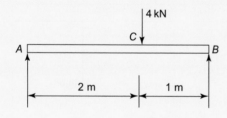

Figure 7.31

[see Figure 7.45(b) on page 100)]

3.

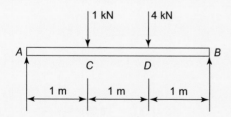

Figure 7.32

[see Figure 7.45(c) on page 100)]

4.

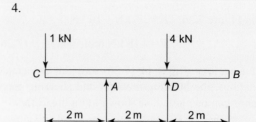

Figure 7.33

[see Figure 7.45(d) on page 100)]

5.

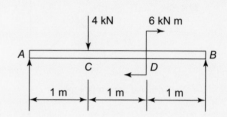

Figure 7.34

[see Figure 7.45(e) on page 101)]

6.

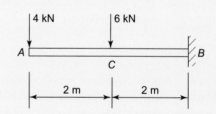

Figure 7.35

[see Figure 7.45(f) on page 101)]

7.

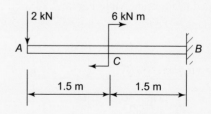

Figure 7.36

[see Figure 7.45(g) on page 101)]

8. A horizontal beam of negligible mass is of length 7 m. The beam is simply-supported at its ends and carries three vertical loads, pointing in a downward direction. The first load is of magnitude 3 kN and acts 2 m from the left end, the second load is of magnitude 2 kN and acts 4 m from the left end, and the third load is of magnitude 4 kN and acts 6 m from the left end. Calculate the bending moment and shearing force at the points of discontinuity, working from the left support to the right support.

[Bending moments (kN m): 0, 7.14, 8.28, 5.42, 0; Shearing forces (kN): 3.57, 3.57/0.57, 0.57/−1.43, −1.43/−5.43, −5.43]

7.4   Uniformly distributed loads

Uniformly distributed loads (UDL) appear as snow loads, self-weight of the beam, uniform pressure loads, and so on. In all cases they are assumed to be spread uniformly over the length of the beam in which they apply. The units for a uniformly distributed load are N/m, kN/m, MN/m, and so on.

Worked problems 7 and 8 involve uniformly distributed loads.

Problem 7. Determine expressions for the bending moment and shearing force distributions for the cantilever shown in Figure 7.37, which is subjected to a uniformly distributed load, acting downwards, and spread over the entire length of the cantilever.

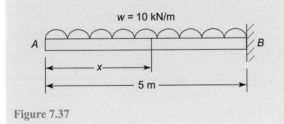

Figure 7.37

Bending moment (BM)

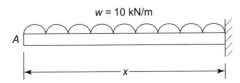

Figure 7.38

At any distance x in Figure 7.38,

$$M = -10 \text{ kN} \times x \times \frac{x}{2} \tag{7.27}$$

i.e. $M = -5x^2$ (a parabola)   (7.28)

In equation (7.27), the weight of the uniformly distributed beam up to the point x is $(10 \times x)$. As the centre of gravity of the *UDL* is at a distance of $\frac{x}{2}$ from the right end of Figure 7.38, $M = -10x \times \frac{x}{2}$

The equation is negative because the beam is hogging.

At $x = 0$, $M = 0$

At $x = 5$ m, $M = -10 \times 5 \times \frac{5}{2} = -125$ kN m

Shearing force (SF)

At any distance x in Figure 7.38, the weight of the UDL is $(10 \times x)$ and this causes the left side to slide down, or alternatively the right side to slide up.

Hence, $F = -10x$ (a straight line)   (7.29)

At $x = 0$, $F = 0$

At $x = 5$ m, $F = -10 \times 5 = -50$ kN

Plotting of equations (7.28) and (7.29) results in the distributions for the bending moment and shearing force diagrams shown in Figure 7.39.

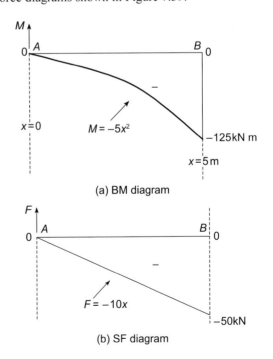

Figure 7.39

Problem 8. Determine expressions for the bending moment and shearing force diagrams for the simply supported beam of Figure 7.40. The beam is subjected to a uniformly distributed load (*UDL*) of 5 kN/m, which acts downwards, and it is spread over the entire length of the beam.

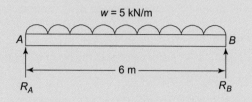

Figure 7.40

Firstly, it will be necessary to calculate the reactions R_A and R_B. As the beam is symmetrically loaded, it is evident that:

$$R_A = R_B \tag{7.30}$$

Taking moments about B gives:

Clockwise moments about B =
anti-clockwise moments about B

i.e. $$R_A \times 6 \text{ m} = 5\, \frac{\text{kN}}{\text{m}} \times 6 \text{ m} \times 3 \text{ m} \tag{7.31}$$

$$= 90 \text{ kN m}$$

from which, $$\boldsymbol{R_A} = \frac{90}{6} = \textbf{15 kN} = \boldsymbol{R_B}$$

On the right hand side of equation (7.31), the term 5 kN/m × 6 m is the weight of the *UDL*, and the length of 3 m is the distance of the centre of gravity of the *UDL* from B.

Bending moment (BM)

At any distance x in Figure 7.41,

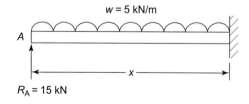

Figure 7.41

$$M = R_A \times x - 5\, \frac{\text{kN}}{\text{m}} \times x \times \frac{x}{2} \tag{7.32}$$

i.e. $$\boldsymbol{M = 15x - 2.5x^2} \text{ (a parabola)} \tag{7.33}$$

On the right hand side of equation (7.32), the term: $(R_A \times x)$ is the bending moment (sagging) caused by

the reaction, and the term $\left(5\, \frac{\text{kN}}{\text{m}} \times x \times \frac{x}{2}\right)$, which is hogging, is caused by the *UDL*, where $(5\, \frac{\text{kN}}{\text{m}} \times x)$ is the weight of the *UDL* up to the point x, and $\frac{x}{2}$ is the distance of the centre of gravity of the *UDL* from the right side of Figure 7.41.

At $x = 0$, $M = 0$

At $x = 3$ m, $M = 15 \times 3 - 2.5 \times 3^2 = 22.5$ kN m

At $x = 6$ m, $M = 15 \times 6 - 2.5 \times 6^2 = 0$

Shearing force (SF)

At any distance x in Figure 7.41,

$$F = R_A - 5\, \frac{\text{kN}}{\text{m}} \times x \tag{7.34}$$

i.e. $$\boldsymbol{F = 15 - 5x} \text{ (a straight line)} \tag{7.35}$$

On the right hand side of equation (7.34), the term $\left(5\, \frac{\text{kN}}{\text{m}} \times x\right)$ is the weight of the *UDL* up to the point x; this causes a negative configuration to the shearing force as it is causing the left side to slide downwards.

At $x = 0$, $F = 15$ kN

At $x = 3$ m, $F = 15 - 5 \times 3 = 0$

At $x = 6$ m, $F = 15 - 5 \times 6 = -15$ kN

Plotting of equations (7.33) and (7.35) results in the bending moment and shearing force diagrams of Figure 7.42.

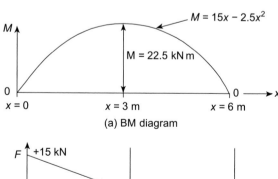

(a) BM diagram

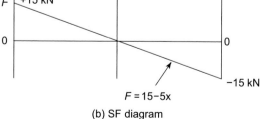

(b) SF diagram

Figure 7.42

Now try the following Practise Exercises

Practise Exercise 42 **Further problems on bending moment and shearing force diagrams**

Determine expressions for the bending moment and shearing force distributions for each of the following simply supported beams; hence, plot the bending moment and shearing force diagrams.

1. Figure 7.43(a)

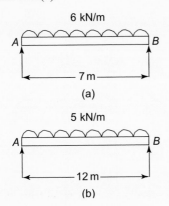

Figure 7.43 Simply supported beams

[see Figure 7.46(a) on page 101)]

2. Figure 7.43(b)
    [see Figure 7.46(b) on page 101)]

Determine expressions for the bending moment and shearing force distributions for each of the following cantilevers; hence, or otherwise, plot the bending moment and shearing force diagrams.

3. Figure 7.44(a)

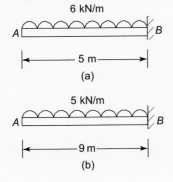

Figure 7.44 Cantilevers

[see Figure 7.47(a) on page 101)]

4. Figure 7.44(b)
    [see Figure 7.47(b) on page 101)]

Practise Exercise 43 **Short-answer questions on bending moment and shearing force diagrams**

1. Define a rigid-jointed framework.

2. Define bending moment.

3. Define sagging and hogging.

4. State two practical examples of uniformly distributed loads.

5. Show that the value of the maximum shearing force for a beam simply supported at its ends, with a centrally placed load of 3 kN, is 1.5 kN.

6. If the beam in question 5 were of span 4 m, show that its maximum bending moment is 3 kN m.

7. Show that the values of maximum bending moment and shearing force for a cantilever of length 4 m, loaded at its free end with a concentrated load of 3 kN, are 12 kN m and 3 kN.

Practise Exercise 44 **Multiple-choice questions on bending moment and shearing force diagrams**

(Answers on page 297)

1. A beam simply supported at its end, carries a centrally placed load of 4 kN. Its maximum shearing force is:

    (a) 4 kN   (b) 2 kN
    (c) 8 kN   (d) 0

2. Instead of the centrally placed load, the beam of question 1 has a uniformly distributed load of 1 kN/m spread over its span of length 4 m. Its maximum shearing force is now:

    (a) 4 kN   (b) 1 kN
    (c) 2 kN   (d) 4 kN

Part Two

3. A cantilever of length 3 m has a load of 4 kN placed on its free end. The magnitude of its maximum bending moment is:
   (a) 3 kN m   (b) 4 kN m
   (c) 12 kN m   (d) 4/3 kN/m

4. The magnitude of the maximum shearing force for the cantilever of question 3 is:
   (a) 4 kN   (b) 12 kN
   (c) 3 kN   (d) zero

5. A cantilever of 3 m length carries a *UDL* of 2 kN/m. The magnitude of the maximum shearing force is:
   (a) 3 kN   (b) 2 kN
   (c) 6 kN   (d) zero

6. In the cantilever of question 5, the magnitude of the maximum bending moment is:
   (a) 6 kN m   (b) 9 kN m
   (c) 2 kN m   (d) 3 kN m

Answers to Exercise 41 (page 96)

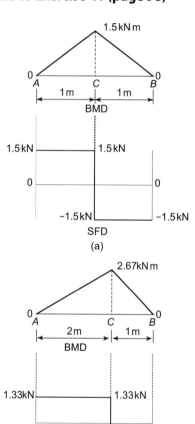

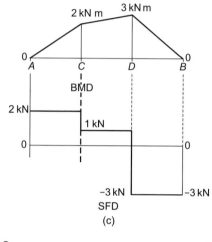

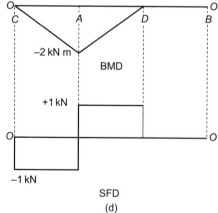

Figure 7.45

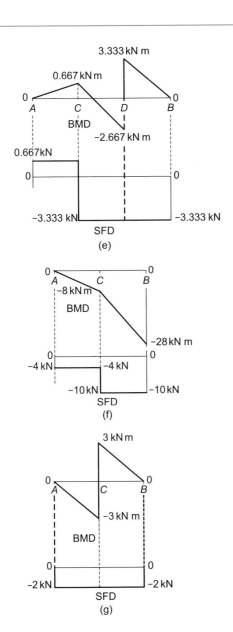

Figure 7.45 *(Continued)*

Answers to Exercise 42 (page 99)

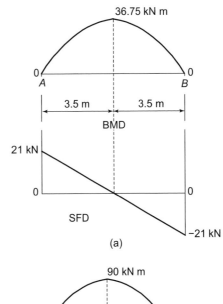

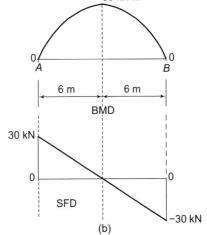

Figure 7.46

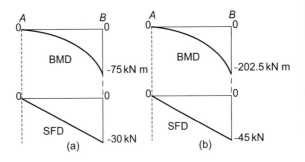

Figure 7.47

Part Two

Chapter 8

First and second moment of areas

The first moment of area is usually used to find the centroid or centre of gravity of a two-dimensional shape or an artefact. The position of the centroid and centre of gravity is fundamental and often it is very important in very many branches of engineering. The second moment of area can be described as the geometric ability of a two-dimensional figure or lamina to resist rotation about an axis through its plane. It is extensively used in structures, where it measures the geometric ability of the structure to resist bending. Likewise in the hydrostatic stability of ships and yachts, the second moment of area of a yacht or a ship's water plane is a measure of the yacht or ship to resist rotations, such as those due to 'heeling' and 'trimming'. This chapter starts by defining the first and second moments of area of two-dimensional figures and laminas, and extends this to shapes that are useful in engineering, such as the shapes of the cross-sections of beams. For example, I beams, Tee beams and channel cross-sections are all very important in structural engineering.

At the end of this chapter you should be able to:

- define a centroid
- define first moment of area
- calculate centroids using integration
- define second moment of area
- define radius of gyration
- state the parallel axis and perpendicular axis theorems
- calculate the second moment of area and radius of gyration of regular sections using a table of standard results
- calculate the second moment of area of I, T and channel bar beam sections

8.1 Centroids

A **lamina** is a thin flat sheet having uniform thickness. The **centre of gravity** of a lamina is the point where it balances perfectly, i.e. the lamina's **centre of moment of mass**. When dealing with an area (i.e. a lamina of negligible thickness and mass) the term **centre of moment of area** or **centroid** is used for the point where the centre of gravity of a lamina of that shape would lie.

8.2 The first moment of area

The **first moment of area** is defined as the product of the area and the perpendicular distance of its centroid from a given axis in the plane of the area. In Figure 8.1, the first moment of area A about axis XX is given by (Ay) cubic units.

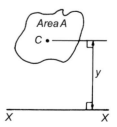

Figure 8.1

Mechanical Engineering Principles, Bird and Ross, ISBN 9780415517850

8.3   Centroid of area between a curve and the *x*-axis

(i)   Figure 8.2 shows an area *PQRS* bounded by the curve $y = f(x)$, the *x*-axis and ordinates $x = a$ and $x = b$. Let this area be divided into a large number of strips, each of width δx. A typical strip is shown shaded drawn at point (x, y) on $f(x)$. The area of the strip is approximately rectangular and is given by $y\delta x$. The centroid, *C*, has coordinates $\left(x, \dfrac{y}{2}\right)$

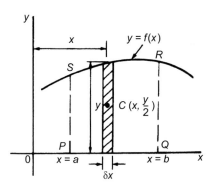

Figure 8.2

(ii) First moment of area of shaded strip about axis
$$Oy = (y\delta x)(x) = xy\delta x$$
Total first moment of area *PQRS* about axis
$$Oy = \lim_{\delta x \to 0} \sum_{x=a}^{x=b} xy\,\delta x = \int_a^b xy\,dx$$

(iii) First moment of area of shaded strip about axis
$$Ox = (y\delta x)\left(\frac{y}{2}\right) = \frac{1}{2}y^2 x$$
Total first moment of area *PQRS* about axis
$$Ox = \lim_{\delta x \to 0} \sum_{x=a}^{x=b} \frac{1}{2}y^2 = \frac{1}{2}\int_a^b y^2\,dx$$

(iv) Area of *PQRS*, $A = \int_a^b y\,dx$ (see *Engineering mathematics, 6th Edition, page 500*)

(v)   Let $\bar{x}$ and $\bar{y}$ be the distances of the centroid of area *A* about *Oy* and *Ox* respectively then:
$$(\bar{x})(A) = \text{total first moment of area } A \text{ about axis}$$
$$Oy = \int_a^b xy\,dx$$

from which, $\bar{x} = \dfrac{\displaystyle\int_a^b xy\,dy}{\displaystyle\int_a^b y\,dy}$

and $(\bar{y})(A) = $ total moment of area *A* about axis

$$Ox = \frac{1}{2}\int_a^b y^2\,dx$$

from which, $\bar{y} = \dfrac{\dfrac{1}{2}\displaystyle\int_a^b y^2\,dx}{\displaystyle\int_a^b y\,dx}$

8.4   Centroid of area between a curve and the *y*-axis

If $\bar{x}$ and $\bar{y}$ are the distances of the centroid of area *EFGH* in Figure 8.3 from *Oy* and *Ox* respectively, then, by similar reasoning as above:

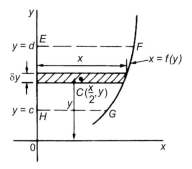

Figure 8.3

$$(\bar{x})(\text{total area}) = \lim_{\delta y \to 0} \sum_{y=c}^{y=d} x\,\delta y\left(\frac{x}{2}\right) = \frac{1}{2}\int_c^d x^2\,dy$$

from which, $\bar{x} = \dfrac{\dfrac{1}{2}\displaystyle\int_c^d x^2\,dy}{\displaystyle\int_c^d x\,dy}$

and $(\bar{y})(\text{total area}) = \lim_{\delta y \to 0} \sum_{y=c}^{y=d} (x\,\delta y)y = \int_c^d xy\,dy$

from which, $\bar{y} = \dfrac{\displaystyle\int_c^d xy\,dy}{\displaystyle\int_c^d x\,dy}$

8.5   Worked problems on centroids of simple shapes

Problem 1.   Show, by integration, that the centroid of a rectangle lies at the intersection of the diagonals.

Let a rectangle be formed by the line $y = b$, the x-axis and ordinates $x = 0$ and $x = L$ as shown in Figure 8.4. Let the coordinates of the centroid C of this area be $(\bar{x}, \bar{y})$.

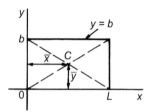

Figure 8.4

By integration, $\bar{x} = \dfrac{\int_0^L xy\,dx}{\int_0^L y\,dx} = \dfrac{\int_0^L (x)(b)\,dx}{\int_0^L b\,dx}$

$$= \dfrac{\left[b\dfrac{x^2}{2}\right]_0^L}{[bx]_0^L} = \dfrac{\dfrac{bL^2}{2}}{bL} = \dfrac{L}{2}$$

and $\bar{y} = \dfrac{\dfrac{1}{2}\int_0^L y^2\,dx}{\int_0^L y\,dx} = \dfrac{\dfrac{1}{2}\int_0^L b^2\,dx}{bL}$

$$= \dfrac{\dfrac{1}{2}[b^2 x]_0^L}{bL} = \dfrac{\dfrac{b^2 L}{2}}{bL} = \dfrac{b}{2}$$

i.e. **the centroid lies at** $\left(\dfrac{L}{2}, \dfrac{b}{2}\right)$ **which is at the inter-**

section of the diagonals.

Problem 2.   Find the position of the centroid of the area bounded by the curve $y = 3x^2$, the x-axis and the ordinates $x = 0$ and $x = 2$.

If $(\bar{x}, \bar{y})$ are the co-ordinates of the centroid of the given area then:

$$\bar{x} = \dfrac{\int_0^2 xy\,dy}{\int_0^2 y\,dx} = \dfrac{\int_0^2 x(3x^2)\,dx}{\int_0^2 3x^2\,dx}$$

$$= \dfrac{\int_0^2 3x^3\,dx}{\int_0^2 3x^2\,dx} = \dfrac{\left[\dfrac{3x^4}{4}\right]_0^2}{[x^3]_0^2} = \dfrac{12}{8} = \mathbf{1.5}$$

$$\bar{y} = \dfrac{\dfrac{1}{2}\int_0^2 y^2\,dy}{\int_0^2 y\,dx} = \dfrac{\dfrac{1}{2}\int_0^2 (3x^2)^2\,dx}{8}$$

$$= \dfrac{\dfrac{1}{2}\int_0^2 9x^4\,dx}{8} = \dfrac{\dfrac{9}{2}\left[\dfrac{x^5}{5}\right]_0^2}{8}$$

$$= \dfrac{\dfrac{9}{2}\left(\dfrac{32}{5}\right)}{8} = \dfrac{18}{5} = \mathbf{3.6}$$

Hence the centroid lies at (1.5, 3.6)

Problem 3.   Determine by integration the position of the centroid of the area enclosed by the line $y = 4x$, the x-axis and ordinates $x = 0$ and $x = 3$.

Let the coordinates of the centroid be $(\bar{x}, \bar{y})$ as shown in Figure 8.5.

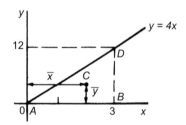

Figure 8.5

Then $\bar{x} = \dfrac{\int_0^3 xy\,dy}{\int_0^3 y\,dx} = \dfrac{\int_0^3 x(4x)\,dx}{\int_0^2 4x\,dx}$

$$= \dfrac{\int_0^3 4x^2\,dx}{\int_0^3 4x\,dx} = \dfrac{\left[\dfrac{4x^3}{3}\right]_0^3}{[2x^2]_0^3} = \dfrac{36}{18} = 2$$

$$\overline{y} = \frac{\frac{1}{2}\int_0^3 y^2\,dy}{\int_0^3 y\,dx} = \frac{\frac{1}{2}\int_0^3 (4x)^2\,dx}{18}$$

$$= \frac{\frac{1}{2}\int_0^3 16x^2\,dx}{18} = \frac{\frac{1}{2}\left[\frac{16x^3}{3}\right]_0^3}{18} = \frac{72}{18} = 4$$

Hence the centroid lies at (2, 4)

In Figure 8.5, ABD is a right-angled triangle. The centroid lies 4 units from AB and 1 unit from BD showing that the centroid of a triangle lies at one-third of the perpendicular height above any side as base.

Now try the following Practise Exercise

> **Practise Exercise 45   Further problems on centroids of simple shapes**
>
> In Problems 1 to 5, find the position of the centroids of the areas bounded by the given curves, the x-axis and the given ordinates.
>
> 1. $y = 2x$; $x = 0$, $x = 3$ $[(2, 2)]$
>
> 2. $y = 3x + 2$; $x = 0$, $x = 4$ $[(2.50, 4.75)]$
>
> 3. $y = 5x^2$; $x = 1$, $x = 4$ $[(3.036, 24.36)]$
>
> 4. $y = 2x^3$; $x = 0$, $x = 2$ $[(1.60, 4.57)]$
>
> 5. $y = x(3x + 1)$; $x = -1$, $x = 0$
>
> $[(-0.833, 0.633)]$

8.6   Further worked problems on centroids of simple shapes

> **Problem 4.**   Determine the co-ordinates of the centroid of the area lying between the curve $y = 5x - x^2$ and the x-axis.

$y = 5x - x^2 = x(5 - x)$. When $y = 0$, $x = 0$ or $x = 5$. Hence the curve cuts the x-axis at 0 and 5 as shown in Figure 8.6. Let the co-ordinates of the centroid be $(\overline{x}, \overline{y})$ then, by integration,

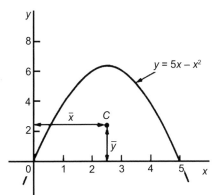

Figure 8.6

$$\overline{x} = \frac{\int_0^5 xy\,dy}{\int_0^5 y\,dx} = \frac{\int_0^5 x(5x - x^2)\,dx}{\int_0^5 (5x - x^2)\,dx}$$

$$= \frac{\int_0^5 (5x^2 - x^3)\,dx}{\int_0^5 (5x - x^2)\,dx} = \frac{\left[\frac{5x^3}{3} - \frac{x^4}{4}\right]_0^5}{\left[\frac{5x^2}{2} - \frac{x^3}{3}\right]_0^5}$$

$$= \frac{\frac{625}{3} - \frac{625}{4}}{\frac{125}{2} - \frac{125}{3}} = \frac{\frac{625}{12}}{\frac{125}{6}}$$

$$= \left(\frac{625}{12}\right)\left(\frac{6}{125}\right) = \frac{5}{2} = \mathbf{2.5}$$

$$\overline{y} = \frac{\frac{1}{2}\int_0^5 y^2\,dy}{\int_0^5 y\,dx} = \frac{\frac{1}{2}\int_0^5 (5x - x^2)^2\,dx}{\int_0^5 (5x - x^2)\,dx}$$

$$= \frac{\frac{1}{2}\int_0^5 (25x^2 - 10x^3 + x^4)\,dx}{\frac{125}{6}}$$

$$= \frac{\frac{1}{2}\left[\frac{25x^3}{3} - \frac{10x^4}{4} + \frac{x^5}{5}\right]_0^5}{\frac{125}{6}}$$

$$= \frac{\frac{1}{2}\left(\frac{25(125)}{3} - \frac{6250}{4} + 625\right)}{\frac{125}{6}} = \mathbf{2.5}$$

Hence the centroid of the area lies at (2.5, 2.5)

(Note from Figure 8.6 that the curve is symmetrical about $x = 2.5$ and thus $\bar{x}$ could have been determined 'on sight').

Problem 5.   Locate the centroid of the area enclosed by the curve $y = 2x^2$, the y-axis and ordinates $y = 1$ and $y = 4$, correct to 3 decimal places.

From Section 8.4, $\bar{x} = \dfrac{\dfrac{1}{2}\displaystyle\int_1^4 x^2\, dy}{\displaystyle\int_1^4 x\, dy} = \dfrac{\dfrac{1}{2}\displaystyle\int_1^4 \dfrac{y}{2}\, dy}{\displaystyle\int_1^4 \sqrt{\dfrac{y}{2}}\, dy}$

$= \dfrac{\dfrac{1}{2}\left[\dfrac{y^2}{4}\right]_1^4}{\left[\dfrac{2y^{3/2}}{3\sqrt{2}}\right]_1^4} = \dfrac{\dfrac{15}{8}}{\dfrac{14}{3\sqrt{2}}} = 0.568$

and $\bar{y} = \dfrac{\displaystyle\int_1^4 xy\, dy}{\displaystyle\int_1^4 x\, dy} = \dfrac{\displaystyle\int_1^4 \sqrt{\dfrac{y}{2}}\,(y)\, dy}{\dfrac{14}{3\sqrt{2}}}$

$= \dfrac{\displaystyle\int_1^4 \dfrac{y^{3/2}}{\sqrt{2}}\, dy}{\dfrac{14}{3\sqrt{2}}} = \dfrac{\dfrac{1}{\sqrt{2}}\left[\dfrac{y^{5/2}}{\dfrac{5}{2}}\right]_1^4}{\dfrac{14}{3\sqrt{2}}}$

$= \dfrac{\dfrac{2}{5\sqrt{2}}(31)}{\dfrac{14}{3\sqrt{2}}} = 2.657$

Hence the position of the centroid is at (0.568, 2.657)

Now try the following Practise Exercise

Practise Exercise 46   **Further problems on centroids of simple shapes**

1.   Determine the position of the centroid of a sheet of metal formed by the curve $y = 4x - x^2$ which lies above the x-axis.   [(2, 1.6)]

2.   Find the coordinates of the centroid of the area that lies between the curve $\dfrac{y}{x} = x - 2$ and the x-axis.   [(1, −0.4)]

3.   Determine the coordinates of the centroid of the area formed between the curve $y = 9 - x^2$ and the x-axis.   [(0, 3.6)]

4.   Determine the centroid of the area lying between $y = 4x^2$, the y-axis and the ordinates $y = 0$ and $y = 4$.   [(0.375, 2.40]

5.   Find the position of the centroid of the area enclosed by the curve $y = \sqrt{5x}$, the x-axis and the ordinate $x = 5$.   [(3.0, 1.875)]

6.   Sketch the curve $y^2 = 9x$ between the limits $x = 0$ and $x = 4$. Determine the position of the centroid of this area.   [(2.4, 0)]

8.7   Second moments of area of regular sections

The **first moment of area** about a fixed axis of a lamina of area A, perpendicular distance y from the centroid of the lamina is defined as Ay cubic units.

The **second moment of area** of the same lamina as above is given by Ay^2, i.e. the perpendicular distance from the centroid of the area to the fixed axis is squared. Second moments of areas are usually denoted by I and have units of mm⁴, cm⁴, and so on.

Several areas, a_1, a_2, a_3,... at distances y_1, y_2, y_3,... from a fixed axis, may be replaced by a single area A, where $A = a_1 + a_2 + a_3 + ..$ at distance k from the axis, such that $Ak^2 = \sum ay^2$.

k is called the **radius of gyration** of area A about the given axis. Since $Ak^2 = \sum ay^2 = I$ then the radius of gyration, $k = \sqrt{\dfrac{I}{A}}$

The second moment of area is a quantity much used in the theory of bending of beams (see Chapter 9), in the torsion of shafts (see Chapter 11), and in calculations involving water planes and centres of pressure (see Chapter 22).

The **procedure to determine the second moment of area of regular sections** about a given axis is (i) to find the second moment of area of a typical element and (ii) to sum all such second moments of area by integrating between appropriate limits.

For example, the second moment of area of the rectangle shown in Figure 8.7 about axis PP is found by initially considering an elemental strip of width δx, parallel to and distance x from axis PP. Area of shaded strip $= b\delta x$. Second moment of area of the shaded strip about $PP = (x^2)(b\delta x)$.

The second moment of area of the whole rectangle about PP is obtained by summing all such strips between $x = 0$ and $x = d$, i.e. $\sum\limits_{x=0}^{x=d} x^2 b \delta x$

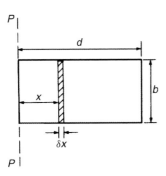

Figure 8.7

It is a fundamental theorem of integration that

$$\lim\limits_{\delta x \to x}\sum\limits_{x=0}^{x=d} x^2 b\,\delta x = \int_0^d x^2 b\,dx$$

Thus the second moment of area of the rectangle about $PP = b\int_0^d x^2\,dx = b\left[\dfrac{x^3}{3}\right]_0^d = \dfrac{bd^3}{3}$

Since the total area of the rectangle, $A = db$,

then $I_{PP} = (db)\left(\dfrac{d^2}{3}\right) = \dfrac{Ad^2}{3}$

$I_{pp} = Ak_{pp}^2$ thus $k_{pp}^2 = \dfrac{d^2}{3}$

i.e. the radius of gyration about axis PP,

$$k_{pp} = \sqrt{\dfrac{d^2}{3}} = \dfrac{d}{\sqrt{3}}$$

Parallel axis theorem

In Figure 8.8, axis GG passes through the centroid C of area A. Axes DD and GG are in the same plane, are parallel to each other and distance H apart. The parallel axis theorem states:

$$I_{DD} = I_{GG} + AH^2$$

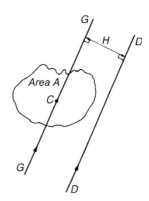

Figure 8.8

Using the parallel axis theorem the second moment of area of a rectangle about an axis through the centroid may be determined. In the rectangle shown in Figure 8.9, $I_{pp} = \dfrac{bd^3}{3}$ (from above)

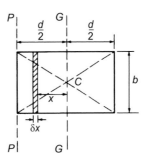

Figure 8.9

From the parallel axis theorem

$$I_{pp} = I_{GG} + (bd)\left(\dfrac{d}{2}\right)^2$$

i.e. $\dfrac{bd^3}{3} = I_{GG} + \dfrac{bd^3}{4}$ from which,

$$I_{GG} = \dfrac{bd^3}{3} - \dfrac{bd^3}{4} = \dfrac{bd^3}{12}$$

Perpendicular axis theorem

In Figure 8.10, axes OX, OY and OZ are mutually perpendicular. If OX and OY lie in the plane of area A then the perpendicular axis theorem states:

$$I_{OZ} = I_{OX} + I_{OY}$$

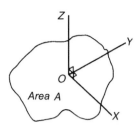

Figure 8.10

A summary of derived standard results for the second moment of area and radius of gyration of regular sections are listed in Table 8.1.

The second moment of area of a **hollow cross-section**, such as that of a tube, can be obtained by subtracting the second moment of area of the hole about its centroid from the second moment of area of the outer circumference about its centroid. This is demonstrated in worked problems 10, 12 and 13 following.

Problem 6.   Determine the second moment of area and the radius of gyration about axes AA, BB and CC for the rectangle shown in Figure 8.11.

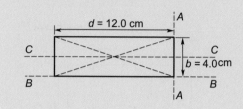

Figure 8.11

Table 8.1 Summary of Standard Results of the Second Moments of Areas of Regular Sections

Shape	Position of axis	Second moment of area, I	Radius of gyration, k
Rectangle length d breadth b	(1) Coinciding with b	$\dfrac{bd^3}{3}$	$\dfrac{d}{\sqrt{3}}$
	(2) Coinciding with d	$\dfrac{db^3}{3}$	$\dfrac{b}{\sqrt{3}}$
	(3) Through centroid, parallel to b	$\dfrac{bd^3}{12}$	$\dfrac{d}{\sqrt{12}}$
	(4) Through centroid, parallel to d	$\dfrac{db^3}{12}$	$\dfrac{b}{\sqrt{12}}$
Triangle Perpendicular height h base b	(1) Coinciding with b	$\dfrac{bh^3}{12}$	$\dfrac{h}{\sqrt{6}}$
	(2) Through centroid, parallel to base	$\dfrac{bh^3}{36}$	$\dfrac{h}{\sqrt{18}}$
	(3) Through vertex, parallel to base	$\dfrac{bh^3}{4}$	$\dfrac{h}{\sqrt{2}}$
Circle radius r diameter d	(1) Through centre perpendicular to plane(i.e. polar axis)	$\dfrac{\pi r^4}{2}$ or $\dfrac{\pi d^4}{32}$	$\dfrac{r}{\sqrt{2}}$
	(2) Coinciding with diameter	$\dfrac{\pi r^4}{4}$ or $\dfrac{\pi d^4}{64}$	$\dfrac{r}{2}$
	(3) About a tangent	$\dfrac{5\pi r^4}{4}$ or $\dfrac{5\pi d^4}{64}$	$\dfrac{\sqrt{5}}{2} r$
Semicircle radius r	Coinciding with diameter	$\dfrac{\pi r^4}{8}$	$\dfrac{r}{2}$

From Table 8.1, the second moment of area about axis AA,

$$I_{AA} = \frac{bd^3}{3} = \frac{(4.0)(12.0)^3}{3} = \textbf{2304 cm}^4$$

Radius of gyration, $k_{AA} = \frac{d}{\sqrt{3}} = \frac{12.0}{\sqrt{3}} = \textbf{6.93 cm}$

Similarly, $\quad I_{BB} = \frac{db^3}{3} = \frac{(12.0)(4.0)^3}{3} = \textbf{256 cm}^4$

and $\quad k_{BB} = \frac{b}{\sqrt{3}} = \frac{4.0}{\sqrt{3}} = \textbf{2.31 cm}$

The second moment of area about the centroid of a rectangle is $\frac{bd^3}{12}$ when the axis through the centroid is parallel with the breadth b. In this case, the axis CC is parallel with the length d

Hence $\quad I_{CC} = \frac{db^3}{12} = \frac{(12.0)(4.0)^3}{12} = \textbf{64 cm}^4$

and $\quad k_{CC} = \frac{b}{\sqrt{12}} = \frac{4.0}{\sqrt{12}} = \textbf{1.15 cm}$

Problem 7. Find the second moment of area and the radius of gyration about axis PP for the rectangle shown in Figure 8.12.

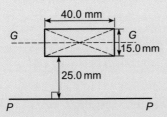

Figure 8.12

$$I_{GG} = \frac{dh^3}{12} \text{ where } d = 40.0 \text{ mm}$$

and $\quad h = 15.0$ mm

Hence $\quad I_{GG} = \frac{(40.0)(15.0)^3}{12} = 11250 \text{ mm}^4$

From the parallel axis theorem,

$$I_{PP} = I_{GG} + AH^2,$$

where $\quad A = 40.0 \times 15.0 = 600$ mm^2 and

$$H = 25.0 + 7.5 = 32.5 \text{ mm},$$

the perpendicular distance between GG and PP.

Hence $\quad \textbf{\textit{I}}_{PP} = 11250 + (600)(32.5)^2 = \textbf{645000 mm}^4$

$I_{PP} = Ak^2_{PP}$, from which,

$$k_{PP} = \sqrt{\frac{I_{PP}}{area}} = \sqrt{\left(\frac{645000}{600}\right)} = \textbf{32.79 mm}$$

Problem 8. Determine the second moment of area and radius of gyration about axis QQ of the triangle BCD shown in Figure 8.13.

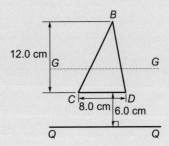

Figure 8.13

Using the parallel axis theorem: $I_{QQ} = I_{GG} + AH^2$, where I_{GG} is the second moment of area about the centroid of the triangle,

i.e. $\quad \frac{bh^3}{36} = \frac{(8.0)(12.0)^3}{36}$

$$= 384 \text{ cm}^4,$$

A is the area of the triangle $= \frac{1}{2}bh$

$$= \frac{1}{2}(8.0)(12.0)$$

$$= 48 \text{ cm}^2$$

and H is the distance between axes GG and QQ

$$= 6.0 + \frac{1}{3}(12.0)$$

$$= 10 \text{ cm}$$

Hence the second moment of area about axis QQ,

$$\textbf{\textit{I}}_{QQ} = 384 + (48)(10)^2$$

$$= \textbf{5184 cm}^4$$

Radius of gyration, $k_{QQ} = \sqrt{\frac{I_{QQ}}{area}} = \sqrt{\left(\frac{5184}{48}\right)}$

$$= \textbf{10.4 cm}$$

Problem 9. Determine the second moment of area and radius of gyration of the circle shown in Figure 8.14 about axis YY.

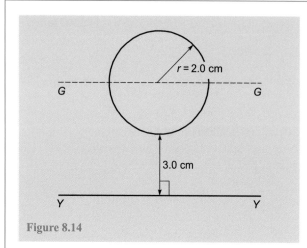

Figure 8.14

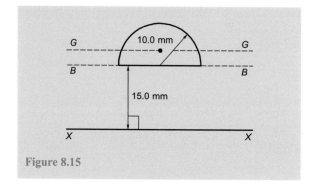

Figure 8.15

In Figure 8.14,

$$I_{GG} = \frac{\pi r^4}{4} = \frac{\pi}{4}(2.0)^4 = 4\pi \text{ cm}^4$$

Using the parallel axis theorem,

$$I_{YY} = I_{GG} + AH^2,$$

where $H = 3.0 + 2.0 = 5.0$ cm.

Hence, $I_{YY} = 4\pi + [\pi(2.0)^2](5.0)^2$

$$= 4\pi + 100\pi = 104\pi = \textbf{327 cm}^4$$

Radius of gyration,

$$k_{YY} = \sqrt{\frac{I_{YY}}{\text{area}}} = \sqrt{\left(\frac{104\pi}{\pi(2.0)^2}\right)} = \sqrt{26} = \textbf{5.10 cm}$$

Problem 10. Determine the second moment of area of an annular section, about its centroidal axis. The outer diameter of the annulus is D_2 and its inner diameter is D_1

Second moment of area of annulus about its centroid, I_{XX}

$= (I_{XX}$ of outer circle about its diameter$)$
$\quad - (I_{XX}$ of inner circle about its diameter$)$

$$= \frac{\pi D_2^4}{64} - \frac{\pi D_1^4}{64} \text{ from Table 8.1}$$

i.e. $I_{XX} = \dfrac{\pi}{64}\left(D_2^4 - D_1^4\right)$

Problem 11. Determine the second moment of area and radius of gyration for the semicircle shown in Figure 8.15 about axis XX.

The centroid of a semicircle lies at $\dfrac{4r}{3\pi}$ from its diameter (see *Engineering Mathematics 6th Edition*, page 523).

Using the parallel axis theorem: $I_{BB} = I_{GG} + AH^2$, where (from Table 8.1)

$$I_{BB} = \frac{\pi r^4}{8} = \frac{\pi(10.0)^4}{8} = 3927 \text{ mm}^4,$$

$$A = \frac{\pi r^2}{2} = \frac{\pi(10.0)^2}{2} = 157.1 \text{ mm}^2$$

and $H = \dfrac{4r}{3\pi} = \dfrac{4(10.0)}{3\pi} = 4.244$ mm

Hence, $3927 = I_{GG} + (157.1)(4.244)^2$

i.e. $3927 = I_{GG} + 2830$, from which,

$$I_{GG} = 3927 - 2830 = 1097 \text{ mm}^4$$

Using the parallel axis theorem again:

$$I_{XX} = I_{GG} + A(15.0 + 4.244)^2$$

i.e. $I_{XX} = 1097 + (157.1)(19.244)^2$

$$= 1097 + 58179 = 59276 \text{ mm}^4$$

or   **59280 mm^4,** correct to 4 significant figures.

Radius of gyration,

$$k_{XX} = \sqrt{\frac{I_{XX}}{\text{area}}} = \sqrt{\left(\frac{59276}{157.1}\right)}$$

$$= \textbf{19.42 mm}$$

Problem 12. Determine the polar second moment of area of an annulus about its centre. The outer diameter of the annulus is D_2 and its inner diameter is D_1

The polar second moment of area is denoted by J.

Hence, for the annulus, $J = (J$ of outer circle about its centre$) - (J$ of inner circle about its centre$)$

$$= \frac{\pi D_2{}^4}{32} - \frac{\pi D_1{}^4}{32} \text{ from Table 8.1}$$

i.e. $$J = \frac{\pi}{32}\left(D_2{}^4 - D_1{}^4\right)$$

Problem 13. Determine the polar second moment of area of the propeller shaft cross-section shown in Figure 8.16.

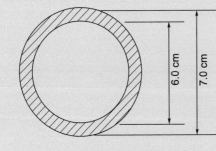

Figure 8.16

The polar second moment of area of a circle, $J = \dfrac{\pi d^4}{32}$

The polar second moment of area of the shaded area is given by the polar second moment of area of the 7.0 cm diameter circle minus the polar second moment of area of the 6.0 cm diameter circle.

Hence, from Problem 12, the polar second moment of area of the cross-section shown

$$= \frac{\pi}{32}\left(7.0^4 - 6.0^4\right) = \frac{\pi}{32}(1105) = \textbf{108.5 cm}^4$$

Problem 14. Determine the second moment of area and radius of gyration of a rectangular lamina of length 40 mm and width 15 mm about an axis through one corner, perpendicular to the plane of the lamina.

The lamina is shown in Figure 8.17.

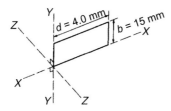

Figure 8.17

From the perpendicular axis theorem: $I_{ZZ} = I_{XX} + I_{YY}$

$$I_{XX} = \frac{db^3}{3} = \frac{(40)(15)^3}{3} = 45000 \text{ mm}^4$$

and $$I_{YY} = \frac{bd^3}{3} = \frac{(15)(40)^3}{3} = 320000 \text{ mm}^4$$

Hence, $I_{ZZ} = 45000 + 320000$
$= \textbf{365000 mm}^4 \text{ or } \textbf{36.5 cm}^4$

Radius of gyration, $k_{ZZ} = \sqrt{\dfrac{I_{ZZ}}{\text{area}}} = \sqrt{\left(\dfrac{365000}{(40)(15)}\right)}$

$$= \textbf{24.7 mm} \text{ or } \textbf{2.47 cm}$$

Problem 15. Determine correct to 3 significant figures, the second moment of area about axis XX for the composite area shown in Figure 8.18.

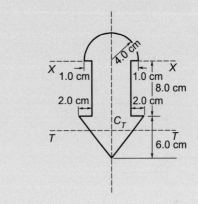

Figure 8.18

For the semicircle, $I_{XX} = \dfrac{\pi r^4}{8} = \dfrac{\pi(4.0)^4}{8}$
$$= 100.5 \text{ cm}^4$$

For the rectangle, $I_{XX} = \dfrac{db^3}{3} = \dfrac{(6.0)(8.0)^3}{3}$
$$= 1024 \text{ cm}^4$$

For the triangle, about axis TT through centroid C_T,

$$I_{TT} = \frac{bh^3}{36} = \frac{(10)(6.0)^3}{36} = 60 \text{ cm}^4$$

By the parallel axis theorem, the second moment of area of the triangle about axis XX

$$= 60 + \left[\frac{1}{2}(10)(6.0)\right]\left[8.0 + \frac{1}{3}(6.0)\right]^2 = 3060 \text{ cm}^4$$

Total second moment of area about $XX = 100.5 + 1024 + 3060 = 4184.5 = \textbf{4180 cm}^4$, correct to 3 significant figures.

Now try the following Practise Exercise

Practise Exercise 47 **Further problems on second moment of areas of regular sections**

1. Determine the second moment of area and radius of gyration for the rectangle shown in Figure 8.19 about (a) axis *AA* (b) axis *BB*, and (c) axis *CC*.

$$\begin{bmatrix} \text{(a) 72 cm}^4\text{, 1.73 cm (b) 128 cm}^4\text{, 2.31 cm} \\ \text{(c) 512 cm}^4\text{, 4.62 cm} \end{bmatrix}$$

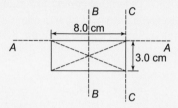

Figure 8.19

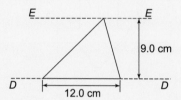

Figure 8.20

2. Determine the second moment of area and radius of gyration for the triangle shown in Figure 8.20 about (a) axis *DD* (b) axis *EE*, and (c) an axis through the centroid of the triangle parallel to axis *DD*.

$$\begin{bmatrix} \text{(a) 729 mm}^4\text{, 3.67 mm (b) 2187 mm}^4\text{,} \\ \text{6.36 mm (c) 243 mm}^4\text{, 2.12 mm} \end{bmatrix}$$

3. For the circle shown in Figure 8.21, find the second moment of area and radius of gyration about (a) axis *FF* and (b) axis *HH*

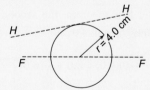

Figure 8.21

[(a) 201 cm⁴, 2.0 cm (b) 1005 cm⁴, 4.47 cm]

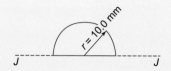

Figure 8.22

4. For the semicircle shown in Figure 8.22, find the second moment of area and radius of gyration about axis *JJ*.

[3927 mm⁴, 5.0 mm]

5. For each of the areas shown in Figure 8.23 determine the second moment of area and radius of gyration about axis *LL*, by using the parallel axis theorem.

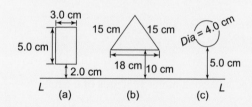

Figure 8.23

$$\begin{bmatrix} \text{(a) 335 cm}^4\text{, 4.73 cm (b) 22030 cm}^4\text{,} \\ \text{14.3 cm (c) 628 cm}^4\text{, 7.07 cm} \end{bmatrix}$$

6. Calculate the radius of gyration of a rectangular door 2.0 m high by 1.5 m wide about a vertical axis through its hinge. [0.866 m]

7. A circular door of a boiler is hinged so that it turns about a tangent. If its diameter is 1.0 m, determine its second moment of area and radius of gyration about the hinge.
[0.245 m⁴, 0.559 m]

8. A circular cover, centre 0, has a radius of 12.0 cm. A hole of radius 4.0 cm and centre *X*, where 0*X* = 6.0 cm, is cut in the cover. Determine the second moment of area and the radius of gyration of the remainder about a diameter through 0 perpendicular to 0*X*. [14280 cm⁴, 5.96 cm]

9. For the sections shown in Figure 8.24, find the second moment of area and the radius of gyration about axis *XX*.

$$\begin{bmatrix} \text{(a) 12190 mm}^4\text{, 10.9 mm} \\ \text{(b) 549.5 cm}^4\text{, 4.18 cm} \end{bmatrix}$$

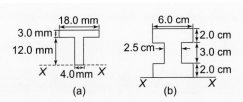

Figure 8.24

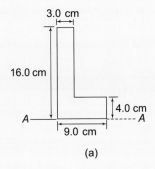

(a)

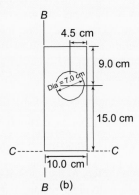

(b)

Figure 8.25

10. Determine the second moments of areas about the given axes for the shapes shown in Figure 8.25 (In Figure 8.25(b), the circular area is removed.)

$$\left[I_{AA} = 4224 \text{ cm}^4, I_{BB} = 6718 \text{ cm}^4, \atop I_{CC} = 37300 \text{ cm}^4 \right]$$

8.8 Second moment of area for 'built-up' sections

The cross-sections of many beams and members of a framework are in the forms of rolled steel joists (RSJ's or *I* beams), tees, and channel bars, as shown in Figure 8.26. These shapes usually afford better bending resistances than solid rectangular or circular sections.

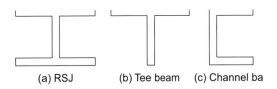

(a) RSJ (b) Tee beam (c) Channel ba

Figure 8.26 Built-up sections

Calculation of the second moments of area and the position of the centroidal, or neutral axes for such sections are demonstrated in the following worked problems.

Problem 16. Determine the second moment of area about a horizontal axis passing through the centroid, for the *I* beam shown in Figure 8.27.

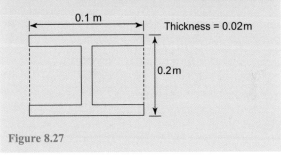

Figure 8.27

The centroid of this beam will lie on the horizontal axis *NA*, as shown in Figure 8.28.

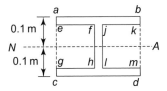

Figure 8.28

The second moment of area of the *I* beam is given by:

$$I_{NA} = (I \text{ of rectangle } abdc) - (I \text{ of rectangle } efhg) - (I \text{ of rectangle } jkml)$$

Hence, from Table 8.1,

$$I_{NA} = \frac{0.1 \times 0.2^3}{12} - \frac{0.04 \times 0.16^3}{12} - \frac{0.04 \times 0.16^3}{12}$$

$$= 6.667 \times 10^{-5} - 1.365 \times 10^{-5} - 1.365 \times 10^{-5}$$

i.e. $I_{NA} = \mathbf{3.937 \times 10^{-5} \ m^4}$

Problem 17. Determine the second moment of area about a horizontal axis passing through the centroid, for the channel section shown in Figure 8.29.

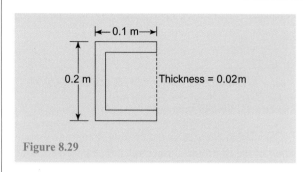

Figure 8.29

The centroid of this beam will be on the horizontal axis *NA*, as shown in Figure 8.30.

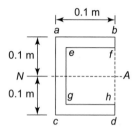

Figure 8.30

The second moment of area of the channel section is given by:

$$I_{NA} = (I \text{ of rectangle } abdc \text{ about } NA) -$$
$$(I \text{ of rectangle } efhg \text{ about } NA)$$

$$= \frac{0.1 \times 0.2^3}{12} - \frac{0.08 \times 0.16^3}{12}$$

$$= 6.667 \times 10^{-5} - 2.731 \times 10^{-5}$$

i.e. $I_{NA} = \mathbf{3.936 \times 10^{-5} \ m^4}$

Problem 18. Determine the second moment of area about a horizontal axis passing through the centroid, for the tee beam shown in Figure 8.31.

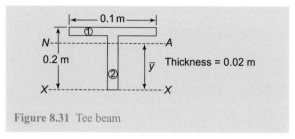

Figure 8.31 Tee beam

In this case, we will first need to find the position of the centroid, i.e. we need to calculate $\overline{y}$ in Figure 8.31. There are several methods of achieving this; the tabular method is as good as any since it can lead to the use of a spreadsheet. The method is explained below with the aid of Table 8.2, below.

First, we divide the tee beam into two rectangles, as shown in Figure 8.31.

In the first column we refer to each of the two rectangles, namely rectangle (1) and rectangle (2). Thus, the second row in Table 8.2 refers to rectangle (1) and the third row to rectangle (2). The fourth row refers to the summation of each column as appropriate. The second column refers to the areas of each individual rectangular element, *a*.

Thus, area of rectangle (1), $a_1 = 0.1 \times 0.02 = 0.002 \ m^2$ and area of rectangle (2), $a_2 = 0.18 \times 0.02 = 0.0036 \ m^2$

Hence, $\sum a = 0.002 + 0.0036 = 0.0056 \ m^2$

The third column refers to the vertical distance of the centroid of each individual rectangular element from the base, namely *XX*.

Thus, $y_1 = 0.2 - 0.01 = 0.19 \ m$

and $y_2 = \dfrac{0.18}{2} = 0.09 \ m$

In the fourth column, the product *ay* is obtained by multiplying the cells of column 2 with the cells of column 3,

Table 8.2

Column	1	2	3	4	5	6
Row 1	Section	a	y	ay	ay^2	i
Row 2	(1)	0.002	0.19	3.8×10^{-4}	7.22×10^{-5}	6.6×10^{-8}
Row 3	(2)	0.0036	0.09	3.24×10^{-4}	2.916×10^{-5}	9.72×10^{-6}
Row 4	$\sum$	0.0056		7.04×10^{-4}	1.014×10^{-4}	9.786×10^{-6}

i.e. $a_1y_1 = 0.002 \times 0.19 = 3.8 \times 10^{-4}\text{m}^3$

$a_2y_2 = 0.0036 \times 0.09 = 3.24 \times 10^{-4}\text{m}^3$

and $\sum ay = 3.8 \times 10^{-4} + 3.24 \times 10^{-4} = 7.04 \times 10^{-4}\text{m}^3$

In the fifth column, the product ay^2 is obtained by multiplying the cells of column 3 by the cells of column 4,

i.e. $\sum ay^2$ is **part** of the second moment of area of the tee beam about XX,

i.e. $a_1y_1^2 = 0.19 \times 3.8 \times 10^{-4} = 7.22 \times 10^{-5}\text{m}^4$

$a_2y_2^2 = 0.09 \times 3.24 \times 10^{-4} = 2.916 \times 10^{-5}\text{m}^4$

and $\sum ay^2 = 7.22 \times 10^{-5} + 2.916 \times 10^{-5}$

$= 1.014 \times 10^{-4}\text{m}^4$

In the sixth column, the symbol i refers to the second moment of area of each individual rectangle about its own local centroid.

Now $i = \dfrac{bd^3}{12}$ from Table 8.1

Hence, $i_1 = \dfrac{0.1 \times 0.02^3}{12} = 6.6 \times 10^{-8}\ \text{m}^4$

$i_2 = \dfrac{0.02 \times 0.18^3}{12} = 9.72 \times 10^{-6}\ \text{m}^4$

and $\sum i = 6.6 \times 10^{-8} + 9.72 \times 10^{-6} = 9.789 \times 10^{-6}\ \text{m}^4$

From the parallel axis theorem: $I_{XX} = \sum i + \sum ay^2$ (8.1)

The cross-sectional area of the tee beam

$= \sum a = \textbf{0.0056 m}^2$ from Table 8.2.

Now the centroidal position, namely $\bar{y}$, is given by:

$$\bar{y} = \frac{\sum ay}{\sum a} = \frac{7.04 \times 10^{-4}}{0.0056} = \textbf{0.1257 m}$$

From equation (8.1), $I_{XX} = \sum i + \sum ay^2$

$= 9.786 \times 10^{-6} + 1.014 \times 10^{-4}$

i.e. $\boldsymbol{I_{XX} = 1.112 \times 10^{-4}\text{m}^4}$

From the parallel axis theorem:

$$I_{NA} = I_{XX} - (\bar{y})^2 \sum a$$

$= 1.112 \times 10^{-4} - (0.1257)^2 \times 0.0056$

$\boldsymbol{I_{NA} = 2.27 \times 10^{-5}\text{m}^4}$

It should be noted that the least second moment of area of a section is always about an axis through its centroid.

Problem 19.   (a) Determine the second moment of area and the radius of gyration about axis XX for the I-section shown in Figure 8.32.

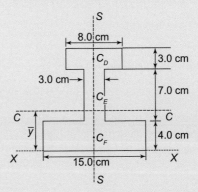

Figure 8.32

(b)   Determine the position of the centroid of the I-section.

(c)   Calculate the second moment of area and radius of gyration about an axis CC through the centroid of the section, parallel to axis XX.

The I-section is divided into three rectangles, D, E and F and their centroids denoted by C_D, C_E and C_F respectively.

(a)   *For rectangle D:*

The second moment of area about C_D (an axis through C_D parallel to XX)

$= \dfrac{bd^3}{12} = \dfrac{(8.0)(3.0)^3}{12} = 18\ \text{cm}^4$

Using the parallel axis theorem: $I_{XX} = 18 + AH^2$

where $A = (8.0)(3.0) = 24\ \text{cm}^2$ and $H = 12.5\ \text{cm}$

Hence $I_{XX} = 18 + 24(12.5)^2 = 3768\ \text{cm}^4$

For rectangle E:

The second moment of area about C_E (an axis through C_E parallel to XX)

$= \dfrac{bd^3}{12} = \dfrac{(3.0)(7.0)^3}{12} = 85.75\ \text{cm}^4$

Using the parallel axis theorem:
$I_{XX} = 85.75 + (7.0)(3.0)(7.5)^2 = 1267\ \text{cm}^4$

For rectangle F: $I_{XX} = \dfrac{bd^3}{3} = \dfrac{(15.0)(4.0)^3}{3}$

$= 320\ \text{cm}^4$

Total second moment of area for the I-section about axis XX,

$$I_{XX} = 3768 + 1267 + 320$$
$$= \textbf{5355 cm}^4$$

Total area of I-section

$$= (8.0)(3.0) + (3.0)(7.0) + (15.0)(4.0)$$
$$= 105 \text{ cm}^2$$

Radius of gyration, $k_{XX} = \sqrt{\dfrac{I_{XX}}{\text{area}}} = \sqrt{\left(\dfrac{5355}{105}\right)}$
$$= \textbf{7.14 cm}$$

(b) The centroid of the I-section will lie on the axis of symmetry, shown as SS in Figure 8.32. Using a tabular approach:

Table 8.3

Part	Area (a cm²)	Distance of centroid from XX (i.e. y cm)	Moment about XX (i.e. ay cm³)
D	24	12.5	300
E	21	7.5	157.5
F	60	2.0	120
$\sum a = A = 105$			$\sum ay = 577.5$

$A\bar{y} = \sum ay$, from which,

$$\bar{y} = \frac{\sum ay}{A} = \frac{577.5}{105} = 5.5 \text{ cm}$$

Thus the centroid is positioned on the axis of symmetry 5.5 cm from axis XX.

(c) From the parallel axis theorem:

$$I_{XX} = I_{CC} + AH^2$$
i.e. $$5355 = I_{CC} + (105)(5.5)^2$$
$$= I_{CC} + 3176$$

from which, **second moment of area about axis CC,** $I_{CC} = 5355 - 3176$
$$= \textbf{2179 cm}^4$$

Radius of gyration, $k_{CC} = \sqrt{\dfrac{I_{CC}}{\text{area}}} = \sqrt{\dfrac{2179}{105}}$
$$= \textbf{4.56 cm}$$

Now try the following Practise Exercises

Practise Exercise 48   Further problems on second moment of area of 'built-up' sections

Determine the second moments of area about a horizontal axis, passing through the centroids, for the 'built-up' sections shown below. All dimensions are in mm and all the thicknesses are 2 mm.

1.

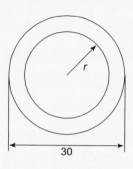

Figure 8.33

[17329 mm⁴]

2.

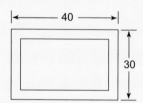

Figure 8.34

[37272 mm⁴]

3.

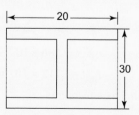

Figure 8.35

[18636 mm⁴]

4.

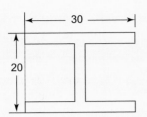

Figure 8.36

[10443 mm^4]

5.

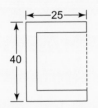

Figure 8.37

[43909 mm^4]

6.

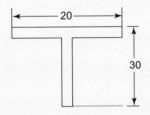

Figure 8.38

[8922 mm^4]

7.

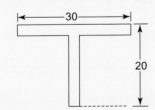

Figure 8.39

[3242 mm^4]

8.

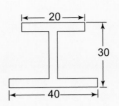

Figure 8.40

[24683 mm^4]

Practise Exercise 49 **Short-answer questions on first and second moment of areas**

1. Define a centroid.

2. Define the first moment of area.

3. Define second moment of area.

4. Define radius of gyration.

5. State the parallel axis theorem.

6. State the perpendicular axis theorem.

Practise Exercise 50 **Multiple-choice questions on first and second moment of areas**

(Answers on page 297)

1. The centroid of the area bounded by the curve $y = 3x$, the x-axis and ordinates $x = 0$ and $x = 3$, lies at:
    (a) (3, 2)   (b) (2, 6)
    (c) (2, 3)   (d) (6, 2)

2. The second moment of area about axis GG of the rectangle shown in Figure 8.41 is:
    (a) 16 cm^4 (b) 4 cm^4
    (c) 36 cm^4 (d) 144 cm^4

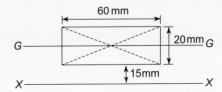

Figure 8.41

Part Two

3. The second moment of area about axis XX of the rectangle shown in Figure 8.41 is:

   (a) 111 cm^4 (b) 31 cm^4
   (c) 63 cm^4 (d) 79 cm^4

4. The radius of gyration about axis GG of the rectangle shown in Figure 8.41 is:

   (a) 5.77 mm   (b) 17.3 mm
   (c) 11.55 mm   (d) 34.64 mm

5. The radius of gyration about axis XX of the rectangle shown in Figure 8.41 is:

   (a) 30.41 mm   (b) 25.66 mm
   (c) 16.07 mm   (d) 22.91 mm

The circumference of a circle is 15.71 mm. Use this fact in questions 6 to 8.

6. The second moment of area of the circle about an axis coinciding with its diameter is:

   (a) 490.9 mm^4 (b) 61.36 mm^4
   (c) 30.69 mm^4 (d) 981.7 mm^4

7. The second moment of area of the circle about a tangent is:

   (a) 153.4 mm^4 (b) 9.59 mm^4
   (c) 2454 mm^4 (d) 19.17 mm^4

8. The polar second moment of area of the circle is:

   (a) 3.84 mm^4 (b) 981.7 mm^4
   (c) 61.36 mm^4 (d) 30.68 mm^4

9. The second moment of area about axis XX of the triangle ABC shown in Figure 8.42 is:

   (a) 24 cm^4 (b) 10.67 cm^4
   (c) 310.67 cm^4 (d) 324 cm^4

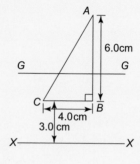

Figure 8.42

10. The radius of gyration about axis GG of the triangle shown in Figure 8.42 is:

    (a) 1.41 cm   (b) 2 cm
    (c) 2.45 cm   (d) 4.24 cm

This Revision Test covers the material contained in Chapters 6 to 8. *The marks for each question are shown in brackets at the end of each question.*

1. Determine the unknown internal forces in the pin-jointed truss of Figure RT3.1.

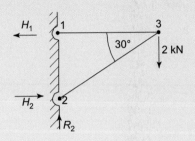

Figure RT3.1

(7)

2. Determine the unknown internal forces in the pin-jointed truss of Figure RT3.2.

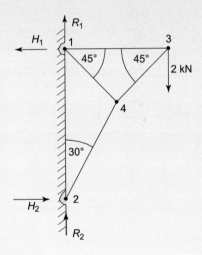

Figure RT3.2

(12)

3. A beam of length 3 m is simply supported at its ends. A clockwise couple of 4 kN m is placed at a distance of 1 m from the left hand support. (a) Determine the end reactions. (b) If the beam now carries an additional downward load of 12 kN at a distance of 1 m from the right hand support, sketch the bending moment and shearing force diagrams. (9)

4. A beam of length 4 m is simply supported at its right extremity and at 1 m from the left extremity. If the beam is loaded with a downward load of 2 kN at its left extremity and with another downward load of 10 kN at a distance of 1 m from its right extremity, sketch its bending moment and shearing force diagrams. (7)

5. (a) Find the second moment of area and radius of gyration about the axis *XX* for the beam section shown in Figure RT3.3.

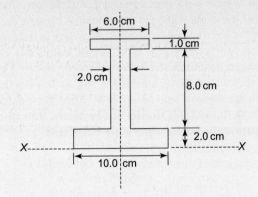

Figure RT3.3

(b) Determine the position of the centroid of the section.

(c) Calculate the second moment of area and radius of gyration about an axis through the centroid parallel to axis *XX*. (25)

Chapter 9

Bending of beams

If a beam of symmetrical cross-section is subjected to a bending moment, then stresses due to bending action will occur. In pure or simple bending, the beam will bend into an arc of a circle. Due to couples, the upper layers of the beam will be in tension, because their lengths have been increased, and the lower layers of the beam will be in compression, because their lengths have been decreased. Somewhere in between these two layers lies a layer whose length has not changed, so that its stress due to bending is zero. This layer is called the neutral layer and its intersection with the beam's cross-section is called the neutral axis. In this chapter stresses in a beam and the radius of curvature due to bending are calculated.

At the end of this chapter you should be able to:

• define neutral layer
• define the neutral axis of a beam's cross-section
• prove that $\dfrac{\sigma}{y} = \dfrac{M}{I} = \dfrac{E}{R}$
• calculate the stresses in a beam due to bending
• calculate the radius of curvature of the neutral layer due to a pure bending moment M

9.1 Introduction

If a beam of symmetrical cross-section is subjected to a bending moment M, then stresses due to bending action will occur. This can be illustrated by the horizontal beam of Figure 9.1, which is of uniform cross-section.

In pure or simple bending, the beam will bend into an arc of a circle as shown in Figure 9.2.

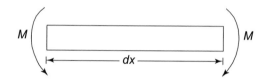

Figure 9.1

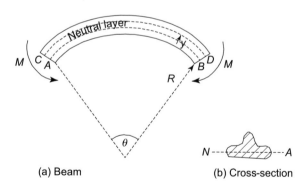

(a) Beam   (b) Cross-section

Figure 9.2

Now in Figure 9.2, it can be seen that due to these couples M, the upper layers of the beam will be in tension, because their lengths have been increased, and the lower layers of the beam will be in compression, because their lengths have been decreased. Somewhere in between these two layers lies a layer whose length has not changed, so that its stress due to bending is zero. This layer is called the **neutral layer** and its intersection with the beam's cross-section is called the **neutral axis (NA).** Later on in this chapter it will

Mechanical Engineering Principles, Bird and Ross, ISBN 9780415517850

be shown that the neutral axis is also the centroidal axis described in Chapter 8.

9.2   To prove that $\dfrac{\sigma}{y} = \dfrac{M}{I} = \dfrac{E}{R}$

In the formula $\dfrac{\sigma}{y} = \dfrac{M}{I} = \dfrac{E}{R}$

σ = the stress due to bending moment M occurring at a distance y from the neutral axis NA,

I = the second moment of area of the beam's cross-section about NA,

E = Young's modulus of elasticity of the beam's material, and

R = radius of curvature of the neutral layer of the beam due to the bending moment M.

Now the original length of the beam element,

$$dx = R\theta \qquad (9.1)$$

At any distance y from NA, the length AB increases its length to:

$$CD = (R + y)\theta \qquad (9.2)$$

Hence, extension of $AB = \delta = (R + y)\theta - R\theta = y\theta$

Now, strain ε = extension/original length,

i.e. $\qquad \varepsilon = \dfrac{y\theta}{R\theta} = \dfrac{y}{R} \qquad (9.3)$

However, $\qquad \dfrac{\text{stress}\,(\sigma)}{\text{strain}\,(\varepsilon)} = E$

or $\qquad \sigma = E\varepsilon \qquad (9.4)$

Substituting equation (9.3) into equation (9.4) gives:

$$\sigma = E\dfrac{y}{R} \qquad (9.5)$$

or $\qquad \dfrac{\sigma}{y} = \dfrac{E}{R} \qquad (9.6)$

Consider now the stresses in the beam's cross-section, as shown in Figure 9.3.

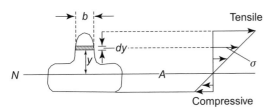

(a) Beam's cross-section   (b) Stress distribution

Figure 9.3

From Figure 9.3, it can be seen that the stress σ causes an elemental couple δM about NA, where:

$$\delta M = \sigma \times (b \times dy) \times y$$

and the total value of the couple caused by all such stresses

$$= M = \sum \delta M = \int \sigma\, b\, y\, dy \qquad (9.7)$$

but from equation (9.5), $\sigma = \dfrac{Ey}{R}$

Therefore, $\qquad M = \int \dfrac{Ey}{R} b\, y\, dy = \int \dfrac{E}{R} y^2\, b\, dy$

Now, E and R are constants, that is, they do not vary with y, hence they can be removed from under the integral sign. Therefore,

$$M = \dfrac{E}{R} \int y^2\, b\, dy$$

However, $\int y^2\, b\, dy = \dfrac{by^3}{3} = I$ = the second moment of area of the beam's cross-section about NA (from Table 8.1, page 108).

Therefore, $\qquad M = \dfrac{E}{R} I$

or $\qquad \dfrac{M}{I} = \dfrac{E}{R} \qquad (9.8)$

Combining equations (9.6) and (9.8) gives:

$$\dfrac{\sigma}{y} = \dfrac{M}{I} = \dfrac{E}{R} \qquad (9.9)$$

Position of NA

From equilibrium considerations, the horizontal force perpendicular to the beam's cross-section, due to the tensile stresses, must equal the horizontal force perpendicular to the beam's cross-section, due to the compressive stresses, as shown in Figure 9.4.

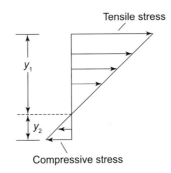

Figure 9.4

Hence,
$$\int_0^{y_1} \sigma\, b\, dy = \int_0^{y_2} \sigma\, b\, dy$$

or
$$\int_0^{y_1} \sigma\, b\, dy - \int_0^{y_2} \sigma\, b\, dy = 0$$

or
$$\int_{-y_2}^{y_1} \sigma\, b\, dy = 0$$

But from equation (9.5), $\sigma = E\dfrac{y}{R}$

Therefore,
$$\int_{-y_2}^{y_1} \frac{Ey}{R} b\, dy = 0$$

Now, E and R are constants, hence
$$\frac{E}{R}\int_{-y_2}^{y_1} y\, b\, dy = 0$$

However, $\int y\, b\, dy$ = the first moment of area about the centroid, and where this is zero, coincides with the centroidal axis, i.e. the **neutral axis** lies on the same axis as the **centroidal axis**.

Moment of resistance (M)

From Figure 9.4, it can be seen that the system of tensile and compressive stresses perpendicular to the beam's cross-section, cause a couple, which resists the applied moment M, where
$$M = \int_{-y_2}^{y_1} \sigma\,(b\, dy)\, y$$

But from equation (9.5), $\sigma = E\dfrac{y}{R}$

Hence,
$$M = \frac{E}{R}\int y^2\, b\, dy = \frac{E}{R}\left(\frac{by^3}{3}\right)$$

or
$$M = \frac{EI}{R} \quad \text{(as required)}$$

9.3 Worked problems on the bending of beams

Problem 1. A solid circular section bar of diameter 20 mm, is subjected to a pure bending moment of 0.3 kN m. If $E = 2 \times 10^{11}\text{N/m}^2$, determine the resulting radius of curvature of the neutral layer of this beam and the maximum bending stress.

From Table 8.1, page 108,
$$I = \frac{\pi d^4}{64} = \frac{\pi \times 20^4}{64} = \textbf{7854 mm}^4$$

Now,
$$M = 0.3 \text{ kN m} \times 1000\frac{\text{N}}{\text{kN}} \times 1000\frac{\text{mm}}{\text{m}}$$
$$= \textbf{3} \times \textbf{10}^5\textbf{N mm}$$

and
$$E = 2 \times 10^{11}\frac{\text{N}}{\text{m}^2} \times 1\frac{\text{m}}{1000\,\text{mm}} \times 1\frac{\text{m}}{1000\,\text{mm}}$$
$$= \textbf{2} \times \textbf{10}^5\textbf{N/mm}^2$$

From equation (9.8), $\dfrac{M}{I} = \dfrac{E}{R}$

hence, radius of curvature,
$$R = \frac{EI}{M} = 2 \times 10^5 \frac{\text{N}}{\text{mm}^2} \times \frac{7854\,\text{mm}^4}{3 \times 10^5\,\text{N mm}}$$

i.e. $\quad$ **R = 5236 mm = 5.24 m**

From equation (9.9), $\dfrac{\sigma}{y} = \dfrac{M}{I}$

and
$$\hat{\sigma} = \frac{M\,\hat{y}}{I}$$

where $\quad \hat{\sigma}$ = maximum stress due to bending

and $\quad \hat{y}$ = outermost fibre from NA $= \dfrac{d}{2} = \dfrac{20}{2}$
$$= 10 \text{ mm}$$

Hence, **maximum bending stress,**
$$\hat{\sigma} = \frac{M\,\hat{y}}{I} = \frac{3 \times 10^5\,\text{N mm} \times 10\,\text{mm}}{7854\,\text{mm}^4}$$
$$= \textbf{382 N/mm}^2 = \textbf{382} \times \textbf{10}^6\textbf{N/m}^2 = \textbf{382 MPa}$$

Problem 2. A beam of length 3 m is simply supported at its ends and has a cross-section, as shown in Figure 9.5. If the beam is subjected to a uniformly distributed load of 2 tonnes/m, determine the maximum stress due to bending and the corresponding value of the radius of curvature of the neutral layer.

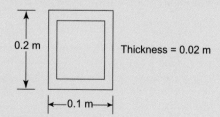

0.2 m $\qquad$ Thickness = 0.02 m

├─0.1 m─┤

Figure 9.5

The total weight on the beam $= wL$, and as the beam is symmetrically loaded, the values of the end reactions, $R = \dfrac{wL}{2}$, as shown in Figure 9.6.

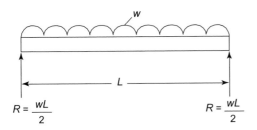

Figure 9.6

Now the maximum bending moment, $\widehat{M}$, occurs at the mid-span, where

$$\widehat{M} = R \times \frac{L}{2} - \frac{wL}{2} \times \frac{L}{4} \text{ but } R = \frac{wL}{2},$$

hence $\widehat{M} = \dfrac{wL}{2} \times \dfrac{L}{2} - \dfrac{wL}{2} \times \dfrac{L}{4}$

$$= \frac{wL^2}{4} - \frac{wL^2}{8} = wL^2\left(\frac{2-1}{8}\right)$$

$$= \frac{wL^2}{8} = \frac{2\dfrac{\text{tonnes}}{\text{m}} \times (3\,\text{m})^2}{8} = 2.25 \text{ tonnes m}$$

$$= 2.25 \text{ tonnes m} \times \frac{1000\,\text{kg}}{\text{tonne}} \times \frac{9.81\,\text{N}}{\text{kg}}$$

i.e. maximum bending moment, $\widehat{M}$ = 22073 N m

Second moment of area, $I = \dfrac{0.1 \times 0.2^3}{12} - \dfrac{0.06 \times 0.16^3}{12}$

$$= 6.667 \times 10^{-5} - 2.048 \times 10^{-5}$$

i.e. $\qquad\qquad$ **$I = 4.619 \times 10^{-5}\text{m}^4$**

The maximum stress $\hat{\sigma}$ occurs in the fibre of the beam's cross-section, which is the furthest distance from NA, namely $\hat{y}$

By inspection, $\qquad \hat{y} = \dfrac{0.2}{2} = 0.1 \text{ m}$

From $\qquad\qquad \dfrac{\hat{\sigma}}{\hat{y}} = \dfrac{M}{I},$

$$\hat{\sigma} = \frac{M\,\hat{y}}{I} = \frac{22073\,\text{N m} \times 0.1\,\text{m}}{4.619 \times 10^{-5}\,\text{m}^4}$$

i.e. **maximum stress, $\hat{\sigma}$ = 47.79 $\times$ 10^6 N/m^2**
$$= \textbf{47.79 MPa}$$

Problem 3. A cantilever beam, whose cross-section is a tube of diameter 0.2 m and wall thickness of 0.02 m, is subjected to a point load, at its free end, of 3 kN, as shown in Figure 9.7. Determine the maximum bending stress in this cantilever.

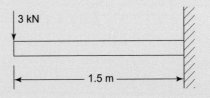

Figure 9.7

From problem 10, page 110, $I = \dfrac{\pi\left(D_2{}^4 - D_1{}^4\right)}{64}$

where D_2 = the external diameter of the tube, and D_1 = the internal diameter of the tube.

Hence $\quad I = \dfrac{\pi\left(0.2^4 - 0.16^4\right)}{64}$

i.e. $\qquad$ **$I = 4.637 \times 10^{-5}\text{m}^4$**

The maximum bending moment, namely $\widehat{M}$, will occur at the built-in end of the beam, i.e. on the extreme right of the beam of Figure 9.7.

Maximum bending moment,

$$\widehat{M} = W \times L = 3 \text{ kN} \times 1.5 \text{ m} \times 1000\,\frac{\text{N}}{\text{kN}}$$

$$= \textbf{4500 N m}$$

The maximum stress occurs at the outermost fibre of the beam's cross-section from NA, namely at $\hat{y}$

By inspection, $\qquad \hat{y} = \dfrac{0.2}{2} = 0.1 \text{ m}$

Hence, $\qquad\qquad \hat{\sigma} = \dfrac{\widehat{M}\,\hat{y}}{I} = \dfrac{4500\,\text{N m} \times 0.1\,\text{m}}{4.637 \times 10^{-5}\,\text{m}^4}$

i.e. **the maximum bending stress,**
$$\hat{\sigma} = \textbf{9.70} \times \textbf{10}^6\textbf{N/m}^2 = \textbf{9.70 MPa}$$

Problem 4. A rectangular plank of wood of length 2.8 m is floating horizontally in still water, as shown in Figure 9.8. If the cross-section of the wooden plank is of rectangular form, of width

$B = 280$ mm and of thickness, $D = 80$ mm, determine the maximum bending stress in the plank, assuming that a concentrated mass of 320 kg is placed at its mid span. Let $g = 9.81$ m/s². Neglect the self-weight of the plank.

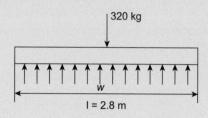

Figure 9.8

Let m = mass of 'weight' = 320 kg,

and W = weight of the mass = mg

$$= 320 \text{ kg} \times 9.81 \text{ m/s}^2 = \mathbf{3139.2 \ N}$$

Let w = load per unit length, caused by the water pressure, acting upwards.

Resolving vertically: upward forces = downward forces

i.e. $w\ell = W$

i.e. $w = \dfrac{W}{\ell} = \dfrac{3139.2}{2.8} = \mathbf{1121.1 \ N/m}$

By inspection, the maximum bending moment, namely $\widehat{M}$, will occur at mid-span, where

$$\widehat{M} = w \times \frac{\ell}{2} \times \frac{\ell}{4} = \frac{w\ell^2}{8}$$

$$= \frac{1121.1 \times (2.8)^2}{8} = \mathbf{1099 \ N \ m}$$

I = second moment of area of the rectangular cross-section about its neutral axis $= \dfrac{BD^3}{12}$,

i.e. $I = \dfrac{280 \times 10^{-3} \times \left(80 \times 10^{-3}\right)^3}{12}$

$$= \mathbf{1.195 \times 10^{-5} m^4}$$

$\hat{y}$ = distance of the fibre of the rectangular cross-section from the neutral axis, i.e.

$$\hat{y} = \frac{D}{2} = \frac{80 \times 10^{-3}}{2} = 40 \times 10^{-3} \text{ m}$$

Maximum stress $= \hat{\sigma} = \dfrac{\widehat{M}\hat{y}}{I} = \dfrac{1099 \times 40 \times 10^{-3}}{1.195 \times 10^{-5}}$

$$= \mathbf{3.68 \ MPa}$$

Now try the following Practise Exercises

Practise Exercise 51   Further problems on the bending of beams

1. A cantilever of solid circular cross-section is subjected to a concentrated load of 30 N at its free end, as shown in Figure 9.9. If the diameter of the cantilever is 10 mm, determine the maximum stress in the cantilever.

Figure 9.9

[367 MPa]

2. If the cantilever of Figure 9.9 were replaced with a tube of the same external diameter, but of wall thickness 2 mm, what would be the maximum stress due to the load shown in Figure 9.9? [421 MPa]

3. A uniform section beam, simply supported at its ends, is subjected to a centrally placed concentrated load of 5 kN. The beam's length is 1 m and its cross-section is a solid circular one.

   If the maximum stress in the beam is limited to 30 MPa, determine the minimum permissible diameter of the beam's cross-section. [75 mm]

4. If the cross-section of the beam of Problem 3 were of rectangular shape, as shown in Figure 9.10, determine its dimensions. Bending can be assumed to take place about the *xx* axis.

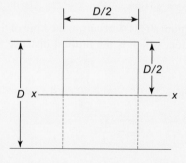

Figure 9.10

[79.4 mm × 39.7 mm]

5. If the cross-section of the beam of Problem 3 is a circular tube of external diameter d and internal diameter $d/2$, determine the value of d. [76.8 mm]

6. A cantilever of length 2 m, carries a uniformly distributed load of 30 N/m, as shown in Figure 9.11. Determine the maximum stress in the cantilever. [39.1 MPa]

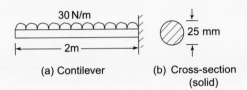

(a) Contilever   (b) Cross-section (solid)

Figure 9.11

7. If the cantilever of Problem 6 were replaced by a uniform section beam, simply supported at its ends and carrying the same uniformly distributed load, determine the maximum stress in the beam. The cross-section of the beam may be assumed to be the same as that of Problem 6. [9.78 MPa]

8. If the load in Problem 7 were replaced by a single concentrated load of 120 N, placed at a distance of 0.75 m from the left support, what would be the maximum stress in the beam due to this concentrated load. [36.7 MPa]

9. If the beam of Figure 9.11 were replaced by another beam of the same length, but which had a cross-section of tee form, as shown in Figure 9.12, determine the maximum stress in the beam.

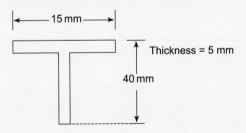

Figure 9.12

[33.9 MPa]

Practise Exercise 52   Short-answer questions on the bending of beams

1. Define neutral layer.

2. Define the neutral axis of a beam's cross-section.

3. Give another name for the neutral axis.

4. Write down the relationship between stress σ and bending moment M.

5. Write down the relationship between stress σ and radius of curvature R.

Practise Exercise 53   Multiple-choice questions on the bending of beams

(Answers on page 297)

1. The maximum stress due to bending occurs:
   (a) at the neutral axis
   (b) at the outermost fibre
   (c) between the neutral axis and the outermost fibre.

2. If the bending moment is increased in a beam, the radius of curvature will:
   (a) increase
   (b) decrease
   (c) stay the same.

3. If the Young's modulus is increased in a beam in bending, due to a constant value of M, the resulting bending stress will:
   (a) increase
   (b) decrease
   (c) stay the same.

Part Two

Chapter 10

Torque

This chapter commences by defining a couple and a torque. It then shows how the energy and work done can be calculated from these terms. It then derives the expression which relates torque to the product of mass moment of inertia and the angular acceleration. The expression for kinetic energy due to rotation is also derived. These expressions are then used for calculating the power transmitted from one shaft to another, via a belt. This work is very important for calculating the power transmitted in rotating shafts and other similar artefacts in many branches of engineering.

At the end of this chapter you should be able to:

- define a couple
- define a torque and state its unit
- calculate torque given force and radius
- calculate work done, given torque and angle turned through
- calculate power, given torque and angle turned through
- appreciate kinetic energy $= \dfrac{I\omega^2}{2}$ where I is the moment of inertia
- appreciate that torque $T = I\alpha$ where α is the angular acceleration
- calculate torque given I and α
- calculate kinetic energy given I and ω
- understand power transmission by means of belt and pulley
- perform calculations involving torque, power and efficiency of belt drives

10.1 Couple and torque

When two equal forces act on a body as shown in Figure 10.1, they cause the body to rotate, and the system of forces is called a **couple**.

Mechanical Engineering Principles, Bird and Ross, ISBN 9780415517850

The turning moment of a couple is called a **torque**, T. In Figure 10.1, torque = magnitude of either force × perpendicular distance between the forces,

i.e. $\qquad T = Fd$

Figure 10.1

The unit of torque is the **newton metre, N m**
When a force F newtons is applied at a radius r metres from the axis of, say, a nut to be turned by a spanner, as shown in Figure 10.2, the torque T applied to the nut is given by:

$$T = Fr \text{ N m}$$

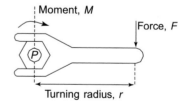

Figure 10.2

Problem 1. Determine the torque when a pulley wheel of diameter 300 mm has a force of 80 N applied at the rim.

Torque $T = Fr$, where force $F = 80$ N

and radius $r = \dfrac{300}{2} = 150$ mm $= 0.15$ m

Hence, **torque,** $\quad T = (80)(0.15) = \textbf{12 N m}$

Problem 2. Determine the force applied tangentially to a bar of a screw jack at a radius of 800 mm, if the torque required is 600 N m.

Torque, T = force × radius, from which

$$\textbf{force} = \frac{\text{torque}}{\text{radius}} = \frac{600\,\text{N m}}{800 \times 10^{-3}\,\text{m}}$$

$$= \textbf{750 N}$$

Problem 3. The circular hand-wheel of a valve of diameter 500 mm has a couple applied to it composed of two forces, each of 250 N. Calculate the torque produced by the couple.

Torque produced by couple, $T = Fd$,

where force $F = 250$ N

and distance between the forces, $d = 500$ mm

$$= 0.5\,\text{m}$$

Hence,   **torque,** $T = (250)(0.5)$

$$= \textbf{125 N m}$$

Now try the following Practise Exercise

Practise Exercise 54   Further problems on torque

1. Determine the torque developed when a force of 200 N is applied tangentially to a spanner at a distance of 350 mm from the centre of the nut.   [70 N m]

2. During a machining test on a lathe, the tangential force on the tool is 150 N. If the torque on the lathe spindle is 12 N m, determine the diameter of the work-piece.   [160 mm]

10.2 Work done and power transmitted by a constant torque

Figure 10.3(a) shows a pulley wheel of radius r metres attached to a shaft and a force F Newton's applied to the rim at point P.

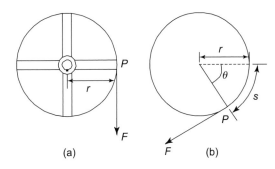

Figure 10.3

Figure 10.3(b) shows the pulley wheel having turned through an angle θ radians as a result of the force F being applied. The force moves through a distance s, where arc length $s = r\theta$

Work done = force × distance moved by the force
$$= F \times r\theta = Fr\theta\,\text{N m} = Fr\theta\,\text{J}$$

However, Fr is the torque T, hence,

$$\textbf{work done} = \textbf{\textit{T}}\boldsymbol{\theta}\ \textbf{joules}$$

Average power = $\dfrac{\text{work done}}{\text{time taken}} = \dfrac{T\theta}{\text{time taken}}$ for a constant torque T

However, (angle θ)/(time taken) = angular velocity, ω rad/s

Hence,   **power,** $P = T\omega$ **watts**   (10.1)

Angular velocity, $\omega = 2\pi n$ rad/s where n is the speed in rev/s

Hence,   **power,** $P = 2\pi nT$ **watts**   (10.2)

Sometimes power is in units of horsepower (hp), where 1 horsepower = 745.7 watts

i.e.   **1 hp = 745.7 watts**

Problem 4. A constant force of 150 N is applied tangentially to a wheel of diameter 140 mm. Determine the work done, in joules, in 12 revolutions of the wheel.

Torque $T = Fr$, where $F = 150$ N

and   radius $r = \dfrac{140}{2} = 70$ mm $= 0.070$ m

Hence,   torque $T = (150)(0.070) = 10.5$ N m

Work done = $T\theta$ joules, where torque, $T = 10.5$ N m and angular displacement,

$\theta = 12$ revolutions $= 12 \times 2\pi$ rad

$$= 24\pi \text{ rad.}$$

Hence,   **work done** $= T\theta = (10.5)(24\pi) = $ **792 J**

Problem 5.   Calculate the torque developed by a motor whose spindle is rotating at 1000 rev/min and developing a power of 2.50 kW.

Power $P = 2\pi n T$ (from above), from which, torque,

$$T = \frac{P}{2\pi n} \text{ N m}$$

where power, $P = 2.50$ kW $= 2500$ W

and speed, $n = 1000/60$ rev/s

Thus, **torque,** $T = \dfrac{P}{2\pi n} = \dfrac{2500}{2\pi\left(\dfrac{1000}{60}\right)} = \dfrac{2500 \times 60}{2\pi \times 1000}$

$$= \textbf{23.87 N m}$$

Problem 6.   An electric motor develops a power of 5 hp and a torque of 12.5 N m. Determine the speed of rotation of the motor in rev/min.

Power, $P = 2\pi n T$, from which,

speed $n = \dfrac{P}{2\pi T}$ rev/s

where power, $P = 5$ hp $= 5 \times 745.7 = 3728.5$ W

and torque $T = 12.5$ N m.

Hence, speed $n = \dfrac{3728.5}{2\pi(12.5)} = 47.47$ rev/s

The speed of rotation of the motor $= 47.47 \times 60$

$$= \textbf{2848 rev/min.}$$

Problem 7.   In a turning-tool test, the tangential cutting force is 50 N. If the mean diameter of the work-piece is 40 mm, calculate (a) the work done per revolution of the spindle (b) the power required when the spindle speed is 300 rev/min.

(a)   Work done $= T\theta$, where $T = Fr$

     Force $F = 50$ N,   radius $r = \dfrac{40}{2} = 20$ mm $= 0.02$ m

     and angular displacement, $\theta = 1$ rev $= 2\pi$ rad.

Hence, **work done per revolution of spindle**

$$= Fr\theta = (50)(0.02)(2\pi) = \textbf{6.28 J}$$

(b)   Power, $P = 2\pi n T$, where

     torque, $T = Fr = (50)(0.02) = 1$ N m   and

     speed, $n = \dfrac{300}{60} = 5$ rev/s.

Hence, **power required,** $P = 2\pi(5)(1) = \textbf{31.42 W}$

Problem 8.   A pulley is 600 mm in diameter and the difference in tensions on the two sides of the driving belt is 1.5 kN. If the speed of the pulley is 500 rev/min, determine (a) the torque developed, and (b) the work done in 3 minutes.

(a)   Torque $T = Fr$, where force $F = 1.5$ kN $= 1500$ N,

     and radius $r = \dfrac{600}{2} = 300$ mm $= 0.3$ m.

Hence,   **torque developed** $= (1500)(0.3)$

$$= \textbf{450 N m}$$

(b)   Work done $= T\theta$, where torque $T = 450$ N m and angular displacement in 3 minutes

$$= (3 \times 500) \text{ rev} = (3 \times 500 \times 2\pi) \text{ rad.}$$

Hence, **work done** $= (450)(3 \times 500 \times 2\pi)$

$$= 4.24 \times 10^6 \text{ J}$$

$$= \textbf{4.24 MJ}$$

Problem 9.   A motor connected to a shaft develops a torque of 5 kN m. Determine the number of revolutions made by the shaft if the work done is 9 MJ.

Work done $= T\theta$, from which, angular displacement,

$$\theta = \frac{\text{work done}}{\text{torque}}$$

Work done $= 9$ MJ $= 9 \times 10^6$ J

and   torque $= 5$ kN m $= 5000$ N m.

Hence, angular displacement, $\theta = \dfrac{9 \times 10^6}{5000}$

$$= 1800 \text{ rad.}$$

2π rad $= 1$ rev, hence, **the number of revolutions**

made by the shaft $= \dfrac{1800}{2\pi} = \textbf{286.5 revs}$

Now try the following Practise Exercise

Practise Exercise 55   Further problems on work done and power transmitted by a constant torque

1. A constant force of 4 kN is applied tangentially to the rim of a pulley wheel of diameter 1.8 m attached to a shaft. Determine the work done, in joules, in 15 revolutions of the pulley wheel.   [339.3 kJ]

2. A motor connected to a shaft develops a torque of 3.5 kN m. Determine the number of revolutions made by the shaft if the work done is 11.52 MJ.   [523.8 rev]

3. A wheel is turning with an angular velocity of 18 rad/s and develops a power of 810 W at this speed. Determine the torque developed by the wheel.   [45 N m]

4. Calculate the torque provided at the shaft of an electric motor that develops an output power of 3.2 hp at 1800 rev/min.   [12.66 N m]

5. Determine the angular velocity of a shaft when the power available is 2.75 kW and the torque is 200 N m.   [13.75 rad/s]

6. The drive shaft of a ship supplies a torque of 400 kN m to its propeller at 400 rev/min. Determine the power delivered by the shaft.   [16.76 MW]

7. A motor is running at 1460 rev/min and produces a torque of 180 N m. Determine the average power developed by the motor.   [27.52 kW]

8. A wheel is rotating at 1720 rev/min and develops a power of 600 W at this speed. Calculate (a) the torque (b) the work done, in joules, in a quarter of an hour.   [(a) 3.33 N m (b) 540 kJ]

9. A force of 60 N is applied to a lever of a scew-jack at a radius of 220 mm. If the lever makes 25 revolutions, determine (a) the work done on the jack (b) the power, if the time taken to complete 25 revolutions is 40 s.   [(a) 2.073 kJ (b) 51.84 W]

10.3 Kinetic energy and moment of inertia

The tangential velocity v of a particle of mass m moving at an angular velocity ω rad/s at a radius r metres (see Figure 10.4) is given by:

$$v = \omega r \text{ m/s}$$

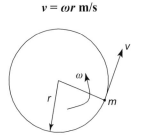

Figure 10.4

The kinetic energy of a particle of mass m is given by:

$$\textbf{Kinetic energy} = \frac{1}{2}mv^2 \text{ (from Chapter 15)}$$

$$= \frac{1}{2}m(\omega r)^2 = \frac{1}{2}m\,\omega^2 r^2 \textbf{ joules}$$

The total kinetic energy of a system of masses rotating at different radii about a fixed axis but with the same angular velocity, as shown in Figure 10.5, is given by:

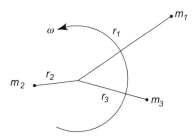

Figure 10.5

$$\text{Total kinetic energy} = \frac{1}{2}\,m_1\omega^2 r_1^2 + \frac{1}{2}\,m_2\omega^2 r_2^2$$

$$+ \frac{1}{2}\,m_3\omega^2 r_3^2$$

$$= (m_1 r_1^2 + m_2 r_2^2 + m_3 r_3^2)\frac{\omega^2}{2}$$

In general, this may be written as:

$$\textbf{Total kinetic energy} = (\Sigma mr^2)\frac{\omega^2}{2} = I\frac{\omega^2}{2}$$

Part Two

where $I (= \Sigma m r^2)$ is called the **moment of inertia** of the system about the axis of rotation and has units of kg m^2.

The moment of inertia of a system is a measure of the amount of work done to give the system an angular velocity of ω rad/s, or the amount of work that can be done by a system turning at ω rad/s.

From Section 10.2, work done $= T\theta$, and if this work is available to increase the kinetic energy of a rotating body of moment of inertia I, then:

$$T\theta = I\left(\frac{\omega_2{}^2 - \omega_1{}^2}{2}\right)$$ where ω_1 and ω_2 are the initial and final angular velocities,

i.e. $$T\theta = I\left(\frac{\omega_2 + \omega_1}{2}\right)(\omega_2 - \omega_1)$$

However, $\left(\dfrac{\omega_2 + \omega_1}{2}\right)$ is the mean angular velocity,

i.e. $\dfrac{\theta}{t}$, where t is the time, and $(\omega_2 - \omega_1)$ is the change in angular velocity, i.e. αt, where α is the angular acceleration

Hence, $T\theta = I\left(\dfrac{\theta}{t}\right)(\alpha t)$ from which,   **torque $T = I\alpha$**

where I is the moment of inertia in kg m^2, α is the angular acceleration in rad/s^2 and T is the torque in N m.

Problem 10.   A shaft system has a moment of inertia of 37.5 kg m^2. Determine the torque required to give it an angular acceleration of 5.0 rad/s^2.

Torque, $T = I\alpha$, where moment of inertia $I = 37.5$ kg m^2 and angular acceleration, $\alpha = 5.0$ rad/s^2.

Hence,   **torque, $T = I\alpha = (37.5)(5.0) = $ 187.5 N m**

Problem 11.   A shaft has a moment of inertia of 31.4 kg m^2. What angular acceleration of the shaft would be produced by an accelerating torque of 495 N m?

Torque, $T = I\alpha$, from which, angular acceleration, $\alpha = \dfrac{T}{I}$, where torque, $T = 495$ N m and moment of inertia $I = 31.4$ kg m^2

Hence,   **angular acceleration, $\alpha = \dfrac{495}{31.4}$**

$$= \textbf{15.76 rad/s}^2$$

Problem 12.   A body of mass 100 g is fastened to a wheel and rotates in a circular path of 500 mm in diameter. Determine the increase in kinetic energy of the body when the speed of the wheel increases from 450 rev/min to 750 rev/min.

From above, kinetic energy $= I\,\dfrac{\omega^2}{2}$

Thus, increase in kinetic energy $= I\left(\dfrac{\omega_2{}^2 - \omega_1{}^2}{2}\right)$

where moment of inertia, $I = mr^2$,

mass, $m = 100$ g $= 0.1$ kg and

radius, $r = \dfrac{500}{2} = 250$ mm $= 0.25$ m.

Initial angular velocity, $\omega_1 = 450$ rev/min

$$= \frac{450 \times 2\pi}{60}\text{ rad/s}$$
$$= 47.12\text{ rad/s},$$

and final angular velocity, $\omega_2 = 750$ rev/min

$$= \frac{750 \times 2\pi}{60}\text{ rad/s}$$
$$= 78.54\text{ rad/s}.$$

Thus, **increase in kinetic energy** $= I\left(\dfrac{\omega_2{}^2 - \omega_1{}^2}{2}\right)$

$$= (mr^2)\left(\frac{\omega_2{}^2 - \omega_1{}^2}{2}\right)$$

$$= (0.1)(0.25^2)\left(\frac{78.54^2 - 47.12^2}{2}\right) = \textbf{12.34 J}$$

Problem 13.   A system consists of three small masses rotating at the same speed about the same fixed axis. The masses and their radii of rotation are: 15 g at 250 mm, 20 g at 180 mm and 30 g at 200 mm. Determine (a) the moment of inertia of the system about the given axis, and (b) the kinetic energy in the system if the speed of rotation is 1200 rev/min.

(a)   Moment of inertia of the system, $I = \Sigma m r^2$

i.e. $I = [(15 \times 10^{-3}\text{kg})(0.25\text{ m})^2]$
$$+ [(20 \times 10^{-3}\text{kg})(0.18\text{ m})^2]$$
$$+ [(30 \times 10^{-3}\text{ kg})(0.20\text{ m})^2]$$
$$= (9.375 \times 10^{-4}) + (6.48 \times 10^{-4}) + (12 \times 10^{-4})$$
$$= 27.855 \times 10^{-4}\text{kg m}^2 = \textbf{2.7855} \times \textbf{10}^{-3}\textbf{kg m}^2$$

(b)   Kinetic energy $= I\dfrac{\omega^2}{2}$, where

moment of inertia, $I = 2.7855 \times 10^{-3}$ kg m^2

and angular velocity, $\omega = 2\pi n = 2\pi\left(\dfrac{1200}{60}\right)$ rad/s

$$= 40\pi \text{ rad/s}$$

Hence, **kinetic energy in the system**

$$= (2.7855 \times 10^{-3})\dfrac{(40\pi)^2}{2} = \textbf{21.99 J}$$

Problem 14.   A shaft with its rotating parts has a moment of inertia of 20 kg m^2. It is accelerated from rest by an accelerating torque of 45 N m. Determine the speed of the shaft in rev/min (a) after 15 s, and (b) after the first 5 revolutions.

(a)   Since torque $T = I\alpha$, then angular acceleration,

$$\alpha = \dfrac{T}{I} = \dfrac{45}{20} = 2.25 \text{ rad/s}^2$$

The angular velocity of the shaft is initially zero, i.e. $\omega_1 = 0$

From chapter 12, page 147, the angular velocity after 15 s,

$$\omega_2 = \omega_1 + \alpha t = 0 + (2.25)(15) = 33.75 \text{ rad/s},$$

i.e. **speed of shaft after 15 s**

$$= (33.75)\left(\dfrac{60}{2\pi}\right) \text{ rev/min}$$

$$= \textbf{322.3 rev/min}$$

(b)   Work done $= T\theta$, where torque $T = 45$ N m and angular displacement, $\theta = 5$ revolutions $= 5 \times 2\pi = 10\pi$ rad.

Hence work done $= (45)(10\pi) = 1414$ J.
This work done results in an increase in kinetic energy, given by $I\dfrac{\omega^2}{2}$, where moment of inertia $I = 20$ kg m^2 and $\omega =$ angular velocity.

Hence, $1414 = (20)\left(\dfrac{\omega^2}{2}\right)$ from which,

$$\omega = \sqrt{\left(\dfrac{1414 \times 2}{20}\right)} = 11.89 \text{ rad/s}$$

i.e. **speed of shaft after the first 5 revolutions**

$$= 11.89 \times \dfrac{60}{2\pi} = \textbf{113.5 rev/min}$$

Problem 15.   The accelerating torque on a turbine rotor is 250 N m.

(a)   Determine the gain in kinetic energy of the rotor while it turns through 100 revolutions (neglecting any frictional and other resisting torques).

(b)   If the moment of inertia of the rotor is 25 kg m^2 and the speed at the beginning of the 100 revolutions is 450 rev/min, determine its speed at the end.

(a)   The kinetic energy gained is equal to the work done by the accelerating torque of 250 N m over 100 revolutions,

i.e. **gain in kinetic energy** = work done

$$= T\theta = (250)(100 \times 2\pi) = \textbf{157.08 kJ}$$

(b)   Initial kinetic energy is given by:

$$I\dfrac{\omega_1^2}{2} = \dfrac{(25)\left(\dfrac{450 \times 2\pi}{60}\right)^2}{2} = 27.76 \text{ kJ}$$

The final kinetic energy is the sum of the initial kinetic energy and the kinetic energy gained,

i.e. $$I\dfrac{\omega_2^2}{2} = 27.76 \text{ kJ} + 157.08 \text{ kJ}$$

$$= 184.84 \text{ kJ}.$$

Hence, $\dfrac{(25)\omega_2^2}{2} = 184840$ from which,

$$\omega_2 = \sqrt{\left(\dfrac{184840 \times 2}{25}\right)}$$

$$= 121.6 \text{ rad/s}.$$

Thus, **speed at end of 100 revolutions**

$$= \dfrac{121.6 \times 60}{2\pi} \text{ rev/min} = \textbf{1161 rev/min}$$

Problem 16.   A shaft with its associated rotating parts has a moment of inertia of 55.4 kg m^2. Determine the uniform torque required to accelerate the shaft from rest to a speed of 1650 rev/min while it turns through 12 revolutions.

From above, $T\theta = I\left(\dfrac{\omega_2^2 - \omega_1^2}{2}\right)$

i.e. $$T = \dfrac{I}{\theta}\left(\dfrac{\omega_2^2 - \omega_1^2}{2}\right)$$

where angular displacement $\theta = 12$ rev $= 12 \times 2\pi$

$$= 24\pi \text{ rad},$$

final speed, $\omega_2 = 1650$ rev/min $= \dfrac{1650}{60} \times 2\pi$

$$= 172.79 \text{ rad/s},$$

initial speed, $\omega_1 = 0$,

and moment of inertia, $I = 55.4$ kg m^2.

Hence, **torque required,**

$$T = \left(\frac{I}{\theta}\right)\left(\frac{\omega_2^2 - \omega_1^2}{2}\right)$$

$$= \left(\frac{55.4}{24\pi}\right)\left(\frac{172.79^2 - 0^2}{2}\right) = \mathbf{10.97 \text{ kN m}}$$

Now try the following Practise Exercise

Practise Exercise 56	Further problems on kinetic energy and moment of inertia

1. A shaft system has a moment of inertia of 51.4 kg m^2. Determine the torque required to give it an angular acceleration of 5.3 rad/s^2. [272.4 N m]

2. A shaft has an angular acceleration of 20 rad/s^2 and produces an accelerating torque of 600 N m. Determine the moment of inertia of the shaft. [30 kg m^2]

3. A uniform torque of 3.2 kN m is applied to a shaft while it turns through 25 revolutions. Assuming no frictional or other resistances, calculate the increase in kinetic energy of the shaft (i.e. the work done). If the shaft is initially at rest and its moment of inertia is 24.5 kg m^2, determine its rotational speed, in rev/min, at the end of the 25 revolutions. [502.65 kJ, 1934 rev/min]

4. An accelerating torque of 30 N m is applied to a motor, while it turns through 10 revolutions. Determine the increase in kinetic energy. If the moment of inertia of the rotor is 15 kg m^2 and its speed at the beginning of the 10 revolutions is 1200 rev/min, determine its speed at the end. [1.885 kJ, 1209.5 rev/min]

5. A shaft with its associated rotating parts has a moment of inertia of 48 kg m^2. Determine the uniform torque required to accelerate the shaft from rest to a speed of 1500 rev/min while it turns through 15 revolutions. [6.283 kN m]

6. A small body, of mass 82 g, is fastened to a wheel and rotates in a circular path of 456 mm diameter. Calculate the increase in kinetic energy of the body when the speed of the wheel increases from 450 rev/min to 950 rev/min. [16.36 J]

7. A system consists of three small masses rotating at the same speed about the same fixed axis. The masses and their radii of rotation are: 16 g at 256 mm, 23 g at 192 mm and 31 g at 176 mm. Determine (a) the moment of inertia of the system about the given axis, and (b) the kinetic energy in the system if the speed of rotation is 1250 rev/min. [(a) 2.857×10^{-3} kg m^2 (b) 24.48 J]

8. A shaft with its rotating parts has a moment of inertia of 16.42 kg m^2. It is accelerated from rest by an accelerating torque of 43.6 N m. Find the speed of the shaft (a) after 15 s, and (b) after the first four revolutions. [(a) 380.3 rev/min (b) 110.3 rev/min]

9. The driving torque on a turbine rotor is 203 N m, neglecting frictional and other resisting torques. (a) What is the gain in kinetic energy of the rotor while it turns through 100 revolutions? (b) If the moment of inertia of the rotor is 23.2 kg m^2 and the speed at the beginning of the 100 revolutions is 600 rev/min, what will be its speed at the end? [(a) 127.55 kJ (b) 1167 rev/min]

10.4 Power transmission and efficiency

A common and simple method of transmitting power from one shaft to another is by means of a **belt** passing over pulley wheels which are keyed to the shafts, as shown in Figure 10.6. Typical applications include an electric motor driving a lathe or a drill, and an engine driving a pump or generator.

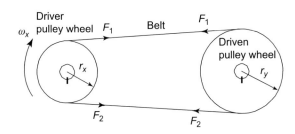

Figure 10.6

For a belt to transmit power between two pulleys there must be a difference in tensions in the belt on either side of the driving and driven pulleys. For the direction of rotation shown in Figure 10.6, $F_2 > F_1$
The torque T available at the driving wheel to do work is given by:

$$T = (F_2 - F_1)r_x \text{ N m}$$

and the available power P is given by:

$$P = T\omega = (F_2 - F_1)r_x\omega_x \text{ watts}$$

From Section 10.3, the linear velocity of a point on the driver wheel, $v_x = r_x\omega_x$
Similarly, the linear velocity of a point on the driven wheel, $v_y = r_y\omega_y$. Assuming no slipping,

$$v_x = v_y \qquad \text{i.e.} \qquad r_x\omega_x = r_y\omega_y$$

Hence $\qquad r_x(2\pi n_x) = r_y(2\pi n_y)$

from which, $\qquad \dfrac{r_x}{r_y} = \dfrac{n_y}{n_x}$

$$\text{Percentage efficiency} = \frac{\text{useful work output}}{\text{energy output}} \times 100$$

or $\qquad$ **efficiency $= \dfrac{\text{power output}}{\text{power input}} \times 100\%$**

Problem 17. An electric motor has an efficiency of 75% when running at 1450 rev/min. Determine the output torque when the power input is 3.0 kW.

$$\text{Efficiency} = \frac{\text{power output}}{\text{power input}} \times 100\%$$

hence $\quad 75 = \dfrac{\text{power output}}{3000} \times 100$

from which, power output $= \dfrac{75}{100} \times 3000$

$$= 2250 \text{ W}$$

From Section 10.2, power output, $P = 2\pi nT$

from which, torque, $T = \dfrac{P}{2\pi n}$

where $n = (14\,50/60)$ rev/s

Hence, **output torque** $= \dfrac{2250}{2\pi\left(\dfrac{1450}{60}\right)} = \textbf{14.82 N m}$

Problem 18. A 15 kW motor is driving a shaft at 1150 rev/min by means of pulley wheels and a belt. The tensions in the belt on each side of the driver pulley wheel are 400 N and 50 N. The diameters of the driver and driven pulley wheels are 500 mm and 750 mm respectively. Determine (a) the efficiency of the motor (b) the speed of the driven pulley wheel.

(a) From above, power output from motor
$= (F_2 - F_1)r_x\omega_x$
Force $F_2 = 400$ N and $F_1 = 50$ N,
hence $(F_2 - F_1) = 350$ N,
radius $r_x = \dfrac{500}{2} = 250$ mm $= 0.25$ m and
angular velocity, $\omega_x = \dfrac{1150 \times 2\pi}{60}$ rad/s
Hence power output from motor $= (F_2 - F_1)r_x\omega_x$

$$= (350)(0.25)\left(\frac{1150 \times 2\pi}{60}\right) = 10.54 \text{ kW}$$

Power input = 15 kW

Hence, **efficiency of the motor** $= \dfrac{\text{power output}}{\text{power input}}$

$$= \frac{10.54}{15} \times 100$$

$$= \textbf{70.27\%}$$

(b) From above, $\dfrac{r_x}{r_y} = \dfrac{n_y}{n_x}$ from which,

speed of driven pulley wheels,

$$n_y = \frac{n_x r_x}{r_y} = \frac{1150 \times 0.25}{\dfrac{0.750}{2}} = \textbf{767 rev/min}$$

Problem 19. A crane lifts a load of mass 5 tonne to a height of 25 m. If the overall efficiency of the crane is 65% and the input power to the hauling motor is 100 kW, determine how long the lifting operation takes.

The increase in potential energy is the work done and is given by mgh (see Chapter 15), where mass, $m = 5$ t $= 5000$ kg, $g = 9.81$ m/s^2 and height $h = 25$ m.

Hence, work done $= mgh = (5000)(9.81)(25)$

$$= 1.226 \text{ MJ.}$$

Input power $= 100$ kW $= 100000$ W

$$\text{Efficiency} = \frac{\text{output power}}{\text{input power}} \times 100$$

hence $\qquad 65 = \dfrac{\text{output power}}{100000} \times 100$

from which, output power $= \dfrac{65}{100} \times 100000 = 65000$ W

$$= \frac{\text{work done}}{\text{time taken}}$$

Thus, **time taken for lifting operation**

$$= \frac{\text{work done}}{\text{output power}} = \frac{1.226 \times 10^6 \text{ J}}{65000 \text{ W}} = \textbf{18.86 } \textit{s}$$

Problem 20. The tool of a shaping machine has a mean cutting speed of 250 mm/s and the average cutting force on the tool in a certain shaping operation is 1.2 kN. If the power input to the motor driving the machine is 0.75 kW, determine the overall efficiency of the machine.

Velocity, $v = 250$ mm/s $= 0.25$ m/s
and force, $F = 1.2$ kN $= 1200$ N

From Chapter 15, power output required at the cutting tool (i.e. power output),

$$P = \text{force} \times \text{velocity} = 1200 \text{ N} \times 0.25 \text{ m/s}$$
$$= 300 \text{ W}$$

Power input $= 0.75$ kW $= 750$ W

Hence, **efficiency of the machine**

$$= \frac{\text{output power}}{\text{input power}} \times 100$$

$$= \frac{300}{750} \times 100 = \textbf{40\%}$$

Problem 21. Calculate the input power of the motor driving a train at a constant speed of 72 km/h on a level track, if the efficiency of the motor is 80% and the resistance due to friction is 20 kN.

Force resisting motion $= 20$ kN $= 20000$ N and

velocity $= 72$ km/h $= \dfrac{72}{3.6} = 20$ m/s

Output power from motor
$=$ resistive force $\times$ velocity of train (from Chapter 15)

$$= 20000 \times 20 = 400 \text{ kW}$$

$$\text{Efficiency} = \frac{\text{power output}}{\text{power input}} \times 100$$

hence $\quad 80 = \dfrac{400}{\text{power input}} \times 100$

from which, **power input** $= 400 \times \dfrac{100}{80} = \textbf{500 kW}$

Now try the following Practise Exercises

Practise Exercise 57   Further problems on power transmission and efficiency

1. A motor has an efficiency of 72% when running at 2600 rev/min. If the output torque is 16 N m at this speed, determine the power supplied to the motor. [6.05 kW]

2. The difference in tensions between the two sides of a belt round a driver pulley of radius 240 mm is 200 N. If the driver pulley wheel is on the shaft of an electric motor running at 700 rev/min and the power input to the motor is 5 kW, determine the efficiency of the motor. Determine also the diameter of the driven pulley wheel if its speed is to be 1200 rev/min. [70.37%, 280 mm]

3. A winch is driven by a 4 kW electric motor and is lifting a load of 400 kg to a height of 5.0 m. If the lifting operation takes 8.6 s, calculate the overall efficiency of the winch and motor. [57.03%]

4. A belt and pulley system transmits a power of 5 kW from a driver to a driven shaft. The driver pulley wheel has a diameter of 200 mm and rotates at 600 rev/min. The diameter of the driven wheel is 400 mm. Determine the speed of the driven pulley and the tension in the slack side of the belt when the tension in the tight side of the belt is 1.2 kN. [404.2 N, 300 rev/min]

5. The average force on the cutting tool of a lathe is 750 N and the cutting speed is 400 mm/s. Determine the power input to the motor driving the lathe if the overall efficiency is 55%.

[545.5 W]

6. A ship's anchor has a mass of 5 tonne. Determine the work done in raising the anchor from a depth of 100 m. If the hauling gear is driven by a motor whose output is 80 kW and the efficiency of the haulage is 75%, determine how long the lifting operation takes.   [4.905 MJ, 1 min 22s]

Practise Exercise 58   Short-answer questions on torque

1. In engineering, what is meant by a couple?

2. Define torque.

3. State the unit of torque.

4. State the relationship between work, torque T and angular displacement θ.

5. State the relationship between power P, torque T and angular velocity ω.

6. Complete the following:
   1 horsepower = ... watts.

7. Define moment of inertia and state the symbol used.

8. State the unit of moment of inertia.

9. State the relationship between torque, moment of inertia and angular acceleration.

10. State one method of power transmission commonly used.

11. Define efficiency.

Practise Exercise 59   Multiple-choice questions on torque

(Answers on page 297)

1. The unit of torque is:
   (a) N   (b) Pa
   (c) N/m   (d) N m

2. The unit of work is:
   (a) N   (b) J
   (c) W   (d) N/m

3. The unit of power is:
   (a) N   (b) J
   (c) W   (d) N/m

4. The unit of the moment of inertia is:
   (a) kg m^2 (b) kg
   (c) kg/m^2 (d) N m

5. A force of 100 N is applied to the rim of a pulley wheel of diameter 200 mm. The torque is:
   (a) 2 N m   (b) 20 kN m
   (c) 10 N m   (d) 20 N m

6. The work done on a shaft to turn it through 5π radians is 25π J. The torque applied to the shaft is:
   (a) 0.2 N m   (b) $125\pi^2$ N m
   (c) 30π N m   (d) 5 N m

7. A 5 kW electric motor is turning at 50 rad/s. The torque developed at this speed is:
   (a) 100 N m   (b) 250 N m
   (c) 0.01 N m   (d) 0.1 N m

8. The force applied tangentially to a bar of a screw-jack at a radius of 500 mm, if the torque required is 1 kN m is:
   (a) 2 N   (b) 2 kN
   (c) 500 N   (d) 0.5 N

9. A 10 kW motor developing a torque of $(200/\pi)$ N m is running at a speed of:
   (a) $(\pi/20)$ rev/s   (b) 50π rev/s
   (c) 25 rev/s   (d) $(20/\pi)$ rev/s

10. A shaft and its associated rotating parts has a moment of inertia of 50 kg m^2. The angular acceleration of the shaft to produce an accelerating torque of 5 kN m is:
    (a) 10 rad/s^2 (b) 250 rad/s^2
    (c) 0.01 rad/s^2 (d) 100 rad/s^2

11. A motor has an efficiency of 25% when running at 3000 rev/min. If the output torque is 10 N m, the power input is:
    (a) 4π kW   (b) 0.25π kW
    (c) 15π kW   (d) 75π kW

12. In a belt-pulley wheel system, the effective tension in the belt is 500 N and the diameter of the driver wheel is 200 mm. If the power output from the driving motor is 5 kW, the driver pulley wheel turns at:
    (a) 50 rad/s   (b) 2500 rad/s
    (c) 100 rad/s   (d) 0.1 rad/s

Chapter 11

Twisting of shafts

The torsion of shafts appears in a number of different branches of engineering, including propeller shafts for ships and aircraft, shafts driving the blades of a helicopter, shafts driving the rear wheels of an automobile and shafts driving food mixers, washing machines, tumble dryers, dishwashers, and so on. If the shaft is overstressed due to a torque, so that the maximum shear stress in the shaft exceeds the yield shear stress of the shaft's material, the shaft can fracture. This is an undesirable phenomenon and normally it should be designed out; hence the need for the theory contained in this chapter.

At the end of this chapter you should be able to:

- appreciate practical applications where torsion of shafts occur
- prove that $\dfrac{\tau}{r} = \dfrac{T}{J} = \dfrac{G\theta}{L}$
- calculate the shearing stress τ, due to a torque, T
- calculate the resulting angle of twist, θ, due to torque, T
- calculate the power that can be transmitted by a shaft

11.1 To prove that $\dfrac{\tau}{r} = \dfrac{T}{J} = \dfrac{G\theta}{L}$

In the formula $\dfrac{\tau}{r} = \dfrac{T}{J} = \dfrac{G\theta}{L}$:

τ = the shear stress at radius r

T = the applied torque

J = polar second moment of area of the shaft (note that for non-circular sections, J is the torsional constant and not the polar second moment of area)

G = rigidity or shear modulus

θ = angle of twist, in radians, over its length L

Prior to proving the above formula, the following assumptions are made for circular section shafts:

(a) the shaft is of circular cross-section
(b) the cross-section of the shaft is uniform along its entire length
(c) the shaft is straight and not bent
(d) the shaft's material is homogeneous (i.e. uniform) and isotropic (i.e. exhibits properties with the same values when measured along different axes) and obeys Hooke's law
(e) the limit of proportionality is not exceeded and the angles of twist due to the torque are small
(f) plane cross-sections remain plane and normal during twisting
(g) radial lines across the shaft's cross-section remain straight and radial during twisting.

Consider a circular section shaft, built-in at one end, namely A, and subjected to a torque T at the other end, namely B, as shown in Figure 11.1.

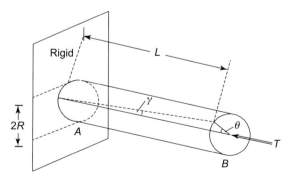

Figure 11.1

Mechanical Engineering Principles, Bird and Ross, ISBN 9780415517850

Let θ be the angle of twist due to this torque T, where the direction of T is according to the right hand screw rule. *N.B.* The direction of a couple, according to the right hand screw rule, is obtained by pointing the right hand in the direction of the double-tailed arrow and rotating the right hand in a clockwise direction.

From Figure 11.1, it can be seen that:

γ = shear strain, and that

$$\gamma L = R\theta \tag{11.1}$$

provided θ is small.

However, from equation (2.1, page 32), $\gamma = \dfrac{\tau}{G}$

Hence, $\qquad \left(\dfrac{\tau}{G}\right)L = R\theta$

or $\qquad\qquad \dfrac{\tau}{R} = \dfrac{G\theta}{L} \tag{11.2}$

From equation (11.2), it can be seen that the shear stress τ is dependent on the value of R and it will be a maximum on the outer surface of the shaft. On the outer surface of the shaft τ will act as shown in Figure 11.2.

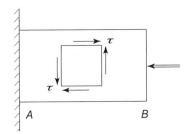

Figure 11.2

For any radius r,

$$\frac{\tau}{r} = \frac{G\theta}{L} \tag{11.3}$$

The shaft in Figure 11.2 is said to be in a state of pure shear on these planes, as these shear stresses will not be accompanied by direct or normal stress.

Consider an annular element of the shaft, as shown in Figure 11.3.

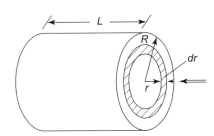

Figure 11.3

The torque T causes constant value shearing stresses on the thin walled annular element shown in Figure 11.4.

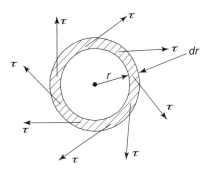

Figure 11.4 Annular element

The elemental torque δT due to these shearing stresses τ at the radius r is given by:

$$\delta T = \tau \times (2\pi r\, dr)r$$

and the total torque $\quad T = \sum \delta T$

or $\qquad\qquad T = \displaystyle\int_0^R \tau(2\pi r^2)\, dr \tag{11.4}$

However, from equation (11.3), $\tau = \dfrac{G\theta}{L}r$

Therefore, $\qquad T = \displaystyle\int_0^R \frac{G\theta}{L}r(2\pi r^2)\, dr$

But G, θ, L and 2π do not vary with r,

hence, $\quad T = \dfrac{G\theta}{L}(2\pi)\displaystyle\int_0^R r^3\, dr$

$$= \frac{G\theta}{L}(2\pi)\left[\frac{r^4}{4}\right]_0^R = \frac{G\theta}{L}\frac{\pi R^4}{2}$$

However, from Table 8.1, page 108, the polar second moment of area of a circle, $J = \dfrac{\pi R^4}{2}$,

hence, $\quad T = \dfrac{G\theta}{L}J$

or $\qquad\qquad \dfrac{T}{J} = \dfrac{G\theta}{L} \tag{11.5}$

Combining equations (11.3) and (11.5) gives:

$$\frac{\tau}{r} = \frac{T}{J} = \frac{G\theta}{L} \tag{11.6}$$

For a solid section of radius R or diameter D,

$$J = \frac{\pi R^4}{2} \text{ or } \frac{\pi D^4}{32} \qquad (11.7)$$

For a hollow tube of circular section and of internal radius R_1 and external radius R_2

$$J = \frac{\pi}{2}\left(R_2{}^4 - R_1{}^4\right) \qquad (11.8)$$

or, in terms of external and internal diameters of D_2 and D_1 respectively, (see Problem 12, Chapter 8, page 110),

$$J = \frac{\pi}{32}\left(D_2{}^4 - D_1{}^4\right) \qquad (11.9)$$

The torsional stiffness of the shaft, k, is defined by:

$$k = \frac{GJ}{L} \qquad (11.10)$$

The next section of worked problems will demonstrate the use of equation (11.6).

11.2 Worked problems on the twisting of shafts

Problem 1. An internal combustion engine of 60 horsepower (hp) transmits power to the car wheels of an automobile at 300 rev/min (rpm). Neglecting any transmission losses, determine the minimum permissible diameter of the solid circular section steel shaft, if the maximum shear stress in the shaft is limited to 50 MPa. What will be the resulting angle of twist of the shaft, due to the applied torque, over a length of 2 m, given that the rigidity modulus, $G = 70$ GPa?
(Note that 1 hp = 745.7 hp).

Power = 60 hp = 60 hp × 745.7 $\frac{W}{hp}$ = 44742 W

From equations (10.1) and (10.2), page 127,

$$\text{power} = T\omega = 2\pi nT \text{ watts,}$$

where n is the speed in rev/s

i.e. $44742 = 2\pi \dfrac{\text{rad}}{\text{rev}} \times \dfrac{300}{60}\dfrac{\text{rev}}{s} \times T$

$$= 31.42 \, T \text{ rad/s}$$

from which, $T = \dfrac{44742 \, W}{31.42 \, \text{rad/s}}$

i.e. **torque** $T = \mathbf{1424 \, N\,m}$ (since 1 W s = 1 N m)

From equation (11.6), $\qquad \dfrac{\tau}{r} = \dfrac{T}{J}$

i.e. $\dfrac{50 \times 10^6}{r} \dfrac{N}{m^2} = \dfrac{1424 \, N\,m}{\dfrac{\pi r^4}{2}}$

and $\dfrac{r^4}{r} = \dfrac{1424 \times 2}{\pi \times 50 \times 10^6} \, m^3$

from which, $r^3 = \dfrac{1424 \times 2}{\pi \times 50 \times 10^6}$

or shaft radius, $r = \sqrt[3]{\left(\dfrac{1424 \times 2}{\pi \times 50 \times 10^6}\right)} = \mathbf{0.0263 \, m}$

Hence, **shaft diameter, d** = 2 × r = 2 × 0.0263

$$= \mathbf{0.0526 \, m}$$

From equation (11.6), $\qquad \dfrac{\tau}{r} = \dfrac{G\theta}{L}$

from which, $\theta = \dfrac{\tau L}{Gr} = \dfrac{50 \times 10^6 \dfrac{N}{m^2} \times 2 \, m}{70 \times 10^9 \dfrac{N}{m^2} \times 0.0263 \, m}$

and $\qquad \boldsymbol{\theta = 0.0543 \, rad}$

$$= 0.0543 \text{ rad} \times \dfrac{360°}{2\pi \text{ rad}}$$

i.e. **angle of twist, $\theta = \mathbf{3.11°}$**

Problem 2. If the shaft in Problem 1 were replaced by a hollow tube of the same external diameter, but of wall thickness 0.005 m, what would be the maximum shear stress in the shaft due to the same applied torque, and the resulting twist of the shaft. The material properties of the shaft may be assumed to be the same as that of Problem 1.

Internal shaft diameter, $D_1 = D_2 - 2 \times$ wall thickness

$$= 0.0525 - 2 \times 0.005$$

i.e. $\qquad \boldsymbol{D_1 = 0.0425 \, m}$

The polar second moment of area for a hollow tube,

$$J = \frac{\pi}{32}\left(D_2{}^4 - D_1{}^4\right) \text{ from Problem 12, page 110.}$$

Hence, $\boldsymbol{J} = \dfrac{\pi}{32}\left(0.0525^4 - 0.0425^4\right)$

$$= \mathbf{4.255 \times 10^{-7} \, m^4}$$

From equation (11.6), $\qquad \dfrac{\tau}{r} = \dfrac{T}{J}$

Hence, **maximum shear stress,**

$$\hat{\tau} = \frac{Tr}{J} = \frac{1424\,\mathrm{N\,m} \times \dfrac{0.0525}{2}\,\mathrm{m}}{4.255 \times 10^{-7}\,\mathrm{m}^4} = \mathbf{87.85\ MPa}$$

From equation (11.6), $\dfrac{\tau}{r} = \dfrac{G\theta}{L}$

from which, $\theta = \dfrac{\tau L}{Gr} = \dfrac{87.85 \times 10^6\, \dfrac{\mathrm{N}}{\mathrm{m}^2} \times 2\,\mathrm{m}}{70 \times 10^9\, \dfrac{\mathrm{N}}{\mathrm{m}^2} \times 0.02625\,\mathrm{m}}$

i.e. $\theta = \mathbf{0.0956\ rad}$

$$= 0.0956\ \mathrm{rad} \times \frac{360°}{2\pi\,\mathrm{rad}}$$

i.e. **angle of twist, $\theta = 5.48°$**

Problem 3.   What would be the maximum shear stress and resulting angle of twist on the shaft of Problem 2, if the wall thickness were 10 mm, instead of 5 mm?

Internal shaft diameter, $D_1 = D_2 - 2 \times$ wall thickness

$$= 0.0525 - 2 \times 10 \times 10^{-3}$$

i.e. $D_1 = \mathbf{0.0325\ m}$

The polar second moment of area for a hollow tube,

$$J = \frac{\pi}{32}\left(D_2{}^4 - D_1{}^4\right) \text{ from Problem 12, page 110.}$$

Hence, $J = \dfrac{\pi}{32}\left(0.0525^4 - 0.0325^4\right)$

$$= \mathbf{6.36 \times 10^{-7}\,m^4}$$

From equation (11.6), $\dfrac{\tau}{r} = \dfrac{T}{J}$

Hence, **maximum shear stress,**

$$\hat{\tau} = \frac{Tr}{J} = \frac{1424\,\mathrm{N\,m} \times \dfrac{0.0525}{2}\,\mathrm{m}}{6.36 \times 10^{-7}\,\mathrm{m}^4} = \mathbf{58.75\ MPa}$$

From equation (11.6), $\dfrac{\tau}{r} = \dfrac{G\theta}{L}$

from which, $\theta = \dfrac{\tau L}{Gr} = \dfrac{58.75 \times 10^6\, \dfrac{\mathrm{N}}{\mathrm{m}^2} \times 2\,\mathrm{m}}{70 \times 10^9\, \dfrac{\mathrm{N}}{\mathrm{m}^2} \times 0.02625\,\mathrm{m}}$

i.e. $\theta = \mathbf{0.06395\ rad}$

$$= 0.06395\ \mathrm{rad} \times \frac{360°}{2\pi\,\mathrm{rad}}$$

i.e. **angle of twist, $\theta = 3.66°$**

N.B. From the calculations in Problems 1 to 3, it can be seen that a hollow shaft is structurally more efficient than a solid section shaft.

Problem 4.   A shaft of uniform circular section is fixed at its ends and it is subjected to an intermediate torque T, as shown in Figure 11.5, where $a > b$. Determine the maximum resulting torque acting on the shaft and then draw the torque diagram.

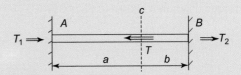

Figure 11.5

From equilibrium considerations, (see Figure 11.5), clockwise torque = the sum of the anticlockwise torques

$$T = T_1 + T_2 \qquad (11.11)$$

Let θ_C = the angle of twist at the point C.

From equation (11.6), $\dfrac{T}{J} = \dfrac{G\theta}{L}$

from which, $T = \dfrac{G\theta J}{L}$

Therefore $T_1 = \dfrac{G\theta_C J}{a} \qquad (11.12)$

and $T_2 = \dfrac{G\theta_C J}{b} \qquad (11.13)$

Dividing equation (11.12) by equation (11.13) gives:

$$\frac{T_1}{T_2} = \frac{b}{a}$$

or $T_1 = \dfrac{bT_2}{a} \qquad (11.14)$

Substituting equation (11.14) into equation (11.11) gives:

$$T = \frac{bT_2}{a} + T_2 = \left(1 + \frac{b}{a}\right)T_2$$

or $T_2 = \dfrac{T}{\left(1 + \dfrac{b}{a}\right)} = \dfrac{T}{\left(\dfrac{a+b}{a}\right)} = \dfrac{T\,a}{(a+b)} \quad (11.15)$

However, $a + b = L$

Therefore $T_2 = \dfrac{T\,a}{L} \qquad (11.16)$

Substituting equation (11.16) into equation (11.14) gives:

$$T_1 = \frac{b\left(\dfrac{Ta}{L}\right)}{a} = \frac{bT}{L} \qquad (11.17)$$

As $a > b$, $T_2 > T_1$; therefore,

$$\text{maximum torque} = T_2 = \frac{Ta}{L}$$

The torque diagram is shown in Figure 11.6.

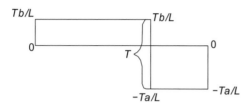

Figure 11.6 Torque diagram

Now try the following Practise Exercises

Practise Exercise 60   Further problems on the twisting of shafts

1. A shaft of uniform solid circular section is subjected to a torque of 1500 N m. Determine the maximum shear stress in the shaft and its resulting angle of twist, if the shaft's diameter is 0.06 m, the shaft's length is 1.2 m, and the rigidity modulus, $G = 77 \times 10^9$ N/m^2. What power can this shaft transmit if it is rotated at 400 rev/min?
    [35.4 MPa, 1.05°, 62.83 kW]

2. If the shaft in Problem 1 were replaced by a similar hollow one of wall thickness 10 mm, but of the same outer diameter, what would be the maximum shearing stress in the shaft and the resulting angle of twist? What power can this shaft transmit if it rotated at 500 rev/min.   [44.1 MPa, 1.31°, 78.5 kW]

3. A boat's propeller shaft transmits 50 hp at 100 rev/min. Neglecting transmission losses, determine the minimum diameter of a solid circular section phosphor bronze shaft, when the maximum permissible shear stress in the shaft is limited to 40 MPa. What will be the resulting angle of twist of this shaft, due to this torque, over a length of 1 m,

given that the rigidity modulus, $G = 40$ GPa.
    [76.8 mm, 1.49°]

4. A shaft to the blades of a helicopter transmits 1000 hp at 200 rev/min. Neglecting transmission losses, determine the minimum external and internal diameters of the hollow circular section aluminium alloy shaft, when the maximum permissible shear stress in the shaft is limited to 30 MPa. It may be assumed that the external diameter of this shaft is twice its internal diameter. What will be the resulting angle of twist of this shaft over a length of 2 m, given that the modulus of rigidity, $G = 26.9$ GPa?
    [182 mm, 91 mm, 1.5°]

5. A solid circular section shaft of diameter d is subjected to a torque of 1000 N m. If the maximum permissible shear stress in this shaft is limited to 30 MPa, determine the minimum value of d.   [55.4 mm]

6. If the shaft in Problem 5 were to be replaced by a hollow shaft of external diameter d_2 and internal diameter $0.5d_2$, determine the minimum value for d_2, the design condition being the same for both shafts. What percentage saving in weight will result by replacing the solid shaft by the hollow one?
    [56.6 mm, 28.3 mm, 21.7%]

7. An internal combustion engine of a ship is of power 3000 hp and it transmits its power through two shafts at 300 rev/min (rpm). If the transmission losses are 10%, determine the minimum possible diameters of the shafts, given that the maximum permissible shear stress in the shafts is 80 MPa. (It may be assumed that 1 hp = 745.7 W).
    [$d = 159$ mm]

Practise Exercise 61   Short-answer questions on the twisting of shafts

1. State three practical examples where the torsion of shafts appears.

2. State the relationship between shear stress τ and torque T for a shaft.

3. State the relationship between torque T and angle of twist θ for a shaft.

4. State whether a solid shaft or a hollow shaft is structurally more efficient.

Practise Exercise 62   Multiple-choice questions on the twisting of shafts

(Answers on page 297)

1. The maximum shear stress for a solid shaft occurs:
    (a) at the outer surface
    (b) at the centre
    (c) in between the outer surface and the centre

2. For a given shaft, if the values of torque T, length L and radius r are kept constant, but rigidity G is increased, the value of shear stress τ:
    (a) increases
    (b) stays the same
    (c) decreases

3. If for a certain shaft, its length is doubled, the angle of twist:

    (a) doubles   (b) halves
    (c) remains the same

4. If a solid shaft is replaced by a hollow shaft of the same external diameter, its angle of twist:
    (a) decreases   (b) stays the same
    (c) increases

5. If a shaft is fixed at its two ends and subjected to an intermediate torque T at mid-length, the maximum resulting torque is equal to:

    (a) T (b) $\dfrac{T}{2}$
    (c) zero

6. If a hollow shaft is subjected to a torque T, the shear stress on the inside surface is:
    (a) a minimum   (b) a maximum
    (c) zero

References

[1] A link to the Twisting of Circular Section Shafts - www.routledge.com/cw/bird
A *video reference on YouTube* which demonstrates a standard experimental test on the twisting of circular cross-section shafts. This experiment has to be carried out to determine the torsional material properties of a metal or a similar material.

Part Two

Revision Test 4   Bending of beams, torque and twisting of shafts

This Revision Test covers the material contained in Chapters 9 to 11. *The marks for each question are shown in brackets at the end of each question.*

1. A beam simply supported at its ends, is of length 1.4 m. If the beam carries a centrally-placed downward concentrated load of 50 kN, determine the minimum permissible dimensions of the beam's cross-section, given that the maximum permissible stress is 40 MPa, and the beam has a solid circular cross-section. (6)

2. Determine the force applied tangentially to a bar of a screw-jack at a radius of 60 cm, if the torque required is 750 N m. (3)

3. Calculate the torque developed by a motor whose spindle is rotating at 900 rev/min and developing a power of 4.20 kW. (5)

4. A motor connected to a shaft develops a torque of 8 kN m. Determine the number of revolutions made by the shaft if the work done is 7.2 MJ. (6)

5. Determine the angular acceleration of a shaft that has a moment of inertia of 32 kg m^2 produced by an accelerating torque of 600 N m. (5)

6. An electric motor has an efficiency of 72% when running at 1400 rev/min. Determine the output torque when the power input is 2.50 kW. (5)

7. A solid circular section shaft is required to transmit 60 hp at 1000 rpm. If the maximum permissible shear stress in the shaft is 35 MPa, determine the minimum permissible diameter of the shaft. Determine the resulting angle of twist of the shaft per metre, assuming that the modulus of rigidity, $G = 70$ GPa and 1 hp = 745.7 W (10)

Part Three

Dynamics

Chapter 12

Linear and angular motion

This chapter commences by defining linear and angular velocity and also linear and angular acceleration. It then derives the well-known relationships, under uniform acceleration, for displacement, velocity and acceleration, in terms of time and other parameters. The chapter then uses elementary vector analysis, similar to that used for forces in Chapter 6, to determine relative velocities. This chapter deals with the basics of kinematics.

At the end of this chapter you should be able to:

- appreciate that 2π radians corresponds to 360°
- define linear and angular velocity
- perform calculations on linear and angular velocity using $\omega = 2\pi n$ and $v = \omega r$
- define linear and angular acceleration
- perform calculations on linear and angular acceleration using $\omega_2 = \omega_1 + \alpha t$ and $a = r\alpha$
- select appropriate equations of motion when performing simple calculations
- appreciate the difference between scalar and vector quantities
- use vectors to determine relative velocities, by drawing and by calculation

12.1   The radian

The unit of angular displacement is the radian, where one radian is the angle subtended at the centre of a circle by an arc equal in length to the radius, as shown in Figure 12.1.

The relationship between angle in radians (θ), arc length (s) and radius of a circle (r) is:

$$s = r\theta \tag{12.1}$$

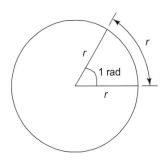

Figure 12.1

Since the arc length of a complete circle is $2\pi r$ and the angle subtended at the centre is 360°, then from equation (12.1), for a complete circle,

$$2\pi r = r\theta \qquad \text{or} \qquad \boldsymbol{\theta = 2\pi \text{ radians}}$$

Thus,   **2π radians corresponds to 360°**   (12.2)

as stated in Chapter 1.

12.2   Linear and angular velocity

Linear velocity v is defined as the rate of change of linear displacement s with respect to time t, and for motion in a straight line:

$$\text{linear velocity} = \frac{\text{change of displacement}}{\text{change of time}}$$

i.e. $$v = \frac{s}{t} \tag{12.3}$$

The unit of linear velocity is metres per second (m/s).

Angular velocity

The speed of revolution of a wheel or a shaft is usually measured in revolutions per minute or revolutions

Mechanical Engineering Principles, Bird and Ross, ISBN 9780415517850

per second, but these units do not form part of a coherent system of units. The basis used in SI units is the angle turned through (in radians) in one second.

Angular velocity is defined as the rate of change of angular displacement θ, with respect to time t, and for an object rotating about a fixed axis at a constant speed:

$$\text{angular velocity} = \frac{\text{angle turned through}}{\text{time taken}}$$

i.e. $$\omega = \frac{\theta}{t} \qquad (12.4)$$

The unit of angular velocity is radians per second (rad/s).

An object rotating at a constant speed of n revolutions per second subtends an angle of $2\pi n$ radians in one second, that is, its angular velocity,

$$\omega = 2\pi n \text{ rad/s} \qquad (12.5)$$

From equation (12.1), $s = r\theta$, and from equation (12.4), $\theta = \omega t$, hence

$$s = r\omega t \quad \text{or} \quad \frac{s}{t} = \omega r$$

However, from equation (12.3), $v = \frac{s}{t}$,

hence $$v = \omega r \qquad (12.6)$$

Equation (12.6) gives the relationship between linear velocity, v, and angular velocity, ω.

> **Problem 1.** A wheel of diameter 540 mm is rotating at $(1500/\pi)$ rev/min. Calculate the angular velocity of the wheel and the linear velocity of a point on the rim of the wheel.

From equation (12.5), angular velocity $\omega = 2\pi n$, where n is the speed of revolution in revolutions per second,

i.e. $n = \dfrac{1500}{60\pi}$ revolutions per second.

Thus, **angular velocity,** $\omega = 2\pi \left(\dfrac{1500}{60\pi}\right) = \textbf{50 rad/s}$

The linear velocity of a point on the rim, $v = \omega r$, where r is the radius of the wheel, i.e.

$$r = 0.54/2 \quad \text{or} \quad 0.27 \text{ m}.$$

Thus, **linear velocity,** $v = \omega r = 50 \times 0.27 = \textbf{13.5 m/s}$

> **Problem 2.** A car is travelling at 64.8 km/h and has wheels of diameter 600 mm.
> (a) Find the angular velocity of the wheels in both rad/s and rev/min.

> (b) If the speed remains constant for 1.44 km, determine the number of revolutions made by a wheel, assuming no slipping occurs.

(a) $64.8 \text{ km/h} = 64.8 \dfrac{\text{km}}{\text{h}} \times 1000 \dfrac{\text{m}}{\text{km}} \times \dfrac{1}{3600} \dfrac{\text{h}}{\text{s}}$

$$= \frac{64.8}{3.6} \text{ m/s}$$

$$= 18 \text{ m/s}$$

i.e. the linear velocity, v, is 18 m/s

The radius of a wheel is $(600/2)$ mm $= 0.3$ m

From equation (12.6), $v = \omega r$, hence $\omega = v/r$

i.e. the **angular velocity,** $\omega = \dfrac{18}{0.3} = \textbf{60 rad/s}$

From equation (12.5), angular velocity, $\omega = 2\pi n$, where n is in revolutions per second. Hence

$n = \omega/2\pi$ and angular speed of a wheel in revolutions per minute is $60\omega/2\pi$; but $\omega = 60$ rad/s,

hence **angular speed** $= \dfrac{60 \times 60}{2\pi} = \textbf{573 revolutions per minute (rpm)}$

(b) From equation (12.3), time taken to travel 1.44 km at a constant speed of 18 m/s is:

$$\frac{1440 \text{ m}}{18 \text{ m/s}} = 80 \text{ s}.$$

Since a wheel is rotating at 573 revolutions per minute, then in 80/60 minutes it makes

$$\frac{573 \times 80}{60} = \textbf{764 revolutions.}$$

Now try the following Practise Exercise

> **Practise Exercise 63 Further problems on linear and angular velocity**
>
> 1. A pulley driving a belt has a diameter of 360 mm and is turning at $2700/\pi$ revolutions per minute. Find the angular velocity of the pulley and the linear velocity of the belt assuming that no slip occurs.
>   [$\omega = 90$ rad/s, $v = 16.2$ m/s]
>
> 2. A bicycle is travelling at 36 km/h and the diameter of the wheels of the bicycle is 500 mm. Determine the angular velocity

of the wheels of the bicycle and the linear velocity of a point on the rim of one of the wheels.

$$[\omega = 40 \text{ rad/s}, v = 10 \text{ m/s}]$$

12.3 Linear and angular acceleration

Linear acceleration, a, is defined as the rate of change of linear velocity with respect to time. For an object whose linear velocity is increasing uniformly:

$$\text{linear acceleration} = \frac{\text{change of linear velocity}}{\text{time taken}}$$

i.e. $a = \dfrac{v_2 - v_1}{t}$ (12.7)

The unit of linear acceleration is metres per second squared (m/s^2). Rewriting equation (12.7) with v_2 as the subject of the formula gives:

$$v_2 = v_1 + at \qquad (12.8)$$

where v_2 = final velocity and v_1 = initial velocity.

Angular acceleration, α, is defined as the rate of change of angular velocity with respect to time. For an object whose angular velocity is increasing uniformly:

$$\text{angular acceleration} = \frac{\text{change of angular velocity}}{\text{time taken}}$$

i.e. $\alpha = \dfrac{\omega_2 - \omega_1}{t}$ (12.9)

The unit of angular acceleration is radians per second squared (rad/s^2). Rewriting equation (12.9) with ω_2 as the subject of the formula gives:

$$\omega_2 = \omega_1 + \alpha t \qquad (12.10)$$

where ω_2 = final angular velocity and ω_1 = initial angular velocity.

From equation (12.6), $v = \omega r$. For motion in a circle having a constant radius r, $v_2 = \omega_2 r$ and $v_1 = \omega_1 r$, hence equation (12.7) can be rewritten as:

$$a = \frac{\omega_2 r - \omega_1 r}{t} = \frac{r(\omega_2 - \omega_1)}{t}$$

But from equation (12.9), $\dfrac{\omega_2 - \omega_1}{t} = \alpha$

Hence $a = r\alpha$ (12.11)

Problem 3. The speed of a shaft increases uniformly from 300 revolutions per minute to 800 revolutions per minute in 10s. Find the angular acceleration, correct to 3 significant figures.

From equation (12.9), $\alpha = \dfrac{\omega_2 - \omega_1}{t}$

Initial angular velocity,

$$\omega_1 = 300 \text{ rev/min} = 300/60 \text{ rev/s}$$

$$= \frac{300 \times 2\pi}{60} \text{ rad/s},$$

final angular velocity,

$$\omega_2 = \frac{800 \times 2\pi}{60} \text{ rad/s and time, } t = 10 \text{ s.}$$

Hence, **angular acceleration,**

$$a = \frac{\dfrac{800 \times 2\pi}{60} - \dfrac{300 \times 2\pi}{60}}{10} \text{ rad/s}^2$$

$$= \frac{500 \times 2\pi}{60 \times 10} = \mathbf{5.24 \text{ rad/s}^2}$$

Problem 4. If the diameter of the shaft in problem 3 is 50 mm, determine the linear acceleration of the shaft on its external surface, correct to 3 significant figures.

From equation (12.11), $a = r\alpha$

The shaft radius is $\dfrac{50}{2}$ mm = 25 mm = 0.025 m, and the angular acceleration, $\alpha = 5.24 \text{ rad/s}^2$,

thus the **linear acceleration,**

$$a = r\alpha = 0.025 \times 5.24 = \mathbf{0.131 \text{ m/s}^2}$$

Now try the following Practise Exercise

Practise Exercise 64   Further problems on linear and angular acceleration

1. A flywheel rotating with an angular velocity of 200 rad/s is uniformly accelerated at a rate of 5 rad/s² for 15 s. Find the final angular velocity of the flywheel both in rad/s and revolutions per minute.

    [275 rad/s, 8250/π rev/min]

Part Three

2. A disc accelerates uniformly from 300 revolutions per minute to 600 revolutions per minute in 25 s. Determine its angular acceleration and the linear acceleration of a point on the rim of the disc, if the radius of the disc is 250 mm.

$$[0.4\pi \text{ rad/s}^2, 0.1\pi \text{ m/s}^2]$$

12.4 Further equations of motion

From equation (12.3), $s = vt$, and if the linear velocity is changing uniformly from v_1 to v_2, then s = mean linear velocity × time

i.e.

$$s = \left(\frac{v_1 + v_2}{2}\right)t \qquad (12.12)$$

From equation (12.4), $\theta = \omega t$, and if the angular velocity is changing uniformly from ω_1 to ω_2, then θ = mean angular velocity × time

i.e.

$$\theta = \left(\frac{\omega_1 + \omega_2}{2}\right)t \qquad (12.13)$$

Two further equations of linear motion may be derived from equations (12.8) and (12.11):

$$s = v_1 t + \frac{1}{2}at^2 \qquad (12.14)$$

and

$$v_2^2 = v_1^2 + 2as \qquad (12.15)$$

Two further equations of angular motion may be derived from equations (12.10) and (12.13):

$$\theta = \omega_1 t + \frac{1}{2}\alpha t^2 \qquad (12.16)$$

and

$$\omega_2^2 = \omega_1^2 + 2\alpha\theta \qquad (12.17)$$

Table 12.1, on page 149, summarises the principal equations of linear and angular motion for uniform changes in velocities and constant accelerations and also gives the relationships between linear and angular quantities.

Problem 5. The speed of a shaft increases uniformly from 300 rev/min to 800 rev/min in 10s. Find the number of revolutions made by the shaft during the 10 s it is accelerating.

From equation (12.13), angle turned through,

$$\theta = \left(\frac{\omega_1 + \omega_2}{2}\right)t$$

$$= \left(\frac{\dfrac{300 \times 2\pi}{60} + \dfrac{800 \times 2\pi}{60}}{2}\right)(10) \text{ rad}$$

However, there are 2π radians in 1 revolution, hence, number of revolutions

$$= \left(\frac{\dfrac{300 \times 2\pi}{60} + \dfrac{800 \times 2\pi}{60}}{2}\right)\left(\frac{10}{2\pi}\right)$$

$$= \frac{1}{2}\left(\frac{1100}{60}\right)(10)$$

$$= \frac{1100}{12}$$

$$= \textbf{91.67 revolutions}$$

Problem 6. The shaft of an electric motor, initially at rest, accelerates uniformly for 0.4 s at 15 rad/s². Determine the angle (in radians) turned through by the shaft in this time.

From equation (12.16), $\theta = \omega_1 t + \dfrac{1}{2}\alpha t^2$

Since the shaft is initially at rest, $\omega_1 = 0$ and $\theta = \dfrac{1}{2}\alpha t^2$, the angular acceleration, $\alpha = 15$ rad/s² and time $t = 0.4$ s.

Hence, **angle turned through,**

$$\theta = 0 + \frac{1}{2} \times 15 \times 0.4^2 = \textbf{1.2 rad}$$

Problem 7. A flywheel accelerates uniformly at 2.05 rad/s² until it is rotating at 1500 rev/min. If it completes 5 revolutions during the time it is accelerating, determine its initial angular velocity in rad/s, correct to 4 significant figures.

Since the final angular velocity is 1500 rev/min,

$$\omega_2 = 1500 \; \frac{\text{rev}}{\text{min}} \times \frac{1\,\text{min}}{60\,\text{s}} \times \frac{2\pi\,\text{rad}}{1\,\text{rev}}$$

$$= 50\pi \text{ rad/s}$$

Table 12.1

s = arc length (m)	r = radius of circle (m)
t = time (s)	θ = angle (rad)
v = linear velocity (m/s)	ω = angular velocity (rad/s)
v_1 = initial linear velocity (m/s)	ω_1 = initial angular velocity (rad/s)
v_2 = final linear velocity (m/s)	ω_2 = final angular velocity (rad/s)
a = linear acceleration (m/s²)	α = angular acceleration (rad/s²)
n = speed of revolution (rev/s)	

Equation number	Linear motion	Angular motion
(12.1)		$s = r\theta$ m
(12.2)		2π rad $= 360°$
(12.3) and (12.4)	$v = \dfrac{s}{t}$	$\omega = \dfrac{\theta}{t}$ rad/s
(12.5)		$\omega = 2\pi n$ rad/s
(12.6)	$v = \omega r$ m/s²	
(12.7) and (12.9)	$a = \dfrac{v_2 - v_1}{t}$	$\alpha = \dfrac{\omega_2 - \omega_1}{t}$
(12.8) and (12.10)	$v_2 = (v_1 + at)$ m/s	$\omega_2 = (\omega_1 + \alpha t)$ rad/s
(12.11)	$a = r\alpha$ m/s²	
(12.12) and (12.13)	$s = \left(\dfrac{v_1 + v_2}{2}\right)t$	$\theta = \left(\dfrac{\omega_1 + \omega_2}{2}\right)t$
(12.14) and (12.16)	$s = v_1 t + \dfrac{1}{2}at^2$	$\theta = \omega_1 t + \dfrac{1}{2}\alpha t^2$
(12.15) and (12.17)	$v_2{}^2 = v_1{}^2 + 2as$	$\omega_2{}^2 = \omega_1{}^2 + 2\alpha\theta$

5 revolutions $= 5$ rev $\times \dfrac{2\pi \text{ rad}}{1\text{ rev}} = 10\pi$ rad

From equation (12.17), $\omega_2{}^2 = \omega_1{}^2 + 2\alpha\theta$

i.e. $(50\pi)^2 = \omega_1{}^2 + (2 \times 2.05 \times 10\pi)$

from which, $\omega_1{}^2 = (50\pi)^2 - (2 \times 2.05 \times 10\pi)$

$= (50\pi)^2 - 41\pi = 24545$

i.e. $\omega_1 = \sqrt{24545} = 156.7$ rad/s

Thus, the initial angular velocity is 156.7 rad/s, correct to 4 significant figures.

Now try the following Practise Exercise

Practise Exercise 65 Further problems on equations of motion

1. A grinding wheel makes 300 revolutions when slowing down uniformly from 1000 rad/s to 400 rad/s. Find the time for this reduction in speed. [2.693 s]

2. Find the angular retardation for the grinding wheel in question 1. [222.8 rad/s²]

3. A disc accelerates uniformly from 300 revolutions per minute to 600 revolutions per minute in 25 s. Calculate the number of revolutions the disc makes during this accelerating period.   [187.5 revolutions]

4. A pulley is accelerated uniformly from rest at a rate of 8 rad/s². After 20 s the acceleration stops and the pulley runs at constant speed for 2 min, and then the pulley comes uniformly to rest after a further 40 s. Calculate:
   (a) the angular velocity after the period of acceleration
   (b) the deceleration
   (c) the total number of revolutions made by the pulley.
   [(a) 160 rad/s (b) 4 rad/s² (c) 12000/π rev]

12.5   Relative velocity

Quantities used in engineering and science can be divided into two groups:

(a) **Scalar quantities** have a size or magnitude only and need no other information to specify them. Thus 20 centimetres, 5 seconds, 3 litres and 4 kilograms are all examples of scalar quantities.

(b) **Vector quantities** have both a size (or magnitude), and a direction, called the line of action of the quantity. Typical vector quantities include velocity, acceleration and force. Thus, a velocity of 30 km/h due west, and an acceleration of 7 m/s² acting vertically downwards, are both vector quantities.

A vector quantity is represented by a straight line lying along the line of action of the quantity, and having a length that is proportional to the size of the quantity, as shown in Chapter 4. Thus **ab** in Figure 12.2 represents a velocity of 20 m/s, whose line of action is due west.

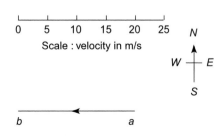

Figure 12.2

The bold letters, **ab**, indicate a vector quantity and the order of the letters indicate that the line of action is from *a* to *b*.

Consider two aircraft *A* and *B* flying at a constant altitude, *A* travelling due north at 200 m/s and *B* travelling 30° east of north, written *N* 30° *E*, at 300 m/s, as shown in Figure 12.3.

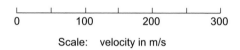

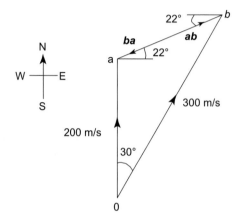

Figure 12.3

Relative to a fixed point 0, **0a** represents the velocity of *A* and **0b** the velocity of *B*. The velocity of *B* relative to *A*, that is the velocity at which *B* seems to be travelling to an observer on *A*, is given by **ab**, and by measurement is 160 m/s in a direction *E* 22° *N*. The velocity of *A* relative to *B*, that is, the velocity at which *A* seems to be travelling to an observer on *B*, is given by **ba** and by measurement is 160 m/s in a direction *W* 22° *S*.

Problem 8. Two cars are travelling on horizontal roads in straight lines, car *A* at 70 km/h at *N* 10° *E* and car *B* at 50 km/h at *W* 60° N. Determine, by drawing a vector diagram to scale, the velocity of car *A* relative to car *B*.

With reference to Figure 12.4(a), **oa** represents the velocity of car *A* relative to a fixed point *o*, and **ob** represents the velocity of car *B* relative to a fixed point *o*. The velocity of car *A* relative to car *B* is given by **ba** and by measurement is **45 km/h in a direction of E 35° N.**

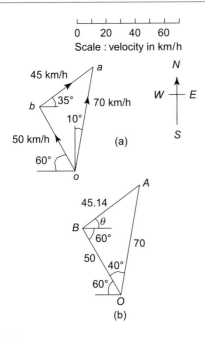

Figure 12.4

Problem 9. Verify the result obtained in Problem 8 by calculation.

The triangle shown in Figure 12.4(b) is similar to the vector diagram shown in Figure 12.4(a). Angle *BOA* is 40°. Using the cosine rule (see chapter 1):

$$BA^2 = 50^2 + 70^2 - 2 \times 50 \times 70 \times \cos 40°$$

from which, **BA** = 45.14

Using the sine rule:

$$\frac{50}{\sin \angle BAO} = \frac{45.14}{\sin 40°} \text{ (also from Chapter 1)}$$

from which, $\sin \angle BAO = \dfrac{50 \sin 40°}{45.14} = 0.7120$

Hence, angle *BAO* = 45.40°

thus, angle $ABO = 180° - (40° + 45.40°)$
$$= 94.60°,$$

and angle $\theta = 94.60° - 60° = 34.60°$

Thus, **ba** is **45.14 km/h in a direction E 34.60° N by calculation.**

Problem 10. A crane is moving in a straight line with a constant horizontal velocity of 2 m/s. At the same time it is lifting a load at a vertical velocity of 5 m/s. Calculate the velocity of the load relative to a fixed point on the earth's surface.

A vector diagram depicting the motion of the crane and load is shown in Figure 12.5. **oa** represents the velocity of the crane relative to a fixed point on the Earth's surface and **ab** represents the velocity of the load relative to the crane. The velocity of the load relative to the fixed point on the Earth's surface is **ob**. By Pythagoras' theorem (from Chapter 1):

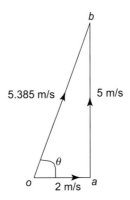

Figure 12.5

$$ob^2 = oa^2 + ab^2$$
$$= 4 + 25 = 29$$

Hence $ob = \sqrt{29} = 5.385$ m/s

$\text{Tan } \theta = \dfrac{5}{2} = 2.5$, hence, $\theta = \tan^{-1} 2.5 = 68.20°$

i.e. the velocity of the load relative to a fixed point on the Earth's surface is **5.385 m/s in a direction 68.20° to the motion of the crane**.

Now try the following Practise Exercises

Practise Exercise 66   Further problems on relative velocity

1. A car is moving along a straight horizontal road at 79.2 km/h and rain is falling vertically downwards at 26.4 km/h. Find the velocity of the rain relative to the driver of the car.

   [83.5 km/h at 71.6° to the vertical]

2. Calculate the time needed to swim across a river 142 m wide when the swimmer can swim at 2 km/h in still water and the river is flowing at 1 km/h. At what angle to the bank should the swimmer swim?

   [4 min 55 s, 60°]

3. A ship is heading in a direction N 60° *E* at a speed which in still water would be 20 km/h. It is carried off course by a current of 8 km/h in a direction of *E* 50° S. Calculate the ship's actual speed and direction.

      [22.79 km/h in a direction *E* 9.78° N]

Practise Exercise 67   **Short-answer questions on linear and angular motion**

1. State and define the unit of angular displacement.

2. Write down the formula connecting an angle, arc length and the radius of a circle.

3. Define linear velocity and state its unit.

4. Define angular velocity and state its unit.

5. Write down a formula connecting angular velocity and revolutions per second in coherent units.

6. State the formula connecting linear and angular velocity.

7. Define linear acceleration and state its unit.

8. Define angular acceleration and state its unit.

9. Write down the formula connecting linear and angular acceleration.

10. Define a scalar quantity and give two examples.

11. Define a vector quantity and give two examples.

Practise Exercise 68   **Multiple-choice questions on linear and angular motion**

(Answers on page 297)

1. Angular displacement is measured in:
   - (a) degrees
   - (b) radians
   - (c) rev/s
   - (d) metres

2. An angle of $\dfrac{3\pi}{4}$ radians is equivalent to:
   - (a) 270°
   - (b) 67.5°
   - (c) 135°
   - (d) 2.356°

3. An angle of 120° is equivalent to:
   - (a) $\dfrac{2\pi}{3}$ rad
   - (b) $\dfrac{\pi}{3}$ rad
   - (c) $\dfrac{3\pi}{4}$ rad
   - (d) $\dfrac{1}{3}$ rad

4. An angle of 2 rad at the centre of a circle subtends an arc length of 40 mm at the circumference of the circle. The radius of the circle is:
   - (a) 40π mm
   - (b) 80 mm
   - (c) 20 mm
   - (d) $(40/\pi)$ mm

5. A point on a wheel has a constant angular velocity of 3 rad/s. The angle turned through in 15 seconds is:
   - (a) 45 rad
   - (b) 10π rad
   - (c) 5 rad
   - (d) 90π rad

6. An angular velocity of 60 revolutions per minute is the same as:
   - (a) $(1/2\pi)$ rad/s
   - (b) 120π rad/s
   - (c) $(30/\pi)$ rad/s
   - (d) 2π rad/s

7. A wheel of radius 15 mm has an angular velocity of 10 rad/s. A point on the rim of the wheel has a linear velocity of:
   - (a) 300π mm/s
   - (b) 2/3 mm/s
   - (c) 150 mm/s
   - (d) 1.5 mm/s

8. The shaft of an electric motor is rotating at 20 rad/s and its speed is increased uniformly to 40 rad/s in 5 s. The angular acceleration of the shaft is:
   - (a) 4000 rad/s^2
   - (b) 4 rad/s^2
   - (c) 160 rad/s^2
   - (d) 12 rad/s^2

9. A point on a flywheel of radius 0.5 m has a uniform linear acceleration of 2 m/s^2. Its angular acceleration is:
   - (a) 2.5 rad/s^2
   - (b) 0.25 rad/s^2
   - (c) 1 rad/s^2
   - (d) 4 rad/s^2

Questions 10 to 13 refer to the following data.
A car accelerates uniformly from 10 m/s to 20 m/s over a distance of 150 m. The wheels of the car each have a radius of 250 mm.

10. The time the car is accelerating is:
    - (a) 0.2 s
    - (b) 15 s
    - (c) 10 s
    - (d) 5 s

11. The initial angular velocity of each of the wheels is:
    - (a) 20 rad/s
    - (b) 40 rad/s
    - (c) 2.5 rad/s
    - (d) 0.04 rad/s

12. The angular acceleration of each of the wheels is:
    (a) 1 rad/s^2 (b) 0.25 rad/s^2
    (c) 400 rad/s^2 (d) 4 rad/s^2

13. The linear acceleration of a point on each of the wheels is:
    (a) 1 m/s^2 (b) 4 m/s^2
    (c) 3 m/s^2 (d) 100 m/s^2

Chapter 13

Linear momentum and impulse

This chapter is of considerable importance in the study of the motion and collision of vehicles, ships, and so on. The chapter commences with defining momentum, impulse, together with Newton's laws of motion. It then applies these laws to solving practical problems in the fields of ballistics, pile drivers, etc.

At the end of this chapter you should be able to:

- define momentum and state its unit
- state Newton's first law of motion
- calculate momentum given mass and velocity
- state Newton's second law of motion
- define impulse and appreciate when impulsive forces occur
- state Newton's third law of motion
- calculate impulse and impulsive force
- use the equation of motion $v^2 = u^2 + 2as$ in calculations

13.1 Linear momentum

The **momentum** of a body is defined as the product of its mass and its velocity, i.e.

$$\text{momentum} = mu$$

where m = mass (in kg) and u = velocity (in m/s). The unit of momentum is kg m/s.

Since velocity is a vector quantity, **momentum is a vector quantity**, i.e. it has both magnitude and direction.

Mechanical Engineering Principles, Bird and Ross, ISBN 9780415517850

Newton's first law of motion states:

a body continues in a state of rest or in a state of uniform motion in a straight line unless acted on by some external force

Hence the momentum of a body remains the same provided no external forces act on it

The principle of conservation of momentum for a closed system (i.e. one on which no external forces act) may be stated as:

the total linear momentum of a system is a constant

The total momentum of a system before collision in a given direction is equal to the total momentum of the system after collision in the same direction. In Figure 13.1, masses m_1 and m_2 are travelling in the same direction with velocity $u_1 > u_2$. A collision will occur, and applying the principle of conservation of momentum:

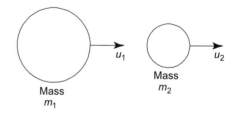

Figure 13.1

total momentum before impact = total momentum after impact

i.e. $m_1 u_1 + m_2 u_2 = m_1 v_1 + m_2 v_2$

where v_1 and v_2 are the velocities of m_1 and m_2 after impact.

Problem 1. Determine the momentum of a pile driver of mass 400 kg when it is moving downwards with a speed of 12 m/s.

Momentum = mass × velocity = 400 kg × 12 m/s

$$= \textbf{4800 kg m/s downwards}$$

Problem 2. A cricket ball of mass 150 g has a momentum of 4.5 kg m/s. Determine the velocity of the ball in km/h.

Momentum = mass × velocity, hence

$$\text{velocity} = \frac{\text{momentum}}{\text{mass}} = \frac{4.5\,\text{kg m/s}}{150 \times 10^{-3}\text{kg}} = 30\text{ m/s}$$

$$30\text{ m/s} = 30\frac{\text{m}}{\text{s}} \times 3600\frac{\text{s}}{\text{h}} \times \frac{1\,\text{km}}{1000\,\text{m}}$$

$$= 30 \times 3.6\text{ km/h} = \textbf{108 km/h}$$

$$= \textbf{velocity of cricket ball.}$$

Problem 3. Determine the momentum of a railway wagon of mass 50 tonnes moving at a velocity of 72 km/h.

Momentum = mass × velocity

Mass = 50 t = 50000 kg (since 1 t = 1000 kg)

and   velocity = 72 km/h = $72\dfrac{\text{km}}{\text{h}} \times \dfrac{1\,\text{h}}{3600\,\text{s}} \times 1000\dfrac{\text{m}}{\text{km}}$

$$= \frac{72}{3.6}\text{ m/s} = 20\text{ m/s}.$$

Hence,   **momentum** = 50000 kg × 20 m/s

$$= 1000000\text{ kg m/s} = \textbf{10}^6\textbf{kg m/s}$$

Problem 4. A wagon of mass 10 t is moving at a speed of 6 m/s and collides with another wagon of mass 15 t, which is stationary. After impact, the wagons are coupled together. Determine the common velocity of the wagons after impact.

Mass m_1 = 10 t = 10000 kg,

m_2 = 15000 kg and velocity u_1 = 6 m/s, u_2 = 0.

Total momentum before impact = $m_1u_1 + m_2u_2$

$$= (10000 \times 6) + (15000 \times 0) = 60000\text{ kg m/s}$$

Let the common velocity of the wagons after impact be v m/s.

Since total momentum before impact = total momentum after impact:

$$60000 = m_1v + m_2v$$
$$= v(m_1 + m_2) = v(25000)$$

Hence $v = \dfrac{60000}{25000} = 2.4\text{ m/s}$

i.e. **the common velocity after impact is 2.4 m/s in the direction in which the 10 t wagon is initially travelling.**

Problem 5. A body has a mass of 30 g and is moving with a velocity of 20 m/s. It collides with a second body which has a mass of 20 g and which is moving with a velocity of 15 m/s. Assuming that the bodies both have the same velocity after impact, determine this common velocity, (a) when the initial velocities have the same line of action and the same sense, and (b) when the initial velocities have the same line of action but are opposite in sense.

Mass m_1 = 30 g = 0.030 kg, m_2 = 20 g = 0.020 kg, velocity u_1 = 20 m/s and u_2 = 15 m/s.

(a)   When the velocities have the same line of action and the same sense, both u_1 and u_2 are considered as positive values.

Total momentum before impact

$$= m_1u_1 + m_2u_2 = (0.030 \times 20) + (0.020 \times 15)$$
$$= 0.60 + 0.30 = 0.90\text{ kg m/s}.$$

Let the common velocity after impact be v m/s.

Total momentum before impact = total momentum after impact

i.e. $0.90 = m_1v + m_2v = v(m_1 + m_2)$

$$0.90 = v(0.030 + 0.020)$$

from which, **common velocity, $v = \dfrac{0.90}{0.050} = $ 18 m/s**

in the direction in which the bodies are initially travelling.

(b)   When the velocities have the same line of action but are opposite in sense, one is considered as positive and the other negative (because velocity is a vector quantity). Taking the direction of mass m_1 as positive gives:

velocity u_1 = +20 m/s and u_2 = −15 m/s.

Total momentum before impact

$$= m_1u_1 + m_2u_2 = (0.030 \times 20) + (0.020 \times -15)$$
$$= 0.60 - 0.30 = +0.30\text{ kg m/s}$$

and since it is positive this indicates a momentum in the same direction as that of mass m_1.

If the common velocity after impact is v m/s then

$$0.30 = v(m_1 + m_2) = v(0.050)$$

from which, **common velocity, $v = \dfrac{0.30}{0.050} = 6$ m/s** **in the direction that the 30 g mass is initially travelling.**

Problem 6. A ball of mass 50 g is moving with a velocity of 4 m/s when it strikes a stationary ball of mass 25 g. The velocity of the 50 g ball after impact is 2.5 m/s in the same direction as before impact. Determine the velocity of the 25 g ball after impact.

Mass $m_1 = 50$ g $= 0.050$ kg, $m_2 = 25$ g $= 0.025$ kg.

Initial velocity $u_1 = 4$ m/s, $u_2 = 0$;

final velocity $v_1 = 2.5$ m/s, v_2 is unknown.

Total momentum before impact
$$= m_1 u_1 + m_2 u_2$$
$$= (0.050 \times 4) + (0.025 \times 0) = 0.20 \text{ kg m/s}$$

Total momentum after impact
$$= m_1 v_1 + m_2 v_2$$
$$= (0.050 \times 2.5) + (0.025\, v_2) = 0.125 + 0.025\, v_2$$

Total momentum before impact = total momentum after impact, hence

$$0.20 = 0.125 + 0.025 v_2$$

from which, **velocity of 25 g ball after impact,**

$$v_2 = \frac{0.20 - 0.125}{0.025} = 3 \text{ m/s}$$

Problem 7. Three masses, P, Q and R lie in a straight line. P has a mass of 5 kg and is moving towards Q at 8 m/s. Q has a mass of 7 kg and a velocity of 4 m/s, and is moving towards R. Mass R is stationary. P collides with Q, and P and Q then collide with R. Determine the mass of R assuming all three masses have a common velocity of 2 m/s after the collision of P and Q with R.

Mass $m_P = 5$ kg, $m_Q = 7$ kg,

velocity $u_P = 8$ m/s and $u_Q = 4$ m/s.

Total momentum before P collides with Q
$$= m_P u_P + m_Q u_Q$$
$$= (5 \times 8) + (7 \times 4) = 68 \text{ kg m/s}.$$

Let P and Q have a common velocity of v_1 m/s after impact.

Total momentum after P and Q collide
$$= m_P v_1 + m_Q v_1$$
$$= v_1 (m_P + m_Q) = 12 v_1$$

Total momentum before impact = total momentum after impact, i.e. $68 = 12 v_1$, from which, common velocity of P and Q, $v_1 = \dfrac{68}{12} = 5\dfrac{2}{3}$ m/s.

Total momentum after P and Q collide with $R = (m_{P+Q} \times 2) + (m_R \times 2)$ (since the common velocity after impact $= 2$ m/s) $= (12 \times 2) + (2\, m_R)$

Total momentum before P and Q collide with R = total momentum after P and Q collide with R, i.e.

$$\left(m_{P+Q} \times 5\frac{2}{3}\right) = (12 \times 2) + 2\, m_R$$

i.e. $$12 \times 5\frac{2}{3} = 24 + 2\, m_R$$

$$68 - 24 = 2\, m_R$$

from which, **mass of R, $m_R = \dfrac{44}{2} = 22$ kg**.

Now try the following Practise Exercise

Practise Exercise 69   Further problems on linear momentum

(Where necessary, take g as 9.81 m/s²)

1. Determine the momentum in a mass of 50 kg having a velocity of 5 m/s. [250 kg m/s]

2. A milling machine and its component have a combined mass of 400 kg. Determine the momentum of the table and component when the feed rate is 360 mm/min. [2.4 kg m/s]

3. The momentum of a body is 160 kg m/s when the velocity is 2.5 m/s. Determine the mass of the body. [64 kg]

4. Calculate the momentum of a car of mass 750 kg moving at a constant velocity of 108 km/h. [22500 kg m/s]

5. A football of mass 200 g has a momentum of 5 kg m/s. What is the velocity of the ball in km/h. [90 km/h]

6. A wagon of mass 8 t is moving at a speed of 5 m/s and collides with another wagon of mass 12 t, which is stationary. After impact, the wagons are coupled together. Determine the common velocity of the wagons after impact. [2 m/s]

7. A car of mass 800 kg was stationary when hit head-on by a lorry of mass 2000 kg travelling at 15 m/s. Assuming no brakes are applied and the car and lorry move as one, determine the speed of the wreckage immediately after collision. [10.71 m/s]

8. A body has a mass of 25 g and is moving with a velocity of 30 m/s. It collides with a second body which has a mass of 15 g and which is moving with a velocity of 20 m/s. Assuming that the bodies both have the same speed after impact, determine their common velocity (a) when the speeds have the same line of action and the same sense, and (b) when the speeds have the same line of action but are opposite in sense.
[(a) 26.25 m/s (b) 11.25 m/s]

9. A ball of mass 40 g is moving with a velocity of 5 m/s when it strikes a stationary ball of mass 30 g. The velocity of the 40 g ball after impact is 4 m/s in the same direction as before impact. Determine the velocity of the 30 g ball after impact. [1.33 m/s]

10. Three masses, X, Y and Z, lie in a straight line. X has a mass of 15 kg and is moving towards Y at 20 m/s. Y has a mass of 10 kg and a velocity of 5 m/s and is moving towards Z. Mass Z is stationary. X collides with Y, and X and Y then collide with Z. Determine the mass of Z assuming all three masses have a common velocity of 4 m/s after the collision of X and Y with Z.
[62.5 kg]

13.2 Impulse and impulsive forces

Newton's second law of motion states:

the rate of change of momentum is directly proportional to the applied force producing the change, and takes place in the direction of this force

In the SI system, the units are such that:

the applied force = rate of change of momentum
$$= \frac{\text{change of momentum}}{\text{time taken}} \quad (13.1)$$

When a force is suddenly applied to a body due to either a collision with another body or being hit by an object such as a hammer, the time taken in equation (13.1) is very small and difficult to measure. In such cases, the total effect of the force is measured by the change of momentum it produces.

Forces that act for very short periods of time are called **impulsive forces**. The product of the impulsive force and the time during which it acts is called the **impulse** of the force and is equal to the change of momentum produced by the impulsive force, i.e.

impulse = applied force × time
= change in linear momentum

Examples where impulsive forces occur include when a gun recoils and when a free-falling mass hits the ground. Solving problems associated with such occurrences often requires the use of the equation of motion: $v^2 = u^2 + 2as$, from Chapter 12.

When a pile is being hammered into the ground, the ground resists the movement of the pile and this resistance is called a **resistive force**.

Newton's third law of motion may be stated as:

for every action there is an equal and opposite reaction

The force applied to the pile is the resistive force; the pile exerts an equal and opposite force on the ground. In practice, when impulsive forces occur, energy is not entirely conserved and some energy is changed into heat, noise, and so on.

Problem 8. The average force exerted on the work-piece of a press-tool operation is 150 kN, and the tool is in contact with the work-piece for 50 ms. Determine the change in momentum.

From above, change of linear momentum = applied force × time (= impulse).

Hence, **change in momentum of work-piece**
$$= 150 \times 10^3 \text{ N} \times 50 \times 10^{-3}\text{s}$$
$$= \textbf{7500 kg m/s (since 1 N = 1 kg m/s}^2\textbf{)}.$$

Problem 9. A force of 15 N acts on a body of mass 4 kg for 0.2 s. Determine the change in velocity.

Impulse = applied force × time

= change in linear momentum

i.e.   15 N × 0.2 s = mass × change in velocity

= 4 kg × change in velocity

from which, **change in velocity**

$$= \frac{15\,N \times 0.2\,s}{4\,kg}$$

= **0.75 m/s** (since 1 N = 1 kg m/s^2).

Problem 10.   A mass of 8 kg is dropped vertically on to a fixed horizontal plane and has an impact velocity of 10 m/s. The mass rebounds with a velocity of 6 m/s. If the mass-plane contact time is 40 ms, calculate (a) the impulse, and (b) the average value of the impulsive force on the plane.

(a)   Impulse = change in momentum = $m(u_1 - v_1)$
where u_1 = impact velocity = 10 m/s and

v_1 = rebound velocity = –6 m/s

(v_1 is negative since it acts in the opposite direction to u_1, and velocity is a vector quantity)

Thus, **impulse** = $m(u_1 - v_1)$ = 8 kg (10 – –6) m/s

= 8 × 16 = **128 kg m/s.**

(b)   **Impulsive force** = $\dfrac{\text{impulse}}{\text{time}} = \dfrac{128\,kg\,m/s}{40 \times 10^{-3}s}$

= **3200 N or 3.2 kN**

Problem 11.   The hammer of a pile-driver of mass 1 t falls a distance of 1.5 m on to a pile. The blow takes place in 25 ms and the hammer does not rebound. Determine the average applied force exerted on the pile by the hammer.

Initial velocity, u = 0,

acceleration due to gravity, g = 9.81 m/s^2

and distance, s = 1.5 m

Using the equation of motion: $v^2 = u^2 + 2gs$

gives: $v^2 = 0^2 + 2(9.81)(1.5)$

from which, impact velocity, $v = \sqrt{(2)(9.81)(1.5)}$

= 5.425 m/s

Neglecting the small distance moved by the pile and hammer after impact,

momentum lost by hammer = the change of momentum

= mv = 1000 kg × 5.425 m/s

Rate of change of momentum

$$= \frac{\text{change of momentum}}{\text{change of time}}$$

$$= \frac{1000 \times 5.425}{25 \times 10^{-3}} = 217000\ N$$

Since the impulsive force is the rate of change of momentum, **the average force exerted on the pile is 217 kN.**

Problem 12.   A mass of 40 g having a velocity of 15 m/s collides with a rigid surface and rebounds with a velocity of 5 m/s. The duration of the impact is 0.20 ms. Determine (a) the impulse, and (b) the impulsive force at the surface.

Mass m = 40 g = 0.040 kg,

initial velocity, u = 15 m/s

and final velocity, v = –5 m/s (negative since the rebound is in the opposite direction to velocity u)

(a)   Momentum before impact = mu = 0.040 × 15

= 0.6 kg m/s

Momentum after impact = mv = 0.040 × – 5

= – 0.2 kg m/s

Impulse = change of momentum

= 0.6 – (–0.2) = **0.8 kg m/s**

(b)   **Impulsive force**

$$= \frac{\text{change of momentum}}{\text{change of time}}$$

$$= \frac{0.8\,kg\,m/s}{0.20 \times 10^{-3}s} = \textbf{4000 N or 4 kN}$$

Problem 13.   A gun of mass 1.5 t fires a shell of mass 15 kg horizontally with a velocity of 500 m/s. Determine (a) the initial velocity of recoil, and (b) the uniform force necessary to stop the recoil of the gun in 200 mm.

Mass of gun, m_g = 1.5 t = 1500 kg,

mass of shell, m_s = 15 kg,

and initial velocity of shell, u_s = 500 m/s.

(a)   Momentum of shell = $m_s u_s$ = 15 × 500

= 7500 kg m/s.

Momentum of gun = $m_g v$ = 1500v

where v = initial velocity of recoil of the gun.

By the principle of conservation of momentum, initial momentum = final momentum, i.e.

$$0 = 7500 + 1500v,$$ from which,

velocity $v = \dfrac{-7500}{1500} = -5$ m/s (the negative sign indicating recoil velocity)

i.e. **the initial velocity of recoil = 5 m/s.**

(b) The retardation of the recoil, a, may be determined using $v_2 = u^2 + 2as$, where v, the final velocity, is zero, u, the initial velocity, is 5 m/s and s, the distance, is 200 mm, i.e. 0.2 m.

Rearranging $v^2 = u^2 + 2as$ for a gives:

$$a = \dfrac{v^2 - u^2}{2s} = \dfrac{0^2 - 5^2}{2(0.2)}$$

$$= \dfrac{-25}{0.4} = -62.5 \text{ m/s}^2$$

Force necessary to stop recoil in 200 mm

$$= \text{mass} \times \text{acceleration}$$
$$= 1500 \text{ kg} \times 62.5 \text{ m/s}^2$$
$$= \textbf{93750 N or 93.75 kN}$$

Problem 14. A vertical pile of mass 100 kg is driven 200 mm into the ground by the blow of a 1 t hammer which falls through 750 mm. Determine (a) the velocity of the hammer just before impact (b) the velocity immediately after impact (assuming the hammer does not bounce), and (c) the resistive force of the ground assuming it to be uniform.

(a) For the hammer, $v^2 = u^2 + 2gs$, where v = final velocity, u = initial velocity = 0, g = 9.81 m/s^2 and s = distance = 750 mm = 0.75 m

Hence, $v^2 = 0^2 + 2(9.81)(0.75)$, from which,

velocity of hammer, just before impact,

$$v = \sqrt{2(9.81)(0.75)} = \textbf{3.84 m/s}$$

(b) Momentum of hammer just before impact

$$= \text{mass} \times \text{velocity}$$
$$= 1000 \text{ kg} \times 3.84 \text{ m/s} = 3840 \text{ kg m/s}$$

Momentum of hammer and pile after impact = momentum of hammer before impact.

Hence, 3840 kg m/s = (mass of hammer and pile) × (velocity immediately after impact)

i.e. 3840 = (1000 + 100)(v), from which,

velocity immediately after impact,

$$v = \dfrac{3840}{1100} = \textbf{3.49 m/s}$$

(c) Resistive force of ground = mass × acceleration. The acceleration is determined using $v^2 = u^2 + 2as$ where v = final velocity = 0, u = initial velocity = 3.49 m/s and s = distance driven in ground = 200 mm = 0.2 m.

Hence, $0^2 = (3.49)^2 + 2(a)(0.2)$, from which,

acceleration, $a = \dfrac{-(3.49)^2}{2(0.2)} = -30.45$ m/s^2 (the minus sign indicates retardation, because acceleration is a vector quantity).

Thus, **resistive force of ground**

$$= \text{mass} \times \text{acceleration}$$
$$= 1100 \text{ kg} \times 30.45 \text{ m/s}^2 = \textbf{33.5 kN}$$

Now try the following Practise Exercises

Practise Exercise 70   Further problems on impulse and impulsive forces

(Where necessary, take g as 9.81 m/s^2)

1. The sliding member of a machine tool has a mass of 200 kg. Determine the change in momentum when the sliding speed is increased from 10 mm/s to 50 mm/s. [8 kg m/s]

2. A force of 48 N acts on a body of mass 8 kg for 0.25 s. Determine the change in velocity. [1.5 m/s]

3. The speed of a car of mass 800 kg is increased from 54 km/h to 63 km/h in 2 s. Determine the average force in the direction of motion necessary to produce the change in speed. [1 kN]

4. A 10 kg mass is dropped vertically on to a fixed horizontal plane and has an impact velocity of 15 m/s. The mass rebounds with a velocity of 5 m/s. If the contact time of mass and plane is 0.025 s, calculate (a) the impulse, and (b) the average value of the impulsive force on the plane. [(a) 200 kg m/s (b) 8 kN]

5. The hammer of a pile driver of mass 1.2 t falls 1.4 m on to a pile. The blow takes place in 20 ms and the hammer does not rebound. Determine the average

applied force exerted on the pile by the hammer.

[314.5 kN]

6. A tennis ball of mass 60 g is struck from rest with a racket. The contact time of ball on racket is 10 ms and the ball leaves the racket with a velocity of 25 m/s. Calculate (a) the impulse, and (b) the average force exerted by a racket on the ball.

[(a) 1.5 kg m/s (b) 150 N]

7. In a press-tool operation, the tool is in contact with the work piece for 40 ms. If the average force exerted on the work piece is 90 kN, determine the change in momentum.

[3600 kg m/s]

8. A gun of mass 1.2 t fires a shell of mass 12 kg with a velocity of 400 m/s. Determine (a) the initial velocity of recoil, and (b) the uniform force necessary to stop the recoil of the gun in 150 mm.

[(a) 4 m/s (b) 64 kN]

9. In making a steel stamping, a mass of 100 kg falls on to the steel through a distance of 1.5 m and is brought to rest after moving through a further distance of 15 mm. Determine the magnitude of the resisting force, assuming a uniform resistive force is exerted by the steel. [98.1 kN]

10. A vertical pile of mass 150 kg is driven 120 mm into the ground by the blow of a 1.1 t hammer, which falls through 800 mm. Assuming the hammer and pile remain in contact, determine (a) the velocity of the hammer just before impact (b) the velocity immediately after impact, and (c) the resistive force of the ground, assuming it to be uniform.

[(a) 3.96 m/s (b) 3.48 m/s (c) 63.08 kN]

Practise Exercise 71 **Short-answer questions on linear momentum and impulse**

1. Define momentum.

2. State Newton's first law of motion.

3. State the principle of the conservation of momentum.

4. State Newton's second law of motion.

5. Define impulse.

6. What is meant by an impulsive force?

7. State Newton's third law of motion.

Practise Exercise 72 **Multiple-choice questions on linear momentum and impulse**

(Answers on page 297)

1. A mass of 100 g has a momentum of 100 kg m/s. The velocity of the mass is:
   (a) 10 m/s   (b) 10^2 m/s
   (c) 10^{-3} m/s   (d) 10^3 m/s

2. A rifle bullet has a mass of 50 g. The momentum when the muzzle velocity is 108 km/h is:
   (a) 54 kg m/s   (b) 1.5 kg m/s
   (c) 15000 kg m/s (d) 21.6 kg m/s

   A body P of mass 10 kg has a velocity of 5 m/s and the same line of action as a body Q of mass 2 kg and having a velocity of 25 m/s. The bodies collide, and their velocities are the same after impact. In questions 3 to 6, select the correct answer from the following:

   (a) 25/3 m/s   (b) 360 kg m/s
   (c) 0   (d) 30 m/s
   (e) 160 kg m/s   (f) 100 kg m/s
   (g) 20 m/s

3. Determine the total momentum of the system before impact when P and Q have the same sense.

4. Determine the total momentum of the system before impact when P and Q have the opposite sense.

5. Determine the velocity of P and Q after impact if their sense is the same before impact.

6. Determine the velocity of P and Q after impact if their sense is opposite before impact.

7. A force of 100 N acts on a body of mass 10 kg for 0.1 s. The change in velocity of the body is:

(a) 1 m/s   (b) 100 m/s
(c) 0.1 m/s   (d) 0.01 m/s.

A vertical pile of mass 200 kg is driven 100 mm into the ground by the blow of a 1 t hammer which falls through 1.25 m. In questions 8 to 12, take g as 10 m/s^2 and select the correct answer from the following:

(a) 25 m/s   (b) 25/6 m/s
(c) 5 kg m/s   (d) 0
(e) 625/6 kN   (f) 5000 kg m/s
(g) 5 m/s   (h) 12 kN

8. Calculate the velocity of the hammer immediately before impact.

9. Calculate the momentum of the hammer just before impact.

10. Calculate the momentum of the hammer and pile immediately after impact assuming they have the same velocity.

11. Calculate the velocity of the hammer and pile immediately after impact assuming they have the same velocity.

12. Calculate the resistive force of the ground, assuming it to be uniform.

Chapter 14

Force, mass and acceleration

When an object is pushed or pulled, a force is applied to the object. The effects of pushing or pulling an object are to cause changes in the motion and shape of the object. If a change occurs in the motion of the object then the object accelerates. Thus, acceleration results from a force being applied to an object. If a force is applied to an object and it does not move, then the object changes shape. Usually the change in shape is so small that it cannot be detected by just watching the object. However, when very sensitive measuring instruments are used, very small changes in dimensions can be detected. A force of attraction exists between all objects. If a person is taken as one object and the Earth as a second object, a force of attraction exists between the person and the Earth. This force is called the gravitational force and is the force that gives a person a certain weight when standing on the Earth's surface. It is also this force that gives freely falling objects a constant acceleration in the absence of other forces. This chapter defines force and acceleration, states Newton's three laws of motion and defines moment of inertia, all demonstrated via practical everyday situations.

At the end of this chapter you should be able to:

- define force and state its unit
- appreciate 'gravitational force'
- state Newton's three laws of motion
- perform calculations involving force $F = ma$
- define 'centripetal acceleration'
- perform calculations involving centripetal force $= \dfrac{mv^2}{r}$
- define 'mass moment of inertia'

14.1 Introduction

As stated above, when an object is pushed or pulled, a **force** is applied to the object. This force is measured in **newtons (N).** The effects of pushing or pulling an object are:

 (i) to cause a change in the motion of the object, and
 (ii) to cause a change in the shape of the object.

If a change occurs in the motion of the object, that is, its velocity changes from u to v, then the object accelerates. Thus, it follows that acceleration results from a force being applied to an object. If a force is applied to an object and it does not move, then the object changes shape, that is, deformation of the object takes place. Usually the change in shape is so small that it cannot be detected by just watching the object. However, when very sensitive measuring instruments are used, very small changes in dimensions can be detected.

A force of attraction exists between all objects. The factors governing the size of this force F are the masses of the objects and the distances between their centres:

$$F \; \alpha \; \frac{m_1 m_2}{d^2}$$

Mechanical Engineering Principles, Bird and Ross, ISBN 9780415517850

Thus, if a person is taken as one object and the Earth as a second object, a force of attraction exists between the person and the Earth. This force is called the **gravitational force,** (first presented by Sir Isaac Newton), and is the force that gives a person a certain weight when standing on the Earth's surface. It is also this force that gives freely falling objects a constant acceleration in the absence of other forces.

14.2 Newton's laws of motion

To make a stationary object move or to change the direction in which the object is moving requires a force to be applied externally to the object. This concept is known as Newton's **first law of motion** and may be stated as:

An object remains in a state of rest, or continues in a state of uniform motion in a straight line, unless it is acted on by an externally applied force

Since a force is necessary to produce a change of motion, an object must have some resistance to a change in its motion. The force necessary to give a stationary pram a given acceleration is far less than the force necessary to give a stationary car the same acceleration on the same surface. The resistance to a change in motion is called the **inertia** of an object and the amount of inertia depends on the mass of the object. Since a car has a much larger mass than a pram, the inertia of a car is much larger than that of a pram.

Newton's **second law of motion** may be stated as:

The acceleration of an object acted upon by an external force is proportional to the force and is in the same direction as the force

Thus, force α acceleration, or force = a constant × acceleration, this constant of proportionality being the mass of the object, i.e.

force = mass × acceleration

The unit of force is the newton (N) and is defined in terms of mass and acceleration. One newton is the force required to give a mass of 1 kilogram an acceleration of 1 metre per second squared. Thus

$$F = ma$$

where F is the force in newtons (N), m is the mass in kilograms (kg) and a is the acceleration in metres per second squared (m/s²), i.e. $1\,N = \dfrac{1\,kg\,m}{s^2}$

It follows that 1 m/s² = 1 N/kg. Hence a gravitational acceleration of 9.8 m/s² is the same as a gravitational field of 9.8 N/kg.

Newton's **third law of motion** may be stated as:

For every force, there is an equal and opposite reacting force

Thus, an object on, say, a table, exerts a downward force on the table and the table exerts an equal upward force on the object, known as a **reaction force** or just a **reaction**.

Problem 1.   Calculate the force needed to accelerate a boat of mass 20 tonne uniformly from rest to a speed of 21.6 km/h in 10 minutes.

The mass of the boat, m, is 20 t, that is 20000 kg.

The law of motion, $v = u + at$ can be used to determine the acceleration a.

The initial velocity, u, is zero, the final velocity,

$$v = 21.6\,km/h = 21.6\frac{km}{h} \times \frac{1\,h}{3600\,s} \times \frac{1000\,m}{1\,km}$$
$$= \frac{21.6}{3.6} = 6\ m/s,$$

and   the time, $t = 10$ min = 600 s.

Thus, $v = u + at$ i.e. $6 = 0 + a \times 600$

from which, $a = \dfrac{6}{600} = 0.01\ m/s^2$

From Newton's second law, $F = ma$

i.e.   **force** = 20000 × 0.01 N = **200 N**

Problem 2.   The moving head of a machine tool requires a force of 1.2 N to bring it to rest in 0.8 s from a cutting speed of 30 m/min. Find the mass of the moving head.

From Newton's second law, $F = ma$, thus $m = \dfrac{F}{a}$, where force is given as 1.2 N. The law of motion

$v = u + at$ can be used to find acceleration a,

where $v = 0$, $u = 30$ m/min $= \dfrac{30}{60}$ m/s = 0.5 m/s,

and $t = 0.8$ s.

Thus, $0 = 0.5 + a \times 0.8$

from which, $a = -\dfrac{0.5}{0.8} = -0.625\ m/s^2$ or a retardation of 0.625 m/s².

Thus the **mass**, $m = \dfrac{F}{a} = \dfrac{1.2}{0.625} = $ **1.92 kg**

Problem 3. A lorry of mass 1350 kg accelerates uniformly from 9 km/h to reach a velocity of 45 km/h in 18 s. Determine (a) the acceleration of the lorry (b) the uniform force needed to accelerate the lorry.

(a) The law of motion $v = u + at$ can be used to determine the acceleration, where final velocity

$$v = 45\frac{km}{h} \times \frac{1h}{3600s} \times \frac{1000m}{1km} = \frac{45}{3.6} \text{ m/s, initial}$$

velocity $u = \frac{9}{3.6}$ m/s and time $t = 18$ s.

Thus $\frac{45}{3.6} = \frac{9}{3.6} + a \times 18$

from which, $a = \frac{1}{18}\left(\frac{45}{3.6} - \frac{9}{3.6}\right) = \frac{1}{18}\left(\frac{36}{3.6}\right)$

$$= \frac{10}{18} = \frac{5}{9} \text{ m/s}^2 \text{ or } \mathbf{0.556 \text{ m/s}^2}$$

(b) From Newton's second law of motion,

force, $F = ma = 1350 \times \frac{5}{9} = \mathbf{750 \text{ N}}$

Problem 4. Find the weight of an object of mass 1.6 kg at a point on the earth's surface where the gravitational field is 9.81 N/kg (or 9.81 m/s^2).

The weight of an object is the force acting vertically downwards due to the force of gravity acting on the object.

Thus: **weight** = force acting vertically downwards
= mass × gravitational field
= $1.6 \times 9.81 = \mathbf{15.696 \text{ N}}$

Problem 5. A bucket of cement of mass 40 kg is tied to the end of a rope connected to a hoist. Calculate the tension in the rope when the bucket is suspended but stationary. Take the gravitational field, g, as 9.81 N/kg (or 9.81 m/s^2).

The **tension** in the rope is the same as the force acting in the rope. The force acting vertically downwards due to the weight of the bucket must be equal to the force acting upwards in the rope, i.e. the tension.

Weight of bucket of cement, $F = mg = 40 \times 9.81$

$$= 392.4 \text{ N}$$

Thus, **the tension in the rope = 392.4 N**

Problem 6. The bucket of cement in Problem 5 is now hoisted vertically upwards with a uniform acceleration of 0.4 m/s^2. Calculate the tension in the rope during the period of acceleration.

With reference to Figure 14.1, the forces acting on the bucket are:

(i) a tension (or force) of T acting in the rope

(ii) a force of mg acting vertically downwards, i.e. the weight of the bucket and cement

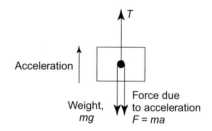

Figure 14.1

The resultant force $F = T - mg$
Hence, $ma = T - mg$

i.e. $40 \times 0.4 = T - 40 \times 9.81$

from which, **tension, $T = 408.4$ N**

By comparing this result with that of Problem 5, it can be seen that there is an increase in the tension in the rope when an object is accelerating upwards.

Problem 7. The bucket of cement in Problem 5 is now lowered vertically downwards with a uniform acceleration of 1.4 m/s^2. Calculate the tension in the rope during the period of acceleration.

With reference to Figure 14.2, the forces acting on the bucket are:

(i) a tension (or force) of T acting vertically upwards

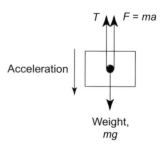

Figure 14.2

(ii) a force of mg acting vertically downwards, i.e. the weight of the bucket and cement

The resultant force, $F = mg - T$

Hence, $ma = mg - T$

from which, **tension,** $T = m(g - a)$

$$= 40(9.81 - 1.4)$$

$$= \textbf{336.4 N}$$

By comparing this result with that of Problem 5, it can be seen that there is a decrease in the tension in the rope when an object is accelerating downwards.

Now try the following Practise Exercise

Practise Exercise 73	**Further problems on Newton's laws of motion**

(Take g as 9.81 m/s², and express answers to three significant figure accuracy)

1. A car initially at rest, accelerates uniformly to a speed of 55 km/h in 14 s. Determine the accelerating force required if the mass of the car is 800 kg.   [873 N]

2. The brakes are applied on the car in Question 1 when travelling at 55 km/h and it comes to rest uniformly in a distance of 50 m. Calculate the braking force and the time for the car to come to rest.   [1.87 kN, 6.55 s]

3. The tension in a rope lifting a crate vertically upwards is 2.8 kN. Determine its acceleration if the mass of the crate is 270 kg.   [0.560 m/s²]

4. A ship is travelling at 18 km/h when it stops its engines. It drifts for a distance of 0.6 km and its speed is then 14 km/h. Determine the value of the forces opposing the motion of the ship, assuming the reduction in speed is uniform and the mass of the ship is 2000 t.   [16.5 kN]

5. A cage having a mass of 2 t is being lowered down a mineshaft. It moves from rest with an acceleration of 4 m/s², until it is travelling at 15 m/s. It then travels at constant speed for 700 m and finally comes to rest in 6 s.

Calculate the tension in the cable supporting the cage during

(a) the initial period of acceleration
(b) the period of constant speed travel
(c) the final retardation period.
   [(a) 11.6 kN (b) 19.6 kN (c) 24.6 kN]

6. A miner having a mass of 80 kg is standing in the cage of Problem 5. Determine the reaction force between the man and the floor of the cage during (a) the initial period of acceleration (b) the period of constant speed travel, and (c) the final retardation period.
   [(a) 464.8 N (b) 784.8 N (c) 984.8 N]

7. During an experiment, masses of 4 kg and 5 kg are attached to a thread and the thread is passed over a pulley so that both masses hang vertically downwards and are at the same height. When the system is released, find (a) the acceleration of the system, and (b) the tension in the thread, assuming no losses in the system.
   [(a) 1.09 m/s² (b) 43.6 N]

14.3   Centripetal acceleration

When an object moves in a circular path at constant speed, its direction of motion is continually changing and hence its velocity (which depends on both magnitude and direction) is also continually changing. Since acceleration is the (change in velocity)/(time taken) the object has an acceleration.

Let the object be moving with a constant angular velocity of ω and a tangential velocity of magnitude v and let the change of velocity for a small change of angle of $\theta \, (= \omega t)$ be V (see Figure 14.3(a)). Then, $v_2 - v_1 = V$.

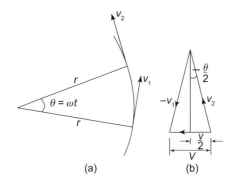

(a)   (b)

Figure 14.3

The vector diagram is shown in Figure 14.3(b) and since the magnitudes of v_1 and v_2 are the same, i.e. v, the vector diagram is also an isosceles triangle. Bisecting the angle between v_2 and v_1 gives:

$$\sin \frac{\theta}{2} = \frac{V/2}{v_2} = \frac{V}{2v}$$

i.e. $$V = 2v \sin \frac{\theta}{2} \qquad (14.1)$$

Since $$\theta = \omega t, \text{ then } t = \frac{\theta}{\omega} \qquad (14.2)$$

Dividing (14.1) by (14.2) gives:

$$\frac{V}{t} = \frac{2v \sin \dfrac{\theta}{2}}{\dfrac{\theta}{\omega}} = \frac{v\omega \sin \dfrac{\theta}{2}}{\dfrac{\theta}{2}}$$

For small angles, $\dfrac{\sin \dfrac{\theta}{2}}{\dfrac{\theta}{2}}$ is very nearly equal to unity

Therefore, $$\frac{V}{t} = v\omega$$

or, $$\frac{V}{t} = \frac{\text{change of velocity}}{\text{change of time}}$$
$$= \text{acceleration}, a = v\omega$$

But, $\omega = v/r$, thus $v\omega = v \times \dfrac{v}{r} = \dfrac{v^2}{r} = \text{acceleration}$

That is, **the acceleration a is** $\dfrac{v^2}{r}$ and is towards the centre of the circle of motion (along V). It is called the **centripetal acceleration**. If the mass of the rotating object is m, then by Newton's second law, the **centripetal force** is $\dfrac{mv^2}{r}$, and its direction is towards the centre of the circle of motion.

> **Problem 8.** A vehicle of mass 750 kg travels round a bend of radius 150 m, at 50.4 km/h. Determine the centripetal force acting on the vehicle.

The centripetal force is given by $\dfrac{mv^2}{r}$ and its direction is towards the centre of the circle.

$$m = 750 \text{ kg},$$
$$v = 50.4 \text{ km/h} = \frac{50.4}{3.6} \text{ m/s} = 14 \text{ m/s}$$
and $$r = 150 \text{ m}$$

Thus, **centripetal force** $= \dfrac{750 \times 14^2}{150}$
$$= \textbf{980 N}$$

> **Problem 9.** An object is suspended by a thread 250 mm long and both object and thread move in a horizontal circle with a constant angular velocity of 2.0 rad/s. If the tension in the thread is 12.5 N, determine the mass of the object.

Centripetal force (i.e. tension in thread) $= \dfrac{mv^2}{r}$
$$= 12.5 \text{ N}$$

The angular velocity, $\omega = 2.0$ rad/s
and radius, $r = 250$ mm $= 0.25$ m.

Since linear velocity $v = \omega r$, $v = 2.0 \times 0.25$
$$= 0.5 \text{ m/s},$$

and since $F = \dfrac{mv^2}{r}$, then $m = \dfrac{Fr}{v^2}$

i.e. **mass of object, m** $= \dfrac{12.5 \times 0.25}{0.5^2} = \textbf{12.5 kg}$

> **Problem 10.** An aircraft is turning at constant altitude, the turn following the arc of a circle of radius 1.5 km. If the maximum allowable acceleration of the aircraft is 2.5 g, determine the maximum speed of the turn in km/h. Take g as 9.8 m/s^2.

The acceleration of an object turning in a circle is $\dfrac{v^2}{r}$.

Thus, to determine the maximum speed of turn $\dfrac{v^2}{r} = 2.5$ g.

Hence, **speed of turn, v** $= \sqrt{2.5gr} = \sqrt{2.5 \times 9.8 \times 1500}$
$$= \sqrt{36750} = 191.7 \text{ m/s}$$
$$= 191.7 \times 3.6 \text{ km/h}$$
$$= \textbf{690 km/h}$$

Now try the following Practise Exercise

Practise Exercise 74   Further problems on centripetal acceleration

1. Calculate the centripetal force acting on a vehicle of mass 1 tonne when travelling round a bend of radius 125 m at 40 km/h. If

this force should not exceed 750 N, determine the reduction in speed of the vehicle to meet this requirement.   [988 N, 34.86 km/h]

2. A speed-boat negotiates an S-bend consisting of two circular arcs of radii 100 m and 150 m. If the speed of the boat is constant at 34 km/h, determine the change in acceleration when leaving one arc and entering the other.
    [0.3 m/s^2]

3. An object is suspended by a thread 400 mm long and both object and thread move in a horizontal circle with a constant angular velocity of 3.0 rad/s. If the tension in the thread is 36 N, determine the mass of the object.   [10 kg]

14.4   Rotation of a rigid body about a fixed axis

A rigid body is said to be a body that does not change its shape or size during motion. Thus, any two particles on a rigid body will remain the same distance apart during motion.

Consider the rigidity of Figure 14.4, which is rotating about the fixed axis O.

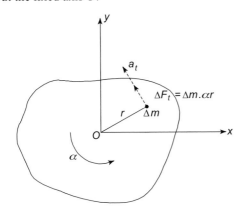

Figure 14.4

In Figure 14.4,

α = the constant angular acceleration

Δm = the mass of a particle

r = the radius of rotation of Δm

a_t = the tangential acceleration of Δm

ΔF_t = the elemental force on the particle

Now,   force $F = ma$

or $\Delta F_t = \Delta m\, a_t$

$= \Delta m\,(\alpha r)$

Multiplying both sides of the above equation by r, gives:

$$\Delta F_t r = \Delta m\,\alpha r^2$$

Since α is a constant

$$\sum \Delta F_t r = \alpha \sum \Delta m r^2$$

But $\Delta T = \Delta F_t r$

and $\Delta m r^2 = \Delta I_o$

Hence, $\boldsymbol{T = I_o \alpha}$ (14.3)

where T = the total turning moment exerted on the rigid body = $\sum \Delta F_t r$

and I_o = the mass moment of inertia (or second moment) about O (in kg m^2).

Equation (14.3) can be seen to be the rotational equivalent of $F = ma$ (Newton's second law of motion).

Problem 11.   Determine the angular acceleration that occurs when a circular disc of mass moment of inertia of 0.5 kg m^2 is subjected to a torque of 6 N m. Neglect friction and other losses.

From equation (14.3), torque $T = I\alpha$,

from which, **angular acceleration**, $\alpha = \dfrac{T}{I} = \dfrac{6\,\text{N m}}{0.5\,\text{kg m}^2}$

$$= \textbf{12 rad/s}^2$$

14.5   Moment of inertia (*I*)

The moment of inertia is required for analysing problems involving the rotation of rigid bodies. It is defined as:

$I = mk^2$ = mass moment of inertia (kg/m^2) or moment of inertia,

where m = the mass of the rigid body

k = its radius of gyration about the point of rotation (see Chapter 8).

In general, $I = \sum \Delta m r^2$ where the definitions of Figure 14.4 apply.

Some typical values of mass and the radius of gyration are given in Table 14.1, where

$$A = \text{cross-sectional area}$$

$$L = \text{length}$$

$$t = \text{disc thickness}$$

$$R = \text{radius of the solid disc}$$

$$R_1 = \text{internal radius}$$

$$R_2 = \text{external radius}$$

$$\rho = \text{density}$$

Table 14.1

Component	mass	k^2
Rod, about mid-point	ρAL	$\dfrac{L^2}{12}$
Rod, about an end	ρAL	$\dfrac{L^2}{3}$
Flat disc	$\rho \pi R^2 t$	$\dfrac{R^2}{2}$
Annulus	$\rho \pi (R_2{}^2 - R_1{}^2)t$	$\dfrac{(R_1{}^2 + R_2{}^2)}{2}$

Parallel axis theorem

This is of similar form to the parallel axis theorem of Chapter 8, where

$$I_{xx} = I_G + mh^2$$

I_{xx} = the mass moment of inertia about the xx axis, which is parallel to an axis passing through the centre of gravity of the rigid body, namely at G

I_G = the mass moment of inertia of the rigid body about an axis passing through G and parallel to the xx axis

h = the perpendicular distance between the above two parallel axes.

Problem 12. Determine the mass moment of inertia about its centroid for a solid uniform thickness disc. For the disc, its radius is 0.2 m, its thickness is 0.05 m, and its density is 7860 kg/m³.

From Table 14.1, for a disc,

mass, $m = \rho \pi R^2 t$

$$= 7860 \, \frac{\text{kg}}{\text{m}^3} \times \pi \times (0.2 \text{ m})^2 \times 0.05 \text{ m}$$

$$= 49.386 \text{ kg}$$

Mass moment of inertia about its centroid,

$$\boldsymbol{I_o = \frac{mR^2}{2} = \frac{49.386 \times 0.2^2}{2} \, \text{kg m}^2}$$

$$= \textbf{0.988 kg m}^2$$

Now try the following Practise Exercises

Practise Exercise 75   Further problems on rotation and moment of inertia

1. Calculate the mass moment of inertia of a thin rod, of length 0.5 m and mass 0.2 kg, about its centroid.   [0.004167 kg m²]

2. Calculate the mass moment of inertia of the thin rod of Problem 1, about an end.
          [0.01667 kg m²]

3. Calculate the mass moment of inertia of a solid disc of uniform thickness about its centroid. The diameter of the disc is 0.3 m and its thickness is 0.08 m. The density of its material of construction is 7860 kg/m³.
          [0.50 kg m²]

4. If a hole of diameter 0.2 m is drilled through the centre of the disc of Problem 3, what will be its mass moment of inertia about its centroid?   [0.401 kg m²]

Practise Exercise 76   Short-answer questions on force, mass and acceleration

1. Force is measured in ....

2. The two effects of pushing or pulling an object are .... or ....

3. A gravitational force gives free-falling objects a .... in the absence of all other forces.

4. State Newton's first law of motion.

5. Describe what is meant by the inertia of an object.

6. State Newton's second law of motion.

7. Define the newton.

8. State Newton's third law of motion.

9. Explain why an object moving round a circle at a constant angular velocity has an acceleration.

10. Define centripetal acceleration in symbols.

11. Define centripetal force in symbols.

12. Define mass moment of inertia.

13. A rigid body has a constant angular acceleration α when subjected to a torque T. The mass moment of inertia, I_o = ...

Practise Exercise 77 **Multiple-choice questions on force, mass and acceleration**

(Answers on page 297)

1. The unit of force is the:
   (a) watt   (b) kelvin
   (c) newton   (d) joule

2. If a = acceleration and F = force, then mass m is given by:
   (a) $m = a - F$ (b) $m = \dfrac{F}{a}$
   (c) $m = F - a$ (d) $m = \dfrac{a}{F}$

3. The weight of an object of mass 2 kg at a point on the earth's surface when the gravitational field is 10 N/kg is:
   (a) 20 N   (b) 0.2 N
   (c) 20 kg   (d) 5 N

4. The force required to accelerate a loaded barrow of 80 kg mass up to 0.2 m/s² on friction-less bearings is:
   (a) 400 N   (b) 3.2 N
   (c) 0.0025 N   (d) 16 N

5. A bucket of cement of mass 30 kg is tied to the end of a rope connected to a hoist. If the gravitational field g = 10 N/kg, the tension in the rope when the bucket is suspended but stationary is:
   (a) 300 N   (b) 3 N
   (c) 300 kg   (d) 0.67 N

A man of mass 75 kg is standing in a lift of mass 500 kg. Use this data to determine the answers to questions 6 to 9. Take g as 10 m/s².

6. The tension in a cable when the lift is moving at a constant speed vertically upward is:
   (a) 4250 N   (b) 5750 N
   (c) 4600 N   (d) 6900 N

7. The tension in the cable supporting the lift when the lift is moving at a constant speed vertically downwards is:
   (a) 4250 N   (b) 5750 N
   (c) 4600 N   (d) 6900 N

8. The reaction force between the man and the floor of the lift when the lift is travelling at a constant speed vertically upwards is:
   (a) 750 N   (b) 900 N
   (c) 600 N   (d) 475 N

9. The reaction force between the man and the floor of the lift when the lift is travelling at a constant speed vertically downwards is:
   (a) 750 N   (b) 900 N
   (c) 600 N   (d) 475 N

A ball of mass 0.5 kg is tied to a thread and rotated at a constant angular velocity of 10 rad/s in a circle of radius 1 m. Use this data to determine the answers to questions 10 and 11

10. The centripetal acceleration is:
   (a) 50 m/s²   (b) $\dfrac{100}{2\pi}$ m/s²
   (c) $\dfrac{50}{2\pi}$ m/s²   (d) 100 m/s²

11. The tension in the thread is:
   (a) 25 N   (b) $\dfrac{50}{2\pi}$ N
   (c) $\dfrac{25}{2\pi}$ N   (d) 50 N

12. Which of the following statements is false?
   (a) An externally applied force is needed to change the direction of a moving object.
   (b) For every force, there is an equal and opposite reaction force.
   (c) A body travelling at a constant velocity in a circle has no acceleration.
   (d) Centripetal acceleration acts towards the centre of the circle of motion.

13. An angular acceleration of 10 rad/s² occurs when a circular disc of mass moment of inertia of 0.5 kg m² is subjected to a torque. The value of the torque is:
   (a) 25 N m   (b) 5 N m
   (c) 20 N m   (d) 0.05 N m

Part Three

Chapter 15

Work, energy and power

This chapter commences by defining work, power and energy. It also provides the mid-ordinate rule, together with an explanation on how to apply it to calculate the areas of irregular figures, such as the areas of ship's water planes. It can also be used for calculating the work done in a force-displacement or similar relationship, which may result in the form of an irregular two-dimensional shape. This chapter is fundamental to the study and application of dynamics to practical problems.

At the end of this chapter you should be able to:

- define work and state its unit
- perform simple calculations on work done
- appreciate that the area under a force/distance graph gives work done
- perform calculations on a force/distance graph to determine work done
- define energy and state its unit
- state several forms of energy
- state the principle of conservation of energy and give examples of conversions
- define and calculate efficiency of systems
- define power and state its unit
- understand that power = force × velocity
- perform calculations involving power, work done, energy and efficiency
- define potential energy
- perform calculations involving potential energy = mgh
- define kinetic energy
- perform calculations involving kinetic energy = $\frac{1}{2}mv^2$
- distinguish between elastic and inelastic collisions

- perform calculations involving kinetic energy in rotation = $\frac{1}{2}I\omega^2$

15.1 Work

If a body moves as a result of a force being applied to it, the force is said to do work on the body. The amount of work done is the product of the applied force and the distance, i.e.

work done = force × distance moved in the direction of the force

The unit of work is the **joule, J**, which is defined as the amount of work done when a force of 1 newton acts for a distance of 1 m in the direction of the force. Thus, **1 J = 1 N m**

If a graph is plotted of experimental values of force (on the vertical axis) against distance moved (on the horizontal axis) a force/distance graph or work diagram is produced. **The area under the graph represents the work done.**

For example, a constant force of 20 N used to raise a load a height of 8 m may be represented on a force/distance graph as shown in Figure 15.1. The area

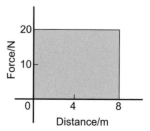

Figure 15.1

Mechanical Engineering Principles, Bird and Ross, ISBN 9780415517850

under the graph shown shaded represents the work done. Hence

work done = 20 N × 8 m = **160 J**

Similarly, a spring extended by 20 mm by a force of 500 N may be represented by the work diagram shown in Figure 15.2, where

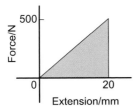

Figure 15.2

$$\textbf{work done} = \text{shaded area} = \frac{1}{2} \times \text{base} \times \text{height}$$

$$= \frac{1}{2} \times (20 \times 10^{-3}) \text{ m} \times 500 \text{ N} = \textbf{5 J}$$

It is shown in Chapter 14 that force = mass × acceleration, and that if an object is dropped from a height it has a constant acceleration of around 9.81 m/s². Thus if a mass of 8 kg is lifted vertically 4 m, the work done is given by:

$$\begin{aligned} \text{work done} &= \text{force} \times \text{distance} \\ &= (\text{mass} \times \text{acceleration}) \times \text{distance} \\ &= (8 \times 9.81) \times 4 = 313.92 \text{ J} \end{aligned}$$

The work done by a variable force may be found by determining the area enclosed by the force/distance graph using an approximate method such as the **mid-ordinate rule**.

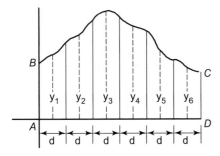

Figure 15.3

To determine the area *ABCD* of Figure 15.3 using the mid-ordinate rule:

(i) Divide base *AD* into any number of equal intervals, each of width *d* (the greater the number of intervals, the greater the precision)

(ii) Erect ordinates in the middle of each interval (shown by broken lines in Figure 15.3)

(iii) Accurately measure ordinates y_1, y_2, y_3, etc.

(iv) Area $ABCD = d(y_1 + y_2 + y_3 + y_4 + y_5 + y_6)$

In general, the mid-ordinate rule states:

Area = (width of interval) (sum of mid-ordinates)

> **Problem 1.** Calculate the work done when a force of 40 N pushes an object a distance of 500 m in the same direction as the force.

Work done = force × distance moved in the direction of the force

$$\begin{aligned} &= 40 \text{ N} \times 500 \text{ m} \\ &= 20000 \text{ J (since 1 J} = 1 \text{ N m)} \end{aligned}$$

i.e. **work done = 20 kJ**

> **Problem 2.** Calculate the work done when a mass is lifted vertically by a crane to a height of 5 m, the force required to lift the mass being 98 N.

When work is done in lifting then:

$$\begin{aligned} \text{work done} &= (\text{weight of the body}) \\ &\quad \times (\text{vertical distance moved}) \end{aligned}$$

Weight is the downward force due to the mass of an object. Hence

work done = 98 N × 5 m = **490 J**

> **Problem 3.** A motor supplies a constant force of 1 kN which is used to move a load a distance of 5 m. The force is then changed to a constant 500 N and the load is moved a further 15 m. Draw the force/distance graph for the operation and from the graph determine the work done by the motor.

The force/distance graph or work diagram is shown in Figure 15.4. Between points *A* and *B* a constant force of 1000 N moves the load 5 m; between points *C* and *D* a constant force of 500 N moves the load from 5 m to 20 m.

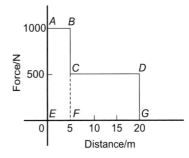

Figure 15.4

Total work done = area under the force/distance graph

$$= \text{area } ABFE + \text{area } CDGF$$

$$= (1000 \text{ N} \times 5 \text{ m}) + (500 \text{ N} \times 15 \text{ m})$$

$$= 5000 \text{ J} + 7500 \text{ J} = 12500 \text{ J} = \textbf{12.5 kJ}$$

Problem 4. A spring, initially in a relaxed state, is extended by 100 mm. Determine the work done by using a work diagram if the spring requires a force of 0.6 N per mm of stretch.

Force required for a 100 mm extension = 100 mm $\times$ 0.6 N/mm = 60 N. Figure 15.5 shows the force/extension graph or work diagram representing the increase in extension in proportion to the force, as the force is increased from 0 to 60 N. The work done is the area under the graph, hence

$$\textbf{work done} = \frac{1}{2} \times \text{base} \times \text{height}$$

$$= \frac{1}{2} \times 100 \text{ mm} \times 60 \text{ N}$$

$$= \frac{1}{2} \times 100 \times 10^{-3}\text{m} \times 60 \text{ N} = \textbf{3 J}$$

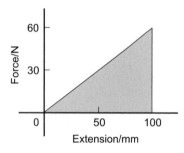

Figure 15.5

(Alternatively, average force during extension

$$= \frac{(60 - 0)}{2} = 30 \text{ N}$$

and total extension = 100 mm = 0.1 m, hence

$$\text{work done} = \text{average force} \times \text{extension}$$

$$= 30 \text{ N} \times 0.1 \text{ m} = 3 \text{ J})$$

Problem 5. A spring requires a force of 10 N to cause an extension of 50 mm. Determine the work done in extending the spring (a) from zero to 30 mm, and (b) from 30 mm to 50 mm.

Figure 15.6 shows the force/extension graph for the spring.

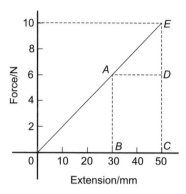

Figure 15.6

(a) Work done in extending the spring from zero to 30 mm is given by area $AB0$ of Figure 15.6,

i.e. **work done** $= \dfrac{1}{2} \times \text{base} \times \text{height}$

$$= \frac{1}{2} \times 30 \times 10^{-3}\text{m} \times 6 \text{ N}$$

$$= 90 \times 10^{-3}\text{J} = \textbf{0.09 J}$$

(b) Work done in extending the spring from 30 mm to 50 mm is given by area $ABCE$ of Figure 15.6, i.e.

work done = area $ABCD$ + area ADE

$$= (20 \times 10^{-3}\text{m} \times 6 \text{ N}) + \frac{1}{2}(20 \times 10^{-3}\text{m})(4 \text{ N})$$

$$= 0.12 \text{ J} + 0.04 \text{ J} = \textbf{0.16 J}$$

Problem 6. Calculate the work done when a mass of 20 kg is lifted vertically through a distance of 5.0 m. Assume that the acceleration due to gravity is 9.81 m/s².

The force to be overcome when lifting a mass of 20 kg vertically upwards is mg,

i.e. $20 \times 9.81 = 196.2$ N (see Chapter 14).

Work done = force $\times$ distance = $196.2 \times 5.0 = \textbf{981 J}$

Problem 7. Water is pumped vertically upwards through a distance of 50.0 m and the work done is 294.3 kJ. Determine the number of litres of water pumped. (1 litre of water has a mass of 1 kg).

Work done = force $\times$ distance

i.e. $294300 = \text{force} \times 50.0$

from which, force $= \dfrac{294300}{50.0} = 5886$ N

The force to be overcome when lifting a mass m kg vertically upwards is mg

i.e. $(m \times 9.81)$ N (see Chapter 14).

Thus, $5886 = m \times 9.81$, from which

mass, $m = \dfrac{5886}{9.81} = 600$ kg.

Since 1 litre of water has a mass of 1 kg, **600 litres of water are pumped**.

Problem 8. The force on a cutting tool of a shaping machine varies over the length of cut as follows:

Distance (mm)	0	20	40	60	80	100
Force (kN)		60	72	65	53	44	50

Determine the work done as the tool moves through a distance of 100 mm.

The force/distance graph for the given data is shown in Figure 15.7. The work done is given by the area under the graph; the area may be determined by an approximate method. Using the mid-ordinate rule, with each strip of width 20 mm, mid-ordinates y_1, y_2, y_3, y_4 and y_5 are erected as shown, and each is measured.

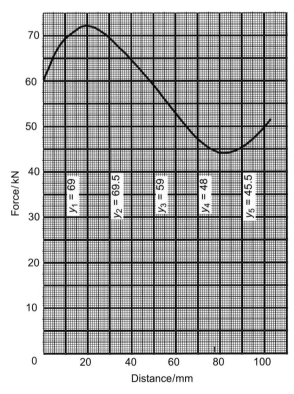

Figure 15.7

Area under curve = (width of each strip) (sum of mid-ordinate values)

$= (20)(69 + 69.5 + 59 + 48 + 45.5)$

$= (20)(291)$

$= 5820$ kN mm $= 5820$ N m

$= 5820$ J

Hence the work done as the tool moves through 100 mm is **5.82 kJ**

Now try the following Practise Exercise

Practise Exercise 78 Further problems on work

1. Determine the work done when a force of 50 N pushes an object 1.5 km in the same direction as the force. [75 kJ]

2. Calculate the work done when a mass of weight 200 N is lifted vertically by a crane to a height of 100 m. [20 kJ]

3. A motor supplies a constant force of 2 kN to move a load 10 m. The force is then changed to a constant 1.5 kN and the load is moved a further 20 m. Draw the force/distance graph for the complete operation, and, from the graph, determine the total work done by the motor. [50 kJ]

4. A spring, initially relaxed, is extended 80 mm. Draw a work diagram and hence determine the work done if the spring requires a force of 0.5 N/mm of stretch. [1.6 J]

5. A spring requires a force of 50 N to cause an extension of 100 mm. Determine the work done in extending the spring (a) from 0 to 100 mm, and (b) from 40 mm to 100 mm.
   [(a) 2.5 J (b) 2.1 J]

6. The resistance to a cutting tool varies during the cutting stroke of 800 mm as follows: (i) the resistance increases uniformly from an initial 5000 N to 10000 N as the tool moves 500 mm, and (ii) the resistance falls uniformly from 10000 N to 6000 N as the tool moves 300 mm. Draw the work diagram and calculate the work done in one cutting stroke. [6.15 kJ]

15.2 Energy

Energy is the capacity, or ability, to do work. The unit of energy is the joule, the same as for work. Energy is expended when work is done. There are several forms of energy and these include:

(i) Mechanical energy
(ii) Heat or thermal energy
(iii) Electrical energy
(iv) Chemical energy
(v) Nuclear energy
(vi) Light energy
(vii) Sound energy

Energy may be converted from one form to another. **The principle of conservation of energy** states that the total amount of energy remains the same in such conversions, i.e. energy cannot be created or destroyed. Some examples of energy conversions include:

(i) Mechanical energy is converted to electrical energy by a generator
(ii) Electrical energy is converted to mechanical energy by a motor
(iii) Heat energy is converted to mechanical energy by a steam engine
(iv) Mechanical energy is converted to heat energy by friction
(v) Heat energy is converted to electrical energy by a solar cell
(vi) Electrical energy is converted to heat energy by an electric fire
(vii) Heat energy is converted to chemical energy by living plants
(viii) Chemical energy is converted to heat energy by burning fuels
(ix) Heat energy is converted to electrical energy by a thermocouple
(x) Chemical energy is converted to electrical energy by batteries
(xi) Electrical energy is converted to light energy by a light bulb
(xii) Sound energy is converted to electrical energy by a microphone
(xiii) Electrical energy is converted to chemical energy by electrolysis.

Efficiency is defined as the ratio of the useful output energy to the input energy. The symbol for efficiency is η (Greek letter eta). Hence

$$\textbf{efficiency, } \eta = \frac{\textbf{useful output energy}}{\textbf{input energy}}$$

Efficiency has no units and is often stated as a percentage. A perfect machine would have an efficiency of 100%. However, all machines have an efficiency lower than this due to friction and other losses. Thus, if the input energy to a motor is 1000 J and the output energy is 800 J then the efficiency is $\frac{800}{1000} \times 100\% = \textbf{80\%}$

Problem 9. A machine exerts a force of 200 N in lifting a mass through a height of 6 m. If 2 kJ of energy are supplied to it, what is the efficiency of the machine?

Work done in lifting mass = force × distance moved
$$= \text{weight of body} \times \text{distance moved}$$
$$= 200 \text{ N} \times 6 \text{ m} = 1200 \text{ J}$$
$$= \text{useful energy output}$$

Energy input = 2 kJ = 2000 J

$$\textbf{Efficiency, } \eta = \frac{\text{useful output energy}}{\text{input energy}}$$

$$= \frac{1200}{2000} = \textbf{0.6 or 60\%}$$

Problem 10. Calculate the useful output energy of an electric motor which is 70% efficient if it uses 600 J of electrical energy.

$$\text{Efficiency, } \eta = \frac{\text{useful output energy}}{\text{input energy}}$$

thus $\frac{70}{100} = \frac{\text{output energy}}{600 \text{ J}}$

from which, **output energy** $= \frac{70}{100} \times 600 = \textbf{420 J}$

Problem 11. 4 kJ of energy are supplied to a machine used for lifting a mass. The force required is 800 N. If the machine has an efficiency of 50%, to what height will it lift the mass?

$$\text{Efficiency, } \eta = \frac{\text{useful output energy}}{\text{input energy}}$$

i.e. $\frac{50}{100} = \frac{\text{output energy}}{4000 \text{ J}}$

from which, output energy $= \frac{50}{100} \times 4000 = 2000 \text{ J}$

Work done = force × distance moved

hence 2000 J = 800 N × height

from which,   **height** $= \dfrac{2000\,\text{J}}{800\,\text{N}} = \textbf{2.5 m}$

Problem 12. A hoist exerts a force of 500 N in raising a load through a height of 20 m. The efficiency of the hoist gears is 75% and the efficiency of the motor is 80%. Calculate the input energy to the hoist.

The hoist system is shown diagrammatically in Figure 15.8.

Figure 15.8

Output energy = work done = force × distance

$$= 500\,\text{N} \times 20\,\text{m} = 10000\,\text{J}$$

For the gearing, efficiency $= \dfrac{\text{output energy}}{\text{input energy}}$

i.e. $\dfrac{75}{100} = \dfrac{10000}{\text{input energy}}$

from which, the input energy to the gears

$$= 10000 \times \dfrac{100}{75} = 13333\,\text{J}$$

The input energy to the gears is the same as the output energy of the motor. Thus, for the motor,

efficiency $= \dfrac{\text{output energy}}{\text{input energy}}$

i.e. $\dfrac{80}{100} = \dfrac{13333}{\text{input energy}}$

Hence, **input energy to the hoist** $= 13333 \times \dfrac{100}{80}$

$$= 16667\,\text{J} = \textbf{16.67 kJ}$$

Now try the following Practise Exercise

Practise Exercise 79   Further problems on energy

1. A machine lifts a mass of weight 490.5 N through a height of 12 m when 7.85 kJ of energy is supplied to it. Determine the efficiency of the machine.   [75%]

2. Determine the output energy of an electric motor which is 60% efficient if it uses 2 kJ of electrical energy.   [1.2 kJ]

3. A machine that is used for lifting a particular mass is supplied with 5 kJ of energy. If the machine has an efficiency of 65% and exerts a force of 812.5 N to what height will it lift the mass?   [4 m]

4. A load is hoisted 42 m and requires a force of 100 N. The efficiency of the hoist gear is 60% and that of the motor is 70%. Determine the input energy to the hoist.   [10 kJ]

15.3   Power

Power is a measure of the rate at which work is done or at which energy is converted from one form to another.

$$\textbf{Power } \boldsymbol{P} = \dfrac{\textbf{energy used}}{\textbf{time taken}} \text{ or } \boldsymbol{P} = \dfrac{\textbf{work done}}{\textbf{time taken}}$$

The unit of power is the **watt, W**, where 1 watt is equal to 1 joule per second. The watt is a small unit for many purposes and a larger unit called the kilowatt, kW, is used, where 1 kW = 1000 W.

The power output of a motor, which does 120 kJ of work in 30 s, is thus given by

$$P = \dfrac{120\,\text{kJ}}{30\,\text{s}} = 4\,\text{kW}$$

Since work done = force × distance, then

$$\text{Power} = \dfrac{\text{work done}}{\text{time taken}} = \dfrac{\text{force} \times \text{distance}}{\text{time taken}}$$

$$= \text{force} \times \dfrac{\text{distance}}{\text{time taken}}$$

However, $\dfrac{\text{distance}}{\text{time taken}} = \text{velocity}$

Hence,   **power = force × velocity**

Problem 13. The output power of a motor is 8 kW. How much work does it do in 30 s?

$$\text{Power} = \dfrac{\text{work done}}{\text{time taken}}$$

from which, **work done** = power × time = 8000 W × 30 s

$$= 240000\,\text{J} = \textbf{240 kJ}$$

Problem 14. Calculate the power required to lift a mass through a height of 10 m in 20 s if the force required is 3924 N.

Work done = force × distance moved
$$= 3924 \text{ N} \times 10 \text{ m} = 39240 \text{ J}$$

$$\textbf{Power} = \frac{\text{work done}}{\text{time taken}} = \frac{39240 \text{ J}}{20 \text{ s}}$$

$$= \textbf{1962 W or 1.962 kW}$$

Problem 15. 10 kJ of work is done by a force in moving a body uniformly through 125 m in 50 s. Determine (a) the value of the force, and (b) the power.

(a) Work done = force × distance

hence 10000 J = force × 125 m

from which, $\qquad$ **force** $= \dfrac{10000 \text{ J}}{125 \text{ m}} = \textbf{80 N}$

(b) **Power** $= \dfrac{\text{work done}}{\text{time taken}} = \dfrac{10000 \text{ J}}{50 \text{ s}} = \textbf{200 W}$

Problem 16. A car hauls a trailer at 90 km/h when exerting a steady pull of 600 N. Calculate (a) the work done in 30 minutes and (b) the power required.

(a) Work done = force × distance moved.

The distance moved in 30 min, i.e. $\dfrac{1}{2}$ h, at 90 km/h = 45 km.

Hence, **work done** = 600 N × 45000 m

$$= \textbf{27000 kJ or 27 MJ}$$

(b) **Power required** $= \dfrac{\text{work done}}{\text{time taken}} = \dfrac{27 \times 10^6 \text{ J}}{30 \times 60 \text{ s}}$

$$= \textbf{15000 W or 15 kW}$$

Problem 17. To what height will a mass of weight 981 N be raised in 40 s by a machine using a power of 2 kW?

Work done = force × distance. Hence,
work done = 981 N × height.

$\text{Power} = \dfrac{\text{work done}}{\text{time taken}}$, from which,

work done = power × time taken

$$= 2000 \text{ W} \times 40 \text{ s} = 80000 \text{ J}$$

Hence, 80000 = 981 N × height, from which,

$$\textbf{height} = \frac{80000 \text{ J}}{981 \text{ N}} = \textbf{81.55 m}$$

Problem 18. A planing machine has a cutting stroke of 2 m and the stroke takes 4 seconds. If the constant resistance to the cutting tool is 900 N, calculate for each cutting stroke (a) the power consumed at the tool point, and (b) the power input to the system if the efficiency of the system is 75%.

(a) Work done in each cutting stroke
$$= \text{force} \times \text{distance} = 900 \text{ N} \times 2 \text{ m} = 1800 \text{ J}$$

Power consumed at tool point

$$= \frac{\text{work done}}{\text{time taken}} = \frac{1800 \text{ J}}{4 \text{ s}} = \textbf{450 W}$$

(b) Efficiency $= \dfrac{\text{output energy}}{\text{input energy}} = \dfrac{\text{output power}}{\text{input power}}$

Hence, $\dfrac{75}{100} = \dfrac{450}{\text{input power}}$ from which, **input**

$$\textbf{power} = 450 \times \frac{100}{75} = \textbf{600 W}$$

Problem 19. An electric motor provides power to a winding machine. The input power to the motor is 2.5 kW and the overall efficiency is 60%. Calculate (a) the output power of the machine, (b) the rate at which it can raise a 300 kg load vertically upwards.

(a) Efficiency, $\eta = \dfrac{\text{power output}}{\text{power input}}$

i.e. $\dfrac{60}{100} = \dfrac{\text{power output}}{2500}$

from which, **power output** $= \dfrac{60}{100} \times 2500$

$$= \textbf{1500 W or 1.5 kW}$$

(b) Power output = force × velocity, from which,

velocity $= \dfrac{\text{power output}}{\text{force}}$

Force acting on the 300 kg load due to gravity
$$= 300 \text{ kg} \times 9.81 \text{ m/s}^2 = 2943 \text{ N}$$

Hence, **velocity** $= \dfrac{1500}{2943}$

$$= \textbf{0.510 m/s or 510 mm/s}$$

Problem 20.   A lorry is travelling at a constant velocity of 72 km/h. The force resisting motion is 800 N. Calculate the tractive power necessary to keep the lorry moving at this speed.

$$\text{Power} = \text{force} \times \text{velocity}.$$

The force necessary to keep the lorry moving at constant speed is equal and opposite to the force resisting motion, i.e. 800 N

$$\text{Velocity} = 72 \text{ km/h} = \frac{72 \times 1000}{60 \times 60} \text{ m/s} = 20 \text{ m/s}.$$

Hence, power $= 800 \text{ N} \times 20 \text{ m/s}$
$$= 16000 \text{ N m/s} = 16000 \text{ J/s}$$
$$= 16000 \text{ W or } 16 \text{ kW}$$

Thus the tractive power needed to keep the lorry moving at a constant speed of 72 km/h is 16 kW

Problem 21.   The variation of tractive force with distance for a vehicle which is accelerating from rest is:

force (kN)	8.0	7.4	5.8	4.5	3.7	3.0
distance (m)	0	10	20	30	40	50

Determine the average power necessary if the time taken to travel the 50 m from rest is 25 s.

The force/distance diagram is shown in Figure 15.9. The work done is determined from the area under the curve. Using the mid-ordinate rule with five intervals gives:

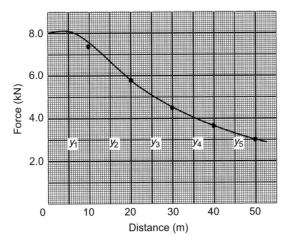

Figure 15.9

area = (width of interval)(sum of mid-ordinate)
$$= (10)[y_1 + y_2 + y_3 + y_4 + y_5]$$
$$= (10)[7.8 + 6.6 + 5.1 + 4.0 + 3.3]$$
$$= (10)[26.8] = 268 \text{ kN m}$$

i.e. work done = 268 kJ

Average power $= \dfrac{\text{work done}}{\text{time taken}} = \dfrac{268000 \text{ J}}{25 \text{ s}}$

$$= \textbf{10720 W or 10.72 kW}$$

Now try the following Practise Exercise

Practise Exercise 80   Further problems on power

1. The output power of a motor is 10 kW. How much work does it do in 1 minute?   [600 kJ]

2. Determine the power required to lift a load through a height of 20 m in 12.5 s if the force required is 2.5 kN.   [4 kW]

3. 25 kJ of work is done by a force in moving an object uniformly through 50 m in 40 s. Calculate (a) the value of the force, and (b) the power.   [(a) 500 N (b) 625 W]

4. A car towing another at 54 km/h exerts a steady pull of 800 N. Determine (a) the work done in $\dfrac{1}{4}$ hr, and (b) the power required.   [(a) 10.8 MJ (b) 12 kW]

5. To what height will a mass of weight 500 N be raised in 20 s by a motor using 4 kW of power?   [160 m]

6. The output power of a motor is 10 kW. Determine (a) the work done by the motor in 2 hours, and (b) the energy used by the motor if it is 72% efficient.
      [(a) 72 MJ (b) 100 MJ]

7. A car is travelling at a constant speed of 81 km/h. The frictional resistance to motion is 0.60 kN. Determine the power required to keep the car moving at this speed.   [13.5 kW]

8. A constant force of 2.0 kN is required to move the table of a shaping machine when

a cut is being made. Determine the power required if the stroke of 1.2 m is completed in 5.0 s. [480 W]

9. The variation of force with distance for a vehicle that is decelerating is as follows:

Distance (m)	600	500	400	300	200	100	0
Force (kN)	24	20	16	12	8	4	0

If the vehicle covers the 600 m in 1.2 minutes, find the power needed to bring the vehicle to rest. [100 kW]

10. A cylindrical bar of steel is turned in a lathe. The tangential cutting force on the tool is 0.5 kN and the cutting speed is 180 mm/s. Determine the power absorbed in cutting the steel. [90 W]

15.4 Potential and kinetic energy

Mechanical engineering is concerned principally with two kinds of energy, potential energy and kinetic energy.

Potential energy is energy due to the position of the body. The force exerted on a mass of m kg is mg N (where $g = 9.81$ m/s^2, the acceleration due to gravity). When the mass is lifted vertically through a height h m above some datum level, the work done is given by:

$$\text{force} \times \text{distance} = (mg)(h) \text{ J}$$

This work done is stored as potential energy in the mass.

Hence, **potential energy = mgh joules**

(the potential energy at the datum level being taken as zero).

Kinetic energy is the energy due to the motion of a body. Suppose a force F acts on an object of mass m originally at rest (i.e. $u = 0$) and accelerates it to a velocity v in a distance s:

$$\text{work done} = \text{force} \times \text{distance}$$
$$= Fs = (ma)(s) \text{ (if no energy is lost)}$$

where a is the acceleration

Since $v^2 = u^2 + 2as$ (see Chapter 12)

and $u = 0$, $v^2 = 2as$, from which $a = \dfrac{v^2}{2s}$

hence, work done $= (ma)(s) = (m)\left(\dfrac{v^2}{2s}\right)(s) = \dfrac{1}{2}mv^2$

This energy is called the kinetic energy of the mass m,

i.e. **kinetic energy $= \dfrac{1}{2}mv^2$ joules**

As stated in Section 15.2, energy may be converted from one form to another. The **principle of conservation of energy** states that the total amount of energy remains the same in such conversions, i.e. energy cannot be created or destroyed.

In mechanics, the potential energy possessed by a body is frequently converted into kinetic energy, and vice versa. When a mass is falling freely, its potential energy decreases as it loses height, and its kinetic energy increases as its velocity increases. Ignoring air frictional losses, at all times:

Potential energy + kinetic energy = a constant

If friction is present, then work is done overcoming the resistance due to friction and this is dissipated as heat. Then,

**Initial energy = final energy
+ work done overcoming frictional resistance**

Kinetic energy is not always conserved in collisions. Collisions in which kinetic energy is conserved (i.e. stays the same) are called **elastic collisions**, and those in which it is not conserved are termed **inelastic collisions.**

Problem 22. A car of mass 800 kg is climbing an incline at 10° to the horizontal. Determine the increase in potential energy of the car as it moves a distance of 50 m up the incline.

With reference to Figure 15.10,

$$\sin 10° = \frac{\text{opposite}}{\text{hypotenuse}} = \frac{h}{50}$$

from which, $h = 50 \sin 10° = 8.682$ m

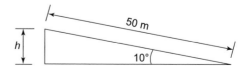

Figure 15.10

Hence, increase in potential energy $= mgh$

$$= 800 \text{ kg} \times 9.81 \text{ m/s}^2 \times 8.682 \text{ m}$$

$$= \mathbf{68140\ J}\ or\ \mathbf{68.14\ kJ}$$

Problem 23.   At the instant of striking, a hammer of mass 30 kg has a velocity of 15 m/s. Determine the kinetic energy in the hammer.

Kinetic energy $= \dfrac{1}{2}mv^2 = \dfrac{1}{2}(30 \text{ kg})(15 \text{ m/s})^2$

i.e. **kinetic energy in hammer** = **3375 J** or **3.375 kJ**

Problem 24.   A lorry having a mass of 1.5 t is travelling along a level road at 72 km/h. When the brakes are applied, the speed decreases to 18 km/h. Determine how much the kinetic energy of the lorry is reduced.

Initial velocity of lorry,

$$v_1 = 72 \text{ km/h} = 72\ \frac{\text{km}}{\text{h}} \times 1000\ \frac{\text{m}}{\text{km}} \times \frac{1\text{h}}{3600\text{s}}$$

$$= \frac{72}{3.6} = 20 \text{ m/s},$$

final velocity of lorry,

$$v_2 = \frac{18}{3.6} = 5 \text{ m/s and mass of lorry,}$$

$$m = 1.5 \text{ t} = 1500 \text{ kg}$$

Initial kinetic energy of the lorry $= \dfrac{1}{2}m\,v_1{}^2$

$$= \frac{1}{2}(1500)(20)^2$$

$$= 300 \text{ kJ}$$

Final kinetic energy of the lorry $= \dfrac{1}{2}m\,v_2{}^2$

$$= \frac{1}{2}(1500)(5)^2$$

$$= 18.75 \text{ kJ}$$

Hence, **the change in kinetic energy** $= 300 - 18.75$

$$= \mathbf{281.25\ kJ}$$

(Part of this reduction in kinetic energy is converted into heat energy in the brakes of the lorry and is hence dissipated in overcoming frictional forces and air friction).

Problem 25.   A canister containing a meteorology balloon of mass 4 kg is fired vertically upwards from a gun with an initial velocity of 400 m/s. Neglecting the air resistance, calculate (a) its initial

kinetic energy (b) its velocity at a height of 1 km (c) the maximum height reached.

(a)   **Initial kinetic energy** $= \dfrac{1}{2}\text{m}v^2$

$$= \frac{1}{2}(4)(400)^2 = \mathbf{320\ kJ}$$

(b)   At a height of 1 km, potential energy
$$= mgh = 4 \times 9.81 \times 1000 = 39.24 \text{ kJ}$$

By the principle of conservation of energy:

potential energy + kinetic energy at 1 km = initial kinetic energy.

Hence $39240 + \dfrac{1}{2}mv^2 = 320000$

from which, $\dfrac{1}{2}(4)v^2 = 320000 - 39240 = 280760$

Hence, $v = \sqrt{\left(\dfrac{2 \times 280760}{4}\right)} = 374.7 \text{ m/s}$

i.e. **the velocity of the canister at a height of 1 km is 374.7 m/s**

(c)   At the maximum height, the velocity of the canister is zero and all the kinetic energy has been converted into potential energy. Hence,

$$\text{potential energy} = \text{initial kinetic energy}$$

$$= 320000 \text{ J (from part (a))}$$

Then, $320000 = mgh = (4)(9.81)(h)$

from which, height $h = \dfrac{320000}{(4)(9.81)}$

$$= 8155 \text{ m}$$

i.e. **the maximum height reached is 8155 m or 8.155 km**

Problem 26.   A pile-driver of mass 500 kg falls freely through a height of 1.5 m on to a pile of mass 200 kg. Determine the velocity with which the driver hits the pile. If, at impact, 3 kJ of energy are lost due to heat and sound, the remaining energy being possessed by the pile and driver as they are driven together into the ground a distance of 200 mm, determine (a) the common velocity immediately after impact (b) the average resistance of the ground.

The potential energy of the pile-driver is converted into kinetic energy.

Thus potential energy = kinetic energy, i.e.

$$mgh = \frac{1}{2}mv^2$$

from which, velocity $v = \sqrt{2gh}$

$$= \sqrt{(2)(9.81)(1.5)} = 5.42 \text{ m/s}.$$

Hence, **the pile-driver hits the pile at a velocity of 5.42 m/s**

(a) Before impact, kinetic energy of pile driver

$$= \frac{1}{2}mv^2 = \frac{1}{2}(500)(5.42)^2 = 7.34 \text{ kJ}$$

Kinetic energy after impact = 7.34 − 3 = 4.34 kJ.
Thus the pile-driver and pile together have a

mass of 500 + 200 = 700 kg and possess kinetic energy of 4.34 kJ

Hence, $4.34 \times 10^3 = \frac{1}{2}mv^2 = \frac{1}{2}(700)v^2$

from which, velocity $v = \sqrt{\left(\dfrac{2 \times 4.34 \times 10^3}{700}\right)}$

$$= 3.52 \text{ m/s}$$

Thus, **the common velocity after impact is 3.52 m/s.**

(b) The kinetic energy after impact is absorbed in overcoming the resistance of the ground, in a distance of 200 mm.

Kinetic energy = work done

$$= \text{resistance} \times \text{distance}$$

i.e. $4.34 \times 10^3 = \text{resistance} \times 0.200$

from which, resistance $= \dfrac{4.34 \times 10^3}{0.200} = 21700 \text{ N}$

Hence, **the average resistance of the ground is 21.7 kN**

Problem 27. A car of mass 600 kg reduces speed from 90 km/h to 54 km/h in 15 s. Determine the braking power required to give this change of speed.

Change in kinetic energy of car $= \frac{1}{2}m\,v_1^2 - \frac{1}{2}m\,v_2^2$

where m = mass of car = 600 kg,

v_1 = initial velocity = 90 km/h

$$= \frac{90}{3.6} \text{ m/s} = 25 \text{ m/s},$$

and v_2 = final velocity = 54 km/h

$$= \frac{54}{3.6} \text{ m/s} = 15 \text{ m/s}.$$

Hence, change in kinetic energy $= \frac{1}{2}m(v_1^2 - v_2^2)$

$$= \frac{1}{2}(600)(25^2 - 15^2)$$

$$= 120000 \text{ J}$$

Braking power $= \dfrac{\text{change in energy}}{\text{time taken}}$

$$= \frac{120000 \text{ J}}{15 \text{ s}} = \mathbf{8000 \text{ W} \text{ or } 8 \text{ kW}}$$

Now try the following Practise Exercise

Practise Exercise 81   Further problems on potential and kinetic energy

(Assume the acceleration due to gravity, $g = 9.81 \text{ m/s}^2$)

1. An object of mass 400 g is thrown vertically upwards and its maximum increase in potential energy is 32.6 J. Determine the maximum height reached, neglecting air resistance.   [8.31 m]

2. A ball bearing of mass 100 g rolls down from the top of a chute of length 400 m inclined at an angle of 30° to the horizontal. Determine the decrease in potential energy of the ball bearing as it reaches the bottom of the chute.   [196.2 J]

3. A vehicle of mass 800 kg is travelling at 54 km/h when its brakes are applied. Find the kinetic energy lost when the car comes to rest.   [90 kJ]

4. A body of mass 15 kg has its speed reduced from 30 km/h to 18 km/h in 4.0 s. Calculate the power required to effect this change of speed.   [83.33 W]

5. Supplies of mass 300 kg are dropped from a helicopter flying at an altitude of 60 m. Determine the potential energy of the supplies relative to the ground at the instant of release, and its kinetic energy as it strikes the ground.   [176.6 kJ, 176.6 kJ]

6. A shell of mass 10 kg is fired vertically upwards with an initial velocity of 200 m/s. Determine its initial kinetic energy and the maximum height reached, correct to the nearest metre, neglecting air resistance.

[200 kJ, 2039 m]

7. The potential energy of a mass is increased by 20.0 kJ when it is lifted vertically through a height of 25.0 m. It is now released and allowed to fall freely. Neglecting air resistance, find its kinetic energy and its velocity after it has fallen 10.0 m. [8 kJ, 14.0 m/s]

8. A pile-driver of mass 400 kg falls freely through a height of 1.2 m on to a pile of mass 150 kg. Determine the velocity with which the driver hits the pile. If, at impact, 2.5 kJ of energy are lost due to heat and sound, the remaining energy being possessed by the pile and driver as they are driven together into the ground a distance of 150 mm, determine (a) the common velocity after impact (b) the average resistance of the ground.

[4.85 m/s (a) 2.83 m/s (b) 14.70 kN]

15.5 Kinetic energy of rotation

When **linear motion** takes place,

kinetic energy $= \sum \dfrac{\Delta m}{2} v^2$

but when **rotational motion** takes place,

kinetic energy $= \dfrac{1}{2} \sum \Delta m (\omega r)^2$

Since ω is a constant, kinetic energy $= \omega^2 \dfrac{1}{2} \sum \Delta m \, r^2$

But $\sum \Delta m \, r^2 = I$

Therefore, **kinetic energy (in rotation)**

$$= \dfrac{1}{2} I \omega^2 \text{ joules}$$

where $I =$ the mass moment of inertia about the point of rotation

and $\omega =$ angular velocity.

Problem 28. Calculate the kinetic energy of a solid flat disc of diameter 0.5 m and of a uniform thickness of 0.1 m, rotating about its centre at 40 rpm. Take the density of the material as 7860 kg/m³.

Angular velocity, $\omega = 2\pi \dfrac{\text{rad}}{\text{rev}} \times 40 \dfrac{\text{rev}}{\text{min}} \times \dfrac{1 \min}{60 \text{ s}}$

$= 4.189$ rad/s

From Table 14.1, page 168,

$$I = \rho \times \pi R^2 \times t \times \dfrac{R^2}{2}$$

$$= 7860 \, \dfrac{\text{kg}}{\text{m}^3} \times \pi \times 0.25^2 \text{ m}^2 \times 0.1 \text{ m} \times \dfrac{0.25^2 \text{ m}^2}{2}$$

i.e. $I = 4.823$ kg m²

Hence,

kinetic energy $= \dfrac{1}{2} I \omega^2$

$$= \dfrac{1}{2} \times 4.823 \, \text{kg m}^2 \times (4.189)^2 \dfrac{1}{\text{s}^2}$$

$$= \textbf{42.32 J}$$

Now try the following Practise Exercises

Practise Exercise 82 Further problems on kinetic energy in rotation

1. Calculate the kinetic energy of a solid flat disc of diameter 0.6 m and of uniform thickness of 0.1 m rotating about its centre at 50 rpm. Take the density of the disc material as 7860 kg/m³. [137.1 J]

2. If the disc of Problem 1 had a hole in its centre of 0.2 m diameter, what would be its kinetic energy? [135.4 J]

3. If an annulus of external diameter 0.4 m and internal diameter 0.2 m were rotated about its centre at 100 rpm, what would be its kinetic energy? Assume the uniform thickness of the annulus is 0.08 m and the density of the material is 7860 kg/m³.

[81.2 J]

Practise Exercise 83 Short-answer questions on work, energy and power

1. Define work in terms of force applied and distance moved.

2. Define energy, and state its unit.

Part Three

3. Define the joule.

4. The area under a force/distance graph represents ....

5. Name five forms of energy.

6. State the principle of conservation of energy.

7. Give two examples of conversion of heat energy to other forms of energy.

8. Give two examples of conversion of electrical energy to other forms of energy.

9. Give two examples of conversion of chemical energy to other forms of energy.

10. Give two examples of conversion of mechanical energy to other forms of energy.

11. (a) Define efficiency in terms of energy input and energy output.
    (b) State the symbol used for efficiency.

12. Define power and state its unit.

13. Define potential energy.

14. The change in potential energy of a body of mass m kg when lifted vertically upwards to a height h m is given by ....

15. What is kinetic energy?

16. The kinetic energy of a body of mass m kg and moving at a velocity of v m/s is given by ....

17. State the principle of conservation of energy.

18. Distinguish between elastic and inelastic collisions.

19. The kinetic energy of rotation of a body of moment of inertia I kg m^2 and moving at an angular velocity of ω rad/s is given by ....

Practise Exercise 84 **Multiple-choice questions on work, energy and power**

(Answers on page 298)

1. State which of the following is incorrect:
   (a) 1 W = 1 J/s
   (b) 1 J = 1 N/m

(c) $\eta = \dfrac{\text{output energy}}{\text{input energy}}$

(d) energy = power × time

2. An object is lifted 2000 mm by a crane. If the force required is 100 N, the work done is:
   (a) $\dfrac{1}{20}$ N m   (b) 200 kN m
   (c) 200 N m   (d) 20 J

3. A motor having an efficiency of 0.8 uses 800 J of electrical energy. The output energy of the motor is:
   (a) 800 J   (b) 1000 J
   (c) 640 J   (d) 6.4 J

4. 6 kJ of work is done by a force in moving an object uniformly through 120 m in 1 minute. The force applied is:
   (a) 50 N   (b) 20 N
   (c) 720 N   (d) 12 N

5. For the object in question 4, the power developed is:
   (a) 6 kW   (b) 12 kW
   (c) 5/6 W   (d) 0.1 kW

6. Which of the following statements is false?
   (a) The unit of energy and work is the same.
   (b) The area under a force/distance graph gives the work done.
   (c) Electrical energy is converted to mechanical energy by a generator.
   (d) Efficiency is the ratio of the useful output energy to the input energy

7. A machine using a power of 1 kW requires a force of 100 N to raise a mass in 10 s. The height the mass is raised in this time is:
   (a) 100 m   (b) 1 km
   (c) 10 m   (d) 1 m

8. A force/extension graph for a spring is shown in Figure 15.11. Which of the following statements is false?

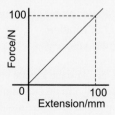

Figure 15.11

The work done in extending the spring:
(a) from 0 to 100 mm is 5 J
(b) from 0 to 50 mm is 1.25 J
(c) from 20 mm to 60 mm is 1.6 J
(d) from 60 mm to 100 mm is 3.75 J

9. A vehicle of mass 1 tonne climbs an incline of 30° to the horizontal. Taking the acceleration due to gravity as 10 m/s^2, the increase in potential energy of the vehicle as it moves a distance of 200 m up the incline is:
(a) 1 kJ   (b) 2 MJ
(c) 1 MJ   (d) 2 kJ

10. A bullet of mass 100 g is fired from a gun with an initial velocity of 360 km/h. Neglecting air resistance, the initial kinetic energy possessed by the bullet is:
(a) 6.48 kJ   (b) 500 J
(c) 500 kJ   (d) 6.48 MJ

11. A small motor requires 50 W of electrical power in order to produce 40 W of mechanical energy output. The efficiency of the motor is:
(a) 10%   (b) 80%
(c) 40%   (d) 90%

12. A load is lifted 4000 mm by a crane. If the force required to lift the mass is 100 N, the work done is:
(a) 400 J   (b) 40 N m
(c) 25 J   (d) 400 kJ

13. A machine exerts a force of 100 N in lifting a mass through a height of 5 m. If 1 kJ of energy is supplied, the efficiency of the machine is:
(a) 10%   (b) 20%
(c) 100%   (d) 50%

14. At the instant of striking an object, a hammer of mass 40 kg has a velocity of 10 m/s. The kinetic energy in the hammer is:
(a) 2 kJ   (b) 1 kJ
(c) 400 J   (d) 8 kJ

15. A machine which has an efficiency of 80% raises a load of 50 N through a vertical height of 10 m. The work input to the machine is:
(a) 400 J   (b) 500 J
(c) 800 J   (d) 625 J

16. The formula for kinetic energy due to rotation is:
(a) mv^2 (b) mgh
(c) $I\dfrac{\omega^2}{2}$ (d) $\omega^2 r$

Revision Test 5   Linear and angular motion, momentum and impulse, force, mass and acceleration, work, energy and power

This Revision Test covers the material contained in Chapters 12 to 15. *The marks for each question are shown in brackets at the end of each question.*

Assume, where necessary, that the acceleration due to gravity, $g = 9.81$ m/s^2

1. A train is travelling at 90 km/h and has wheels of diameter 1600 mm.

   (a) Find the angular velocity of the wheels in both rad/s and rev/min.

   (b) If the speed remains constant for 2 km, determine the number of revolutions made by a wheel, assuming no slipping occurs.   (7)

2. The speed of a shaft increases uniformly from 200 revolutions per minute to 700 revolutions per minute in 12 s. Find the angular acceleration, correct to 3 significant figures.   (5)

3. The shaft of an electric motor, initially at rest, accelerates uniformly for 0.3 *s* at 20 rad/s^2. Determine the angle (in radians) turned through by the shaft in this time.   (4)

4. Determine the momentum of a lorry of mass 10 tonnes moving at a velocity of 81 km/h.   (4)

5. A ball of mass 50 g is moving with a velocity of 4 m/s when it strikes a stationary ball of mass 25 g. The velocity of the 50 g ball after impact is 2.5 m/s in the same direction as before impact. Determine the velocity of the 25 g ball after impact.   (7)

6. A force of 24 N acts on a body of mass 6 kg for 150 ms. Determine the change in velocity.   (4)

7. The hammer of a pile-driver of mass 800 kg falls a distance of 1.0 m on to a pile. The blow takes place in 20 ms and the hammer does not rebound. Determine (a) the velocity of impact (b) the momentum lost by the hammer (c) the average applied force exerted on the pile by the hammer.   (8)

8. Determine the mass of the moving head of a machine tool if it requires a force of 1.5 N to bring it to rest in 0.75 s from a cutting speed of 25 m/min.   (5)

9. Find the weight of an object of mass 2.5 kg at a point on the earth's surface where the gravitational field is 9.8 N/kg.   (4)

10. A van of mass 1200 kg travels round a bend of radius 120 m, at 54 km/h. Determine the centripetal force acting on the vehicle.   (4)

11. A spring, initially in a relaxed state, is extended by 80 mm. Determine the work done by using a work diagram if the spring requires a force of 0.7 N per mm of stretch.   (4)

12. Water is pumped vertically upwards through a distance of 40.0 m and the work done is 176.58 kJ. Determine the number of litres of water pumped. (1 litre of water has a mass of 1 kg).   (4)

13. 3 kJ of energy are supplied to a machine used for lifting a mass. The force required is 1 kN. If the machine has an efficiency of 60%, to what height will it lift the mass?   (4)

14. When exerting a steady pull of 450 N, a lorry travels at 80 km/h. Calculate (a) the work done in 15 minutes and (b) the power required.   (4)

15. An electric motor provides power to a winding machine. The input power to the motor is 4.0 kW and the overall efficiency is 75%. Calculate (a) the output power of the machine (b) the rate at which it can raise a 509.7 kg load vertically upwards   (4)

16. A tank of mass 4800 kg is climbing an incline at 12° to the horizontal. Determine the increase in potential energy of the tank as it moves a distance of 40 m up the incline.   (4)

17. A car of mass 500 kg reduces speed from 108 km/h to 36 km/h in 20 s. Determine the braking power required to give this change of speed.   (4)

Chapter 16

Friction

When an object is placed on a flat surface and sufficient force is applied to the block, the force being parallel to the surface, the block slides across the surface. When the force is removed, motion of the block stops; thus there is a force which resists sliding. In this chapter, both dynamic and static frictions are explained, together with the factors that affect the size and direction of frictional forces. A low coefficient of friction is desirable in bearings, pistons moving within cylinders and on ski runs; however, for a force being transmitted by belt drives and braking systems, a high value of coefficient is necessary. Advantages and disadvantages of frictional forces are discussed and calculations are performed on friction on an inclined plane and screw jack efficiency.

At the end of this chapter you should be able to:

- understand dynamic or sliding friction
- appreciate factors which affect the size and direction of frictional forces
- define coefficient of friction, μ
- perform calculations involving $F = \mu N$
- state practical applications of friction
- state advantages and disadvantages of frictional forces
- understand friction on an inclined plane
- perform calculations on friction on an inclined plane
- calculate the efficiency of a screw jack

16.1 Introduction to friction

When an object, such as a block of wood, is placed on a floor and sufficient force is applied to the block, the force being parallel to the floor, the block slides across the floor. When the force is removed, motion of the block stops; thus there is a force which resists sliding. This force is called **dynamic** or **sliding friction**. A force may be applied to the block, which is insufficient to move it. In this case, the force resisting motion is called the **static friction** or **stiction**. Thus there are two categories into which a frictional force may be split:

(i) dynamic or sliding friction force which occurs when motion is taking place, and

(ii) static friction force which occurs before motion takes place.

There are three factors that affect the size and direction of frictional forces.

(i) The size of the frictional force depends on the type of surface (a block of wood slides more easily on a polished metal surface than on a rough concrete surface).

(ii) The size of the frictional force depends on the size of the force acting at right angles to the surfaces in contact, called the **normal force**; thus, if the weight of a block of wood is doubled, the frictional force is doubled when it is sliding on the same surface.

(iii) The direction of the frictional force is always opposite to the direction of motion. Thus the frictional force opposes motion, as shown in Figure 16.1.

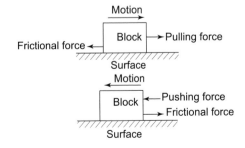

Figure 16.1

Mechanical Engineering Principles, Bird and Ross, ISBN 9780415517850

16.2 Coefficient of friction

The **coefficient of friction, μ,** is a measure of the amount of friction existing between two surfaces. A low value of coefficient of friction indicates that the force required for sliding to occur is less than the force required when the coefficient of friction is high. The value of the coefficient of friction is

given by: $\mu = \dfrac{\text{frictional force}(F)}{\text{normal force}(N)}$

Transposing gives: frictional force = $\mu \times$ normal force

i.e. $$F = \mu N$$

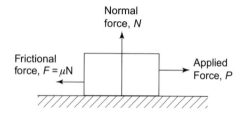

Figure 16.2

The direction of the forces given in this equation is as shown in Figure 16.2.

The coefficient of friction is the ratio of a force to a force, and hence has no units. Typical values for the coefficient of friction when sliding is occurring, i.e. the dynamic coefficient of friction, are:

For polished oiled metal surfaces	less than 0.1
For glass on glass	0.4
For rubber on tarmac	close to 1.0

The coefficient of friction (μ) for dynamic friction is, in general, a little less than that for static friction. However, for dynamic friction, μ increases with speed; additionally, it is dependent on the area of the surface in contact.

Problem 1. A block of steel requires a force of 10.4 N applied parallel to a steel plate to keep it moving with constant velocity across the plate. If the normal force between the block and the plate is 40 N, determine the dynamic coefficient of friction.

As the block is moving at constant velocity, the force applied must be that required to overcome frictional forces, i.e. frictional force, $F = 10.4$ N;

the normal force is 40 N, and since $F = \mu N$,

$$\mu = \frac{F}{N} = \frac{10.4}{40}$$
$$= 0.26$$

i.e. **the dynamic coefficient of friction is 0.26**

Problem 2. The surface between the steel block and plate of Problem 1 is now lubricated and the dynamic coefficient of friction falls to 0.12. Find the new value of force required to push the block at a constant speed.

The normal force depends on the weight of the block and remains unaltered at 40 N. The new value of the dynamic coefficient of friction is 0.12 and since the frictional force $F = \mu N$,

$$F = 0.12 \times 40 = 4.8 \text{ N}$$

The block is sliding at constant speed, thus the force required to overcome the frictional force is also 4.8 N, i.e. **the required applied force is 4.8 N**

Problem 3. The material of a brake is being tested and it is found that the dynamic coefficient of friction between the material and steel is 0.91. Calculate the normal force when the frictional force is 0.728 kN.

The dynamic coefficient of friction, $\mu = 0.91$ and the frictional force, $F = 0.728$ kN $= 728$ N

Since $F = \mu N,$

then normal force, $N = \dfrac{F}{\mu} = \dfrac{728}{0.91} = 800$ N

i.e. **the normal force is 800 N**

Now try the following Practise Exercise

Practise Exercise 85   Further problems on the coefficient of friction

1. The coefficient of friction of a brake pad and a steel disc is 0.82. Determine the normal force between the pad and the disc if the frictional force required is 1025 N.

    [1250 N]

2. A force of 0.12 kN is needed to push a bale of cloth along a chute at a constant speed. If the normal force between the bale and

the chute is 500 N, determine the dynamic coefficient of friction. [0.24]

3. The normal force between a belt and its driver wheel is 750 N. If the static coefficient of friction is 0.9 and the dynamic coefficient of friction is 0.87, calculate
   (a) the maximum force which can be transmitted, and
   (b) maximum force which can be transmitted when the belt is running at a constant speed. [(a) 675 N (b) 652.5 N]

16.3 Applications of friction

In some applications, a low coefficient of friction is desirable, for example, in bearings, pistons moving within cylinders, on ski runs, and so on. However, for such applications as force being transmitted by belt drives and braking systems, a high value of coefficient is necessary.

Problem 4. State three advantages, and three disadvantages of frictional forces.

Instances where frictional forces are an advantage include:

(i) Almost all fastening devices rely on frictional forces to keep them in place once secured, examples being screws, nails, nuts, clips and clamps.

(ii) Satisfactory operation of brakes and clutches rely on frictional forces being present.

(iii) In the absence of frictional forces, most accelerations along a horizontal surface are impossible; for example, a person's shoes just slip when walking is attempted and the tyres of a car just rotate with no forward motion of the car being experienced.

Disadvantages of frictional forces include:

(i) Energy is wasted in the bearings associated with shafts, axles and gears due to heat being generated.

(ii) Wear is caused by friction, for example, in shoes, brake lining materials and bearings.

(iii) Energy is wasted when motion through air occurs (it is much easier to cycle with the wind rather than against it).

Problem 5. Discuss briefly two design implications that arise due to frictional forces and how lubrication may or may not help.

(i) Bearings are made of an alloy called white metal, which has a relatively low melting point. When the rotating shaft rubs on the white metal bearing, heat is generated by friction, often in one spot and the white metal may melt in this area, rendering the bearing useless. Adequate lubrication (oil or grease) separates the shaft from the white metal, keeps the coefficient of friction small and prevents damage to the bearing. For very large bearings, oil is pumped under pressure into the bearing and the oil is used to remove the heat generated, often passing through oil coolers before being re-circulated. Designers should ensure that the heat generated by friction can be dissipated.

(ii) Wheels driving belts, to transmit force from one place to another, are used in many workshops. The coefficient of friction between the wheel and the belt must be high, and it may be increased by dressing the belt with a tar-like substance. Since frictional force is proportional to the normal force, a slipping belt is made more efficient by tightening it, thus increasing the normal and hence the frictional force. Designers should incorporate some belt tension mechanism into the design of such a system.

Problem 6. Explain what is meant by the terms (a) the limiting or static coefficient of friction, and (b) the sliding or dynamic coefficient of friction.

(a) When an object is placed on a surface and a force is applied to it in a direction parallel to the surface, if no movement takes place, then the applied force is balanced exactly by the frictional force. As the size of the applied force is increased, a value is reached such that the object is just on the point of moving. The limiting or static coefficient of friction is given by the ratio of this applied force to the normal force, where the normal force is the force acting at right angles to the surfaces in contact.

(b) Once the applied force is sufficient to overcome the stiction or static friction, its value can be reduced slightly and the object moves across the surface. A particular value of the applied force is then sufficient to keep the object moving at a constant velocity. The sliding or dynamic coefficient of friction is the ratio of the applied force, to maintain constant velocity, to the normal force.

16.4 Friction on an inclined plane

Angle of repose

Consider a mass m lying on an inclined plane, as shown in Figure 16.3. If the direction of motion of this mass is down the plane, then the frictional force F will act up the plane, as shown in Figure 16.3, where $F = \mu mg \cos \theta$.

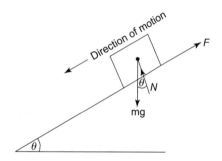

Figure 16.3

Now the weight of the mass is mg and this will cause two other forces to act on the mass, namely N, and the component of the weight down the plane, namely mg sin θ, as shown by the vector diagram of Figure 16.4.

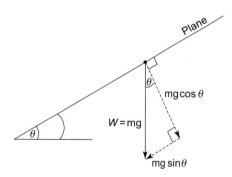

Figure 16.4 Components of mg

It should be noted that N acts normal to the surface.
Resolving forces parallel to the plane gives:

Forces up the plane = forces down the plane

i.e.
$$F = mg \sin \theta \quad (16.1)$$

Resolving force perpendicular to the plane gives:

Forces 'up' = forces 'down'

i.e.
$$N = mg \cos \theta \quad (16.2)$$

Dividing equation (16.1) by (16.2) gives:

$$\frac{F}{N} = \frac{mg \sin \theta}{mg \cos \theta} = \frac{\sin \theta}{\cos \theta} = \tan \theta$$

But
$$\frac{F}{N} = \mu, \text{ hence, } \mathbf{\tan \theta = \mu}$$

where μ = the coefficient of friction, and θ = the **angle of repose.**

If θ is gradually increased until the body starts motion down the plane, then this value of θ is called the **limiting angle of repose**. A laboratory experiment based on the theory is a useful method of obtaining the maximum value of μ for static friction.

16.5 Motion up a plane with the pulling force P parallel to the plane

In this case the frictional force F acts down the plane, opposite to the direction of motion of the body, as shown in Figure 16.5.

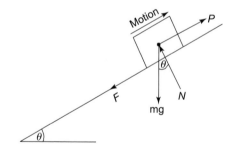

Figure 16.5

The components of the weight mg will be the same as that shown in Figure 16.4.
Resolving forces parallel to the plane gives:

$$P = mg \sin \theta + F \quad (16.3)$$

Resolving forces perpendicular to the plane gives:

$$N = mg \cos \theta \quad (16.4)$$

For limiting friction,

$$F = \mu N \quad (16.5)$$

From equations (16.3) to (16.5), solutions of problems in this category that involve limiting friction can be solved.

> **Problem 7.**   Determine the value of the force P, which will just move the body of mass of 25 kg up the plane shown in Figure 16.6. It may be assumed that the coefficient of limiting friction, $\mu = 0.3$ and $g = 9.81$ m/s².

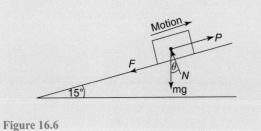

Figure 16.6

From equation (16.4), $N = mg \cos \theta = 25 \times 9.81 \times \cos 15°$

$= 245.3 \times 0.966 = \textbf{236.9 N}$

From equation (16.5), $F = \mu N = 0.3 \times 236.9$

$= 71.1$ N

From equation (16.3), $P = mg \sin \theta + F$

$= 25 \times 9.81 \times \sin 15° + 71.1$

$= 63.48 + 71.1$

i.e.   **force, $P = 134.6$ N**   (16.6)

16.6   Motion down a plane with the pulling force P parallel to the plane

In this case, the frictional force F acts up the plane, opposite to the direction of motion of the plane, as shown in Figure 16.7.

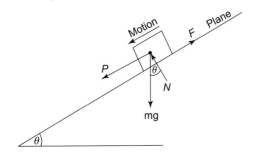

Figure 16.7

The components of the weight mg are shown in Figure 16.4, where it can be seen that the normal reaction,

$N = mg \cos \theta$, and the component of weight parallel to and down the plane $= mg \sin \theta$

Resolving forces perpendicular to the plane gives:

$$N = mg \cos \theta \qquad (16.7)$$

Resolving forces parallel to the plane gives:

$$P + mg \sin \theta = F \qquad (16.8)$$

When the friction is limiting,

$$F = \mu N \qquad (16.9)$$

From equations (16.7) to (16.9), problems arising in this category can be solved.

> **Problem 8.**   If the mass of Problem 7 were subjected to the force P, which acts parallel to and down the plane, as shown in Figure 16.7, determine the value of P to just move the body.

From equation (16.7),

$N = mg \cos \theta = 25 \times 9.81 \cos 15° = \textbf{236.9 N}$

From equation (16.9), $F = \mu N = 0.3 \times 236.9 = \textbf{71.1 N}$

From equation (16.8), $P + mg \sin \theta = F$

i.e. $P + 25 \times 9.81 \sin 15° = 71.1$

i.e. $P + 63.5 = 71.1$

from which, **force,**

$$P = 71.1 - 63.5 = \textbf{7.6 N} \qquad (16.10)$$

From equations (16.6) and (16.10), it can be seen that the force required to move a body down the plane is so much smaller than to move the body up the plane.

16.7   Motion up a plane due to a horizontal force P

This motion, together with the primary forces, is shown in Figure 16.8.

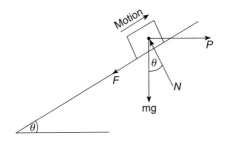

Figure 16.8

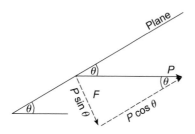

Figure 16.9

In this, the components of mg are as shown in Figure 16.4, and the components of the horizontal force P are shown by the vector diagram of Figure 16.9. Resolving perpendicular to the plane gives:

Forces 'up' = forces 'down'

i.e. $N = mg \cos \theta + P \sin \theta$ (16.11)

Resolving parallel to the plane gives:

$P \cos \theta = F + mg \sin \theta$ (16.12)

and $F = \mu N$ (16.13)

From equations (16.11) to (16.13), problems arising in this category can be solved.

Problem 9.   If the mass of Problem 7 were subjected to a horizontal force P, as shown in Figure 16.8, determine the value of P that will just cause motion up the plane.

Substituting equation (16.13) into equation (16.12) gives:

$P \cos \theta = \mu N + mg \sin \theta$

or $\mu N = P \cos \theta - mg \sin \theta$

i.e. $N = \dfrac{P \cos \theta}{\mu} - \dfrac{mg \sin \theta}{\mu}$ (16.14)

Equating equation (16.11) and equation (16.14) gives:

$mg \cos \theta + P \sin \theta = \dfrac{P \cos \theta}{\mu} - \dfrac{mg \sin \theta}{\mu}$

i.e. $25 \times 9.81 \cos 15° + P \sin 15°$

$= \dfrac{P \cos 15°}{0.3} - \dfrac{25 \times 9.81 \sin 15°}{0.3}$

$245.3 \times 0.966 + P \times 0.259 = \dfrac{P \times 0.966}{0.3} - \dfrac{245.3 \times 0.259}{0.3}$

i.e. $237 + 0.259 P = 3.22 P - 211.8$

$237 + 211.8 = 3.22 P - 0.259 P$

from which, $448.8 = 2.961 P$

and   **force $P = \dfrac{448.8}{2.961} = $ 151.6 N**

Problem 10.   If the mass of Problem 9 were subjected to a horizontal force P, acting down the plane, as shown in Figure 16.10, determine the value of P which will just cause motion down the plane.

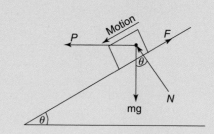

Figure 16.10

The components for mg are shown by the phasor diagram of Figure 16.4, and the components for P are shown by the vector diagram of Figure 16.11.

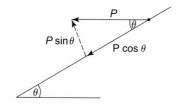

Figure 16.11

Resolving forces down the plane gives:

$P \cos \theta + mg \sin \theta = F$ (16.15)

Resolving forces perpendicular to the plane gives:

Forces up = forces down

$N + P \sin \theta = mg \cos \theta$ (16.16)

and $F = \mu N$ (16.17)

Substituting equation (16.17) into equation (16.15) gives:

$P \cos \theta + mg \sin \theta = \mu N$

from which, $N = \dfrac{P \cos \theta}{\mu} + \dfrac{mg \sin \theta}{\mu}$ (16.18)

From equation (16.16), $N = mg \cos \theta - P \sin \theta$ (16.19)

Equating equations (16.18) and (16.19) gives:

$\dfrac{P \cos \theta}{\mu} + \dfrac{mg \sin \theta}{\mu} = mg \cos \theta - P \sin \theta$

i.e. $\dfrac{P\cos 15°}{0.3} + \dfrac{25 \times 9.81 \sin 15°}{0.3}$

$= 25 \times 9.81 \cos 15° - P \sin 15°$

$3.22\,P + 211.6 = 236.9 - 0.259\,P$

$P(3.22 + 0.259) = 236.9 - 211.6$

$3.479\,P = 25.3$

from which, **force** $P = \dfrac{25.3}{3.479} = \textbf{7.27 N}$

Problem 11. If in Problem 9, the contact surfaces were greased, so that the value of μ decreased and $P = 50$ N, determine the value of μ which will just cause motion down the plane.

The primary forces for this problem are shown in Figure 16.12, where it can be seen that F is opposite to the direction of motion.

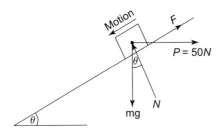

Figure 16.12

Resolving forces perpendicular to the plane gives:

Forces 'up' = forces 'down'

$$N = mg \cos \theta + P \sin \theta \qquad (16.20)$$

Resolving forces parallel to the plane gives:

$$mg \sin \theta = F + P \cos \theta \qquad (16.21)$$

and $$F = \mu N \qquad (16.22)$$

Substituting equation (16.22) into equation (16.21) gives:

$$mg \sin \theta = \mu N + P \cos \theta \qquad (16.23)$$

Substituting equation (16.20) into equation (16.23) gives:

$$mg \sin \theta = \mu(mg \cos \theta + P \sin \theta) + P \cos \theta$$

i.e. $25 \times 9.81 \sin 15° = \mu(25 \times 9.81 \cos 15° + 50 \sin 15°) + 50 \cos 15°$

Hence $63.48 = \mu(236.89 + 12.94) + 48.3$

$63.48 - 48.3 = \mu \times 249.83$

from which, $\mu = \dfrac{15.18}{249.83} = \textbf{0.061}$

Now try the following Practise Exercise

Practise Exercise 86   Further problems on friction on an inclined plane

(Where necessary, take $g = 9.81$ m/s^2)

1. A mass of 40 kg rests on a flat horizontal surface as shown in Figure 16.13. If the coefficient of friction $\mu = 0.2$, determine the minimum value of a horizontal force P which will just cause it to move.
    [78.48 N]

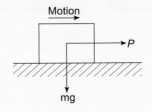

Figure 16.13

2. If the mass of Problem 1 were equal to 50 kg, what will be the value of P?
    [98.1 N]

3. An experiment is required to obtain the static value of μ; this is achieved by increasing the value of θ until the mass just moves down the plane, as shown in Figure 16.14. If the experimentally obtained value for θ was 22.5°, what is the value of μ?
    [$\mu = 0.414$]

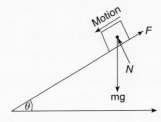

Figure 16.14

4. If in Problem 3, μ was 0.6, what would be the experimental value of θ?   [$\theta = 30.96°$]

5. For a mass of 50 kg just moving up an inclined plane, as shown in Figure 16.5, what would be the value of P, given that $\theta = 20°$ and $\mu = 0.4$?   [$P = 352.1$ N]

6. For a mass of 50 kg, just moving down an inclined plane, as shown in Figure 16.7, what would be the value of P, given that $\theta = 20°$ and $\mu = 0.4$? [$P = 16.6$ N]

7. If in Problem 5, $\theta = 10°$ and $\mu = 0.5$, what would be the value of P? [$P = 326.7$ N]

8. If in Problem 6, $\theta = 10°$ and $\mu = 0.5$, what would be the value of P? [$P = 156.3$ N]

9. Determine P for Problem 5, if it were acting in the direction shown in Figure 16.8. [$P = 438.6$ N]

10. Determine P for Problem 6, if it were acting in the direction shown in Figure 16.10. [$P = 15.43$ N]

11. Determine the value for θ which will just cause motion down the plane, when $P = 250$ N and acts in the direction shown in Figure 16.12. It should be noted that in this problem, motion is down the plane, $\mu = 0.5$ and $m = 50$ kg. [$\theta = 53.58°$]

12. If in Problem 11, $\theta = 30°$, determine the value of μ. [$\mu = 0.052$]

16.8   The efficiency of a screw jack

Screw jacks (see Section 19.4, page 216) are often used to lift weights; one of their most common uses are to raise cars, so that their wheels can be changed. The theory described in Section 16.7 can be used to analyse screw jacks.

Consider the thread of the square-threaded screw jack shown in Figure 16.15.

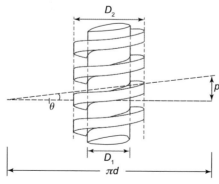

Figure 16.15

Let p be the pitch of the thread, i.e. the axial distance that the weight W is lifted or lowered when the screw is turned through one complete revolution.

From Figure 16.15, the motion of the screw in lifting the weight can be regarded as pulling the weight by a horizontal force P, up an incline θ, where

$$\tan \theta = \frac{p}{\pi d} \qquad \text{as shown in Figure 16.15,}$$

and

$$d = \frac{(D_1 + D_2)}{2}$$

If μ is the coefficient of friction up the slope, then let $\tan \lambda = \mu$

Referring now to Figure 16.16, the screw jack can be analysed.

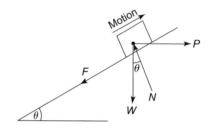

Figure 16.16

Resolving normal to the plane gives:

$$N = W \cos \theta + P \sin \theta \qquad (16.24)$$

Resolving parallel to the plane gives:

$$P \cos \theta = F + W \sin \theta \qquad (16.25)$$

and

$$F = \mu N \qquad (16.26)$$

Substituting equation (16.26) into equation (16.25) gives:

$$P \cos \theta = \mu N + W \sin \theta \qquad (16.27)$$

Substituting equation (16.24) into equation (16.27) gives:

$$P \cos \theta = \mu(W \cos \theta + P \sin \theta) + W \sin \theta$$

Dividing each term by $\cos \theta$ and remembering that $\dfrac{\sin \theta}{\cos \theta} = \tan \theta$ gives:

$$P = \mu(W + P \tan \theta) + W \tan \theta$$

Rearranging gives:

$$P(1 - \mu \tan \theta) = W(\mu + \tan \theta)$$

from which, $P = \dfrac{W(\mu + \tan\theta)}{(1 - \mu\tan\theta)} = \dfrac{W(\tan\lambda + \tan\theta)}{(1 - \tan\lambda\tan\theta)}$

since $\mu = \tan\lambda$

However, from compound angle formulae (see reference [1] on page 195),

$$\tan(\lambda + \theta) = \frac{(\tan\lambda + \tan\theta)}{(1 - \tan\lambda\tan\theta)}$$

Hence, $P = W\tan(\theta + \lambda)$ (16.28)

However, from Figure 16.15,

$$\tan\theta = \frac{p}{\pi d}\ \text{and}\ \tan\lambda = \mu$$

hence $P = \dfrac{W(\tan\lambda + \tan\theta)}{1 - \tan\lambda\tan\theta} = \dfrac{W\left(\mu + \dfrac{p}{\pi d}\right)}{\left(1 - \dfrac{\mu p}{\pi d}\right)}$ (16.29)

Multiplying top and bottom of equation (16.29) by πd gives:

$$P = \frac{W(\mu\pi d + p)}{(\pi d - \mu p)}$$ (16.30)

The **useful work done** in lifting the weight W a distance of $p = Wp$ (16.31)

From Figure 16.15, **the actual work done** = $P \times \pi d$

$$= \frac{W(\mu\pi d + p)}{(\pi d - \mu p)} \times \pi d$$ (16.32)

Efficiency $\eta = \dfrac{\text{useful work done}}{\text{actual work done}}$ which is usually expressed as a percentage

i.e. $\eta = \dfrac{Wp}{\dfrac{W(\mu\pi d + p) \times \pi d}{(\pi d - \mu p)}} = \dfrac{p(\pi d - \mu p)}{(\mu\pi d + p) \times \pi d}$

Dividing throughout by πd gives:

$$\eta = \frac{p\left(1 - \dfrac{\mu p}{\pi d}\right)}{(\mu\pi d + p)} = \frac{p(1 - \tan\lambda\tan\theta)}{\pi d\left(\mu + \dfrac{p}{\pi d}\right)}$$

$$= \frac{p(1 - \tan\lambda\tan\theta)}{\pi d(\tan\lambda + \tan\theta)}$$

However, $\tan(\lambda + \theta) = \dfrac{\tan\lambda + \tan\theta}{(1 - \tan\lambda\tan\theta)}$ from compound angle formulae (see reference [1], page 195)

Hence, $\eta = \dfrac{p}{\pi d}\dfrac{1}{\tan(\lambda + \theta)}$

but $\dfrac{p}{\pi d} = \tan\theta$

hence,   **efficiency, $\eta = \dfrac{\tan\theta}{\tan(\lambda + \theta)}$**   (16.33)

From equations (16.31) and (16.32),

the **work lost in friction**

$$= \frac{W(\mu\pi d + p)}{(\pi d - \mu p)} \times \pi d - Wp$$ (16.34)

Problem 12.   The coefficient of friction on the sliding surface of a screw jack is 0.2. If the pitch equals 1 cm, and $D_1 = 4$ cm and $D_2 = 5$ cm, calculate the efficiency of the screw jack.

Working in millimetres, $d = \dfrac{(D_1 + D_2)}{2} = \dfrac{(40 + 50)}{2}$

$$= 45\ \text{mm},$$

$$p = 1\ \text{cm} = 10\ \text{mm},$$

$$\tan\theta = \frac{p}{\pi d} = \frac{10}{\pi \times 45} = 0.0707,$$

from which, $\theta = \tan^{-1}(0.0707) = 4.05°$

and $\tan\lambda = \mu = 0.2,$

from which, $\lambda = \tan^{-1}(0.2) = 11.31°$

From equation (16.33),

efficiency $\eta = \dfrac{\tan\theta}{\tan(\lambda + \theta)}$

$$= \frac{0.0707}{\tan(11.31 + 4.05)°} = \frac{0.0707}{0.2747}$$

$$= 0.257$$

i.e. $\eta = $ **25.7%**

Now try the following Practise Exercises

Practise Exercise 87   Further problem on the efficiency of a screw jack

1.   The coefficient of friction on the sliding surface of a screw jack whose thread is similar to Figure 16.15, is 0.24. If the pitch equals 12 mm, and $D_1 = 42$ mm and $D_2 = 56$ mm, calculate the efficiency of the screw jack.
      [24.06%]

Practise Exercise 88 Short-answer questions on friction

1. The .... of frictional force depends on the .... of surfaces in contact.

2. The .... of frictional force depends on the size of the .... to the surfaces in contact.

3. The .... of frictional force is always .... to the direction of motion.

4. The coefficient of friction between surfaces should be a .... value for materials concerned with bearings.

5. The coefficient of friction should have a .... value for materials concerned with braking systems.

6. The coefficient of dynamic or sliding friction is given by $\dfrac{.....}{.....}$

7. The coefficient of static or limiting friction is given by $\dfrac{.....}{.....}$ when .... is just about to take place.

8. Lubricating surfaces in contact result in a .... of the coefficient of friction.

9. Briefly discuss the factors affecting the size and direction of frictional forces.

10. Name three practical applications where a low value of coefficient of friction is desirable and state briefly how this is achieved in each case.

11. Name three practical applications where a high value of coefficient of friction is required when transmitting forces and discuss how this is achieved.

12. For an object on a surface, two different values of coefficient of friction are possible. Give the names of these two coefficients of friction and state how their values may be obtained.

13. State the formula for the angle of repose.

14. What theory can be used for calculating the efficiency of a screw jack.

Practise Exercise 89 Multiple-choice questions on friction

(Answers on page 298)

1. A block of metal requires a frictional force F to keep it moving with constant velocity across a surface. If the coefficient of friction is μ, then the normal force N is given by:

    (a) $\dfrac{\mu}{F}$ (b) μF

    (c) $\dfrac{F}{\mu}$ (d) F

2. The unit of the linear coefficient of friction is:
    (a) newtons
    (b) radians
    (c) dimensionless
    (d) newtons/metre

Questions 3 to 7 refer to the statements given below. Select the statement required from each group given.

(a) The coefficient of friction depends on the type of surfaces in contact.

(b) The coefficient of friction depends on the force acting at right angles to the surfaces in contact.

(c) The coefficient of friction depends on the area of the surfaces in contact.

(d) Frictional force acts in the opposite direction to the direction of motion.

(e) Frictional force acts in the direction of motion.

(f) A low value of coefficient of friction is required between the belt and the wheel in a belt drive system.

(g) A low value of coefficient of friction is required for the materials of a bearing.

(h) The dynamic coefficient of friction is given by (normal force)/(frictional force) at constant speed.

(i) The coefficient of static friction is given by (applied force) ÷ (frictional force) as sliding is just about to start.

(j) Lubrication results in a reduction in the coefficient of friction.

3. Which statement is false from (a), (b), (f) and (i)?

4. Which statement is false from (b), (e), (g) and (j) ?

5. Which statement is true from (c), (f), (h) and (i) ?

6. Which statement is false from (b), (c), (e) and (j)?

7. Which statement is false from (a), (d), (g) and (h)?

8. The normal force between two surfaces is 100 N and the dynamic coefficient of friction is 0.4
   The force required to maintain a constant speed of sliding is:

   (a) 100.4 N   (b) 40 N
   (c) 99.6 N   (d) 250 N

9. The normal force between two surfaces is 50 N and the force required to maintain a constant speed of sliding is 25 N. The dynamic coefficient of friction is:

(a) 25   (b) 2
(c) 75   (d) 0.5

10. The maximum force, which can be applied to an object without sliding occurring, is 60 N, and the static coefficient of friction is 0.3. The normal force between the two surfaces is:

   (a) 200 N   (b) 18 N
   (c) 60.3 N   (d) 59.7 N

11. The formula for the angle of repose is:

   (a) $F = \mu N$ (b) $\tan \theta = \mu$
   (c) $\mu = \dfrac{F}{N}$ (d) $\tan \theta = \dfrac{\sin \theta}{\cos \theta}$

Reference

[1] Bird, J O *Engineering Mathematics 6th Edition*, chapter 27, Taylor and Francis Publishers, 2010.

Chapter 17

Motion in a circle

In this chapter, uniform circular motion of particles is considered, and it is assumed that objects such as railway trains and motorcars behave as particles. When a railway train goes round a bend, its wheels will have to produce a centripetal acceleration towards the centre of the turning circle. This in turn will cause the railway tracks to experience a centrifugal thrust, which will tend to cause the track to move outwards. To avoid this unwanted outward thrust on the outer rail, it will be necessary to incline the railway tracks. Although problems involving the motion in a circle are dynamic ones, they can be reduced to static problems through D'Alembert's principle. If a motorcar travels around a bend, its tyres will have to exert centripetal forces to achieve this; this is achieved by the transverse frictional forces acting on the tyres. Problems involving locomotives and cars travelling around bends, a conical pendulum and motion in a vertical circle are solved in this chapter, and the centrifugal clutch is explained.

At the end of this chapter you should be able to:

- understand centripetal force
- understand D'Alembert's principle
- understand centrifugal force
- solve problems involving locomotives and cars travelling around bends
- solve problems involving a conical pendulum
- solve problems involving the motion in a vertical circle
- understand the centrifugal clutch

17.1 Introduction

In this chapter we will restrict ourselves to the uniform circular motion of particles. We will assume that objects such as railway trains and motorcars behave as particles, i.e. rigid body motion is neglected. When a railway train goes round a bend, its wheels will have to produce a centripetal acceleration towards the centre of the turning circle. This in turn will cause the railway tracks to experience a centrifugal thrust (see below), which will tend to cause the track to move outwards. To avoid this unwanted outward thrust on the outer rail, it will be necessary to incline the railway tracks in the manner shown in Figure 17.1.

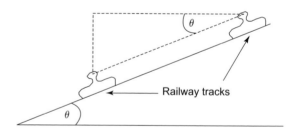

Figure 17.1

From Section 14.3, it can be seen that when a particle moves in a circular path at a constant speed v, its centripetal acceleration,

$$a = 2\,v\,\sin\frac{\theta}{2} \times \frac{1}{t}$$

When θ is small, $\theta \approx \sin\theta$,

hence

$$a = 2\,v\,\frac{\theta}{2} \times \frac{1}{t} = v\,\frac{\theta}{t}$$

However, $\omega = $ uniform angular velocity $= \dfrac{\theta}{t}$

Therefore $a = v\,\omega$

If $r = $ the radius of the turning circle, then

$$v = \omega\,r$$

and

$$a = \omega^2 r = \frac{v^2}{r}$$

Mechanical Engineering Principles, Bird and Ross, ISBN 9780415517850

Now   force = mass × acceleration

Hence,   **centripetal force = $m \omega^2 r = \dfrac{m v^2}{r}$**   (17.1)

D'Alembert's principle

Although problems involving the motion in a circle are dynamic ones, they can be reduced to static problems through D'Alembert's principle. In this principle, the centripetal force is replaced by an imaginary **centrifugal force** which acts equal and opposite to the centripetal force. By using this principle, the dynamic problem is reduced to a static one.

If a motorcar travels around a bend, its tyres will have to exert centripetal forces to achieve this. This is achieved by the transverse frictional forces acting on the tyres, as shown in Figure 17.2.

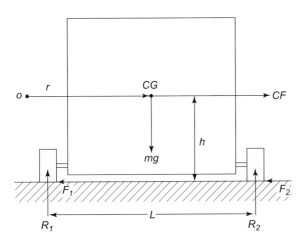

Figure 17.2

In Figure 17.2, the following notation is used:

CG = centre of gravity of the car,

m = mass of car,

R_1 = vertical reaction of 'inner' wheel,

F_1 = frictional force on 'inner' wheel,

h = vertical distance of the centre of gravity of the car from the ground,

L = distance between the centre of the tyres,

μ = coefficient of friction,

CF = centrifugal force $= \dfrac{mv^2}{r}$,

g = acceleration due to gravity,

R_2 = vertical reaction of 'outer' wheel,

F_2 = frictional force on 'outer' wheel,

r = radius of the turning circle.

Problem 1.   Determine expressions for the frictional forces F_1 and F_2 of Figure 17.2. Hence determine the thrust on each tyre.

Resolving forces horizontally gives:

$$F_1 + F_2 = \text{centrifugal force} = CF = \frac{mv^2}{r} \quad (17.2)$$

Resolving forces vertically gives:

$$R_1 + R_2 = mg \quad (17.3)$$

Taking moments about the 'outer' wheel gives:

$$CF \times h + R_1 \times L = mg\frac{L}{2}$$

i.e. $$\frac{mv^2}{r} h + R_1 L = mg\frac{L}{2}$$

or $$R_1 L = mg\frac{L}{2} - \frac{mv^2}{r} h$$

Hence, $$R_1 L = m\left(\frac{gL}{2} - \frac{v^2 h}{r}\right)$$

from which, $$R_1 = \frac{m}{L}\left(\frac{gL}{2} - \frac{v^2 h}{r}\right) \quad (17.4)$$

Also, $$F_1 = \mu R_1 \text{ and } F_2 = \mu R_2 \quad (17.5)$$

Substituting equation (17.4) into equation (17.3) gives:

$$\frac{m}{L}\left(\frac{gL}{2} - \frac{v^2 h}{r}\right) + R_2 = mg$$

Therefore, $$R_2 = mg - \frac{m}{L}\left(\frac{gL}{2} - \frac{v^2 h}{r}\right)$$

$$= mg - \frac{mg}{2} + \frac{m}{L}\frac{v^2 h}{r}$$

i.e. $$R_2 = \frac{m}{L}\left(\frac{gL}{2} + \frac{v^2 h}{r}\right) \quad (17.6)$$

From equations (17.4) to (17.6):

$$\boldsymbol{F_1 = \mu \frac{m}{L}\left(\frac{gL}{2} - \frac{v^2 h}{r}\right)} \quad (17.7)$$

and $$\boldsymbol{F_2 = \mu \frac{m}{L}\left(\frac{gL}{2} + \frac{v^2 h}{r}\right)} \quad (17.8)$$

To calculate the thrust on each tyre:

From Pythagoras' theorem (see Chapter 1),

$$T_1 = \sqrt{F_1^2 + R_1^2} = \sqrt{\mu^2 R_1^2 + R_1^2}$$

i.e. $$T_1 = R_1 \times \sqrt{1 + \mu^2} \quad \text{(see Figure 17.3(a))}$$

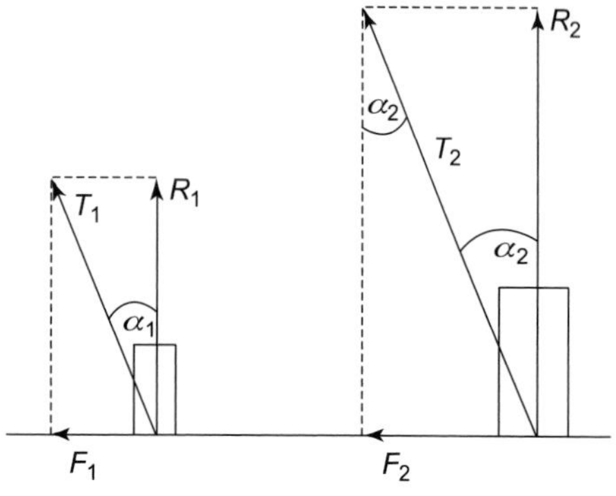

Figure 17.3

Let α_1 = angle of thrust,

i.e. $$\alpha_1 = \tan^{-1}\left(\frac{F_1}{R_1}\right) = \tan^{-1}\mu$$

From Figure 17.3(b), $T_2 = \sqrt{F_2{}^2 + R_2{}^2}$

$$= \sqrt{\mu^2 R_2{}^2 + R_2{}^2}$$

$$= R_2 \times \sqrt{1 + \mu^2}$$

$$\alpha_2 = \tan^{-1}\left(\frac{F_2}{R_2}\right) = \tan^{-1}\mu$$

17.2   Motion on a curved banked track

Problem 2. A railway train is required to travel around a bend of radius r at a uniform speed of v. Determine the amount that the 'outer' rail is to be elevated to avoid an outward centrifugal thrust in these rails, as shown in Figure 17.4.

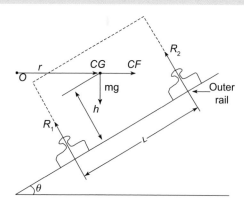

Figure 17.4

To balance the centrifugal force:

$$(R_1 + R_2)\sin\theta = CF = \frac{mv^2}{r}$$

from which, $$\sin\theta = \frac{mv^2}{r(R_1 + R_2)}$$

Let $R = R_1 + R_2$

Then $$\sin\theta = \frac{mv^2}{r\,R} \qquad (17.9)$$

Resolving forces vertically gives:

$$R\cos\theta = mg$$

from which, $$R = \frac{mg}{\cos\theta} \qquad (17.10)$$

Substituting equation (17.10) into equation (17.9) gives:

$$\sin\theta = \frac{mv^2}{r\,mg}\cos\theta$$

Hence $$\tan\theta = \frac{v^2}{r\,g} \quad \left(\text{since } \frac{\sin\theta}{\cos\theta} = \tan\theta\right)$$

Thus, the amount that the outer rail has to be elevated to avoid an outward centrifugal thrust on these rails,

$$\theta = \tan^{-1}\left(\frac{v^2}{rg}\right) \qquad (17.11)$$

Problem 3. A locomotive travels around a curve of 700 m radius. If the horizontal thrust on the outer rail is 1/40th of the locomotive's weight, determine the speed of the locomotive (in km/h). The surface that the rails are on may be assumed to be horizontal and the horizontal force on the inner rail may be assumed to be zero. Take g as 9.81 m/s^2.

Centrifugal force on outer rail $= \dfrac{mg}{40}$

Hence, $$\frac{mv^2}{r} = \frac{mg}{40}$$

from which, $$v^2 = \frac{gr}{40} = \frac{9.81 \times 700}{40}$$

$$= 171.675 \text{ m}^2/\text{s}^2$$

i.e. $$v = \sqrt{171.675} = 13.10 \text{ m/s}$$

$$= (13.10 \times 3.6) \text{ km/h}$$

i.e. **the speed of the locomotive, v = 47.17 km/h**

Problem 4. What angle of banking of the rails is required for Problem 3 above, for the outer rail to have a zero value of thrust? Assume the speed of the locomotive is 40 km/h.

From Problem 2, angle of banking, $\theta = \tan^{-1}\left(\dfrac{v^2}{rg}\right)$

$$v = 40 \text{ km/h}$$

$$= 40\frac{\text{km}}{\text{h}} \times \frac{1\,\text{h}}{3600\,\text{s}} \times \frac{1000\,\text{m}}{1\,\text{km}}$$

$$= \frac{40}{3.6} = 11.11 \text{ m/s}$$

Hence, $\theta = \tan^{-1}\left(\dfrac{11.11^2\,\text{m}^2/\text{s}^2}{700\,\text{m} \times 9.81\,\text{m/s}^2}\right)$

$$= \tan^{-1}(0.01798)$$

i.e. **angle of banking, $\theta = 1.03°$**

Now try the following Practise Exercise

Practise Exercise 90   Further problems on motion in a circle

(Where needed, take $g = 9.81$ m/s²)

1. A locomotive travels around a curve of 500 m radius. If the horizontal thrust on the outer rail is $\dfrac{1}{50}$ of the locomotive weight, determine the speed of the locomotive. The surface that the rails are on may be assumed to be horizontal and the horizontal force on the inner rail may be assumed to be zero.
   [35.64 km/h]

2. If the horizontal thrust on the outer rail of Problem 1 is $\dfrac{1}{100}$ of the locomotive's weight, determine its speed.   [25.2 km/h]

3. What angle of banking of the rails of Problem 1 is required for the outer rail to have a zero value of outward thrust? Assume the speed of the locomotive is 15 km/h.
   [0.203°]

4. What angle of banking of the rails is required for Problem 3, if the speed of the locomotive is 30 km/h ?   [0.811°]

17.3 Conical pendulum

If a mass m were rotated at a constant angular velocity, ω, in a horizontal circle of radius r, by a mass-less taut string of length L, its motion will be in the form of a cone, as shown in Figure 17.5.

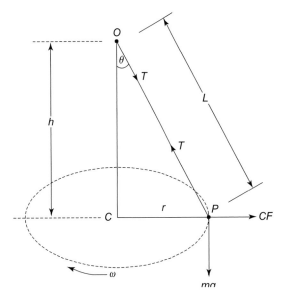

Figure 17.5 Conical pendulum

Let r = radius of horizontal turning circle,
       L = length of string,
       $h = OC$,
       ω = constant angular velocity about C,
       m = mass of particle P,
       T = tension in string,
and θ = cone angle

Problem 5. Determine an expression for the cone angle θ and the tension in the string T, for the conical pendulum of Figure 17.5. Determine also an expression for ω.

Resolving forces horizontally gives:

Centrifugal force = $CF = T \sin\theta$

i.e. $m\,\omega^2 r = T\sin\theta$

from which, $T = \dfrac{m\omega^2 r}{\sin\theta}$ (17.12)

Resolving forces vertically gives:

$$T\cos\theta = mg$$

from which, $T = \dfrac{mg}{\cos\theta}$ (17.13)

Equating equations (17.12) and (17.13) gives:

$$\frac{m\omega^2 r}{\sin\theta} = \frac{mg}{\cos\theta}$$

Rearranging gives: $\dfrac{m\omega^2 r}{mg} = \dfrac{\sin\theta}{\cos\theta}$

i.e. $\tan\theta = \dfrac{\omega^2 r}{g}$

Hence, the cone angle, $\quad \boldsymbol{\theta = \tan^{-1}\left(\dfrac{\omega^2 r}{g}\right)} \quad$ (17.14)

From Figure 17.5, $\quad \sin\theta = \dfrac{r}{L} \quad$ (17.15)

Hence, from equation (17.12), $\quad T = \dfrac{m\omega^2 r}{\dfrac{r}{L}}$

i.e. **the tension in the string, $T = m\,\omega^2 L$** (17.16)

From equation (17.14), $\quad \dfrac{\omega^2 r}{g} = \tan\theta$

But, from Figure 17.5, $\quad \tan\theta = \dfrac{r}{h}$

Hence, $\quad \dfrac{\omega^2 r}{g} = \dfrac{r}{h}$

and $\quad \omega^2 = \dfrac{g}{h}$

Thus, angular velocity about C, $\boldsymbol{\omega = \sqrt{\dfrac{g}{h}}} \quad$ (17.17)

Problem 6. A conical pendulum rotates about a horizontal circle at 90 rpm. If the speed of rotation of the mass increases by 10%, how much does the mass of the pendulum rise (in mm)? Take g as 9.81 m/s².

Angular velocity, $\quad \omega = \dfrac{2\pi n}{60} = \dfrac{2\pi \times 90}{60}$

$= 9.425$ rad/s

From equation (16.17), $\omega = \sqrt{\dfrac{g}{h}}$

or $\quad \omega^2 = \dfrac{g}{h}$

from which, height, $\boldsymbol{h} = \dfrac{g}{\omega^2} = \dfrac{9.81}{9.425^2}$

$= \boldsymbol{0.11044\ m}$ (see Figure 17.5)

When the speed of rotation rises by 10%,

$$n_2 = 90 \times 1.1 = 99 \text{ rpm}$$

Hence, $\quad \omega_2 = \dfrac{2\pi n_2}{60} = \dfrac{2\pi \times 99}{60}$

$= 10.367$ rad/s

From equation (17.17), $\omega_2 = \sqrt{\dfrac{g}{h_2}}$

or $\quad \omega_2^2 = \dfrac{g}{h_2}$

Hence, $\quad h_2 = \dfrac{g}{\omega_2^2} = \dfrac{9.81}{10.367^2}$

i.e. the new value of height, $\boldsymbol{h_2 = 0.09127\ m}$

Rise in height of the pendulum mass = 'old' h – 'new' h

$= h - h_2 = 0.11044 - 0.09127$

$= 0.01917$ m = **19.17 mm**

Problem 7. A conical pendulum rotates at a horizontal angular velocity of 5 rad/s. If the length of the string is 2 m and the pendulum mass is 0.3 kg, determine the tension in the string. Determine also the radius of the turning circle. Take g as 9.81 m/s².

Angular velocity, $\omega = 5$ rad/s

From equation (17.16), tension in the string,

$$T = m\,\omega^2 L$$

$= 0.3 \text{ kg} \times (5 \text{ rad/s})^2 \times 2 \text{ m}$

i.e. $\quad T = 15$ kg m/s²

However, 1 kg m/s² = 1 N, hence,

tension in the string, $T = 15$ N

From equation (17.13), $T = \dfrac{mg}{\cos\theta}$

from which, $\quad \cos\theta = \dfrac{mg}{T}$

$= \dfrac{0.3 \text{ kg} \times 9.81 \text{ m/s}^2}{15 \text{ N}}$

$= 0.1962$

Hence, the cone angle, $\theta = \cos^{-1}(0.1962)$

$= \boldsymbol{78.685°}$

From equation (17.15), $\sin\theta = \dfrac{r}{L}$

from which, **radius of turning circle,**

$r = L\sin\theta = 2 \text{ m} \times \sin 78.685°$

$= \boldsymbol{1.961\ m}$

Now try the following Practise Exercise

**Practise Exercise 91   Further problems on
                                    the conical pendulum**

1. A conical pendulum rotates about a horizontal circle at 100 rpm. If the speed of rotation of the mass increases by 5%, how much does the mass of the pendulum rise?
   [8.32 mm]

2. If the speed of rotation of the mass of Problem 1 decreases by 5%, how much does the mass fall?
   [9.66 mm]

3. A conical pendulum rotates at a horizontal angular velocity of 2 rad/s. If the length of the string is 3 m and the pendulum mass is 0.25 kg, determine the tension in the string. Determine also the radius of the turning circle.
   [3 N, 1.728 m]

17.4   Motion in a vertical circle

This problem is best solved by energy considerations. Consider a particle P rotating in a vertical circle of radius r, about a point O, as shown in Figure 17.6. Neglect losses due to friction.

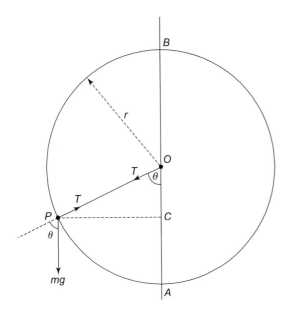

Figure 17.6 Motion in vertical circle

Let T = tension in a mass-less string,
         r = radius of turning circle,
and m = mass of particle.

Problem 8. Determine the minimum tangential velocity at A, namely, v_A, which will just keep the string taut at the point B for the particle moving in the vertical circle of Figure 17.6.

At the point B the **potential energy** $= mg \times 2r$ (17.18)

and the kinetic energy $= \dfrac{m v_B{}^2}{2}$ (17.19)

At the point A, the kinetic energy $(KE) = \dfrac{m v_A{}^2}{2}$ (17.20)

and the potential energy $(PE) = 0$

As there are no energy losses, KE at $A = (KE + PE)$ at B

Hence, from equations (17.18) to (17.20):

$$\frac{m v_A{}^2}{2} = mg \times 2r + \frac{m v_B{}^2}{2}$$

or $$\frac{v_A{}^2}{2} = \frac{v_B{}^2}{2} + 2gr$$

from which, $$v_A{}^2 = v_B{}^2 + 4gr \qquad (17.21)$$

At B, $T = 0$; this is because B is the highest point in the circle, where it can readily be observed that T will be a minimum

Thus,   weight = centrifugal force at B,

or $$mg = \frac{m v_B{}^2}{r}$$

from which, $$v_B{}^2 = gr \qquad (17.22)$$

Substituting equation (17.22) into equation (17.21) gives:

$$v_A{}^2 = gr + 4gr = 5gr$$

Hence, **the minimum tangential velocity at A,**

$$v_A = \sqrt{5gr} \qquad (17.23)$$

Problem 9. A mass of 0.1 kg is being rotated in a vertical circle of radius 0.6 m. If the mass is attached to a mass-less string and the motion is such that the string is just taut when the mass is at the top of the circle, what is the tension in the string when it is horizontal? Neglect losses and take g as 9.81 m/s².

At the top of the circle, potential energy =
$$PE = 2\,mgr$$
and
$$KE = \frac{mv_T{}^2}{2}$$
where v_T = velocity at the top

When the string is **horizontal,**
$$PE = mgr$$
and kinetic energy, $KE = \dfrac{mv_1{}^2}{2}$

where v_1 = velocity of mass at this point.

From the conservation of energy,

$(PE + KE)$ at the top = $(PE + KE)$ when the string is horizontal

i.e.
$$2\,mgr = mgr + \frac{mv_1{}^2}{2} - \frac{mv_T{}^2}{2}$$

or
$$\frac{mv_1{}^2}{2} = 2\,mgr - mgr + \frac{mv_T{}^2}{2}$$

i.e.
$$\frac{v_1{}^2}{2} = gr + \frac{v_T{}^2}{2}$$

but $\quad CF$ at top $= \dfrac{mv_T{}^2}{r} = mg \quad$ or $\quad v_T{}^2 = gr$

Hence, $\quad \dfrac{v_1{}^2}{2} = gr + \dfrac{gr}{2} = \dfrac{3gr}{2}$

i.e.
$$v_1{}^2 = 3gr$$

and $\quad v_1 = \sqrt{3gr} = \sqrt{3 \times 9.81 \times 0.6} = 4.202 \text{ m/s}$

Resolving forces horizontally,

Centrifugal force $= T =$ tension in the string

Therefore, $\quad T = \dfrac{mv_1{}^2}{r} = \dfrac{0.1\,\text{kg} \times (4.202)^2\,\text{m}^2/\text{s}^2}{0.6\,\text{m}}$

i.e. **the tension in the string, $T = 2.943$ N**

Problem 10. What is the tension in the string for Problem 9 when the mass is at the bottom of the circle?

From equation (17.23), the velocity at the bottom of the circle $= v = \sqrt{5gr}$

i.e. $\quad v = \sqrt{5 \times 9.81 \times 0.6} = 5.4249$ m/s.

Resolving forces vertically,

$\quad T$ = tension in the string

$\quad\quad$ = centrifugal force + the weight of the mass

i.e. $\quad T = \dfrac{mv^2}{r} + mg = m\left(\dfrac{v^2}{r} + g\right)$

$$= 0.1 \times \left(\frac{5.4249^2}{0.6} + 9.81\right)$$

$$= 0.1 \times (49.05 + 9.81)$$

$$= 0.1 \times 58.86 \text{ N}$$

i.e. **tension in the string, $T = 5.886$ N**

Problem 11. If the mass of Problem 9 were to rise, so that the string is at 45° to the vertical axis and below the halfway mark, what would be the tension in the string?

At 45°, $\quad PE = \dfrac{mgr}{2}$

and $\quad KE = \dfrac{mv_2{}^2}{2}$

where v_2 = velocity of the mass at this stage.

From the conservation of energy, $(PE + KE)$ at the top $= (PE + KE)$ at this stage

Therefore, $\quad 2\,mgr = \dfrac{mgr}{2} + \dfrac{mv_2{}^2}{2} - \dfrac{mv_T{}^2}{2}$

From Problem 9, $\quad v_T{}^2 = gr$

or $\quad \dfrac{v_2{}^2}{2} = \left(2r - \dfrac{r}{2} + \dfrac{r}{2}\right)g$

$$= 2gr$$

from which, $\quad v_2{}^2 = 4gr$

and $\quad v_2 = \sqrt{4gr}$

$$= \sqrt{4 \times 9.81 \times 0.6}$$

$$= 4.852 \text{ m/s}$$

Resolving forces in a direction along the string,

$\quad T$ = tension in the string = centrifugal force + component of weight at 45° to the vertical

i.e. $\quad T = \dfrac{mv_2{}^2}{r} + mg \cos 45°$

$$= \frac{0.1 \times (4.2852)^2}{0.6} + 0.1 \times 9.81 \times 0.7071$$

$$= 3.924 \text{ N} + 0.6937 \text{ N}$$

i.e. **the tension in the string, $T = 4.618$ N**

Now try the following Practise Exercise

1. A uniform disc of diameter 0.1 m rotates about a vertical plane at 200 rpm. The disc has a mass of 1.5 kg attached at a point on its rim and another mass of 2.5 kg at another point on its rim, where the angle between the two masses is 90° clockwise. Determine the magnitude of the resultant centrifugal force that acts on the axis of the disc, and its position with respect to the 1.5 kg mass.

   [63.94 N at 59° clockwise]

2. If a mass of 4 kg is placed on some position on the disc in Problem 1, determine the position where this mass must be placed to nullify the unbalanced centrifugal force.

   $$\left[\begin{array}{l}\text{At a radius of 36.44 mm,}\\ 121° \text{ anti-clockwise to 1.5 kg mass}\end{array}\right]$$

3. A stone of mass 0.1 kg is whirled in a vertical circle of 1 m radius by a mass-less string, so that the string just remains taut. Determine the velocity and tension in the string at (a) the top of the circle (b) the bottom of the circle (c) midway between (a) and (b).

   $$\left[\begin{array}{l}\text{(a) 3.132 m/s, 0 N   (b) 7 m/s, 5.88 N}\\ \text{(c) 5.42 m/s, 2.94 N}\end{array}\right]$$

17.5   Centrifugal clutch

A clutch is an engineering device used for transferring motion from an engine to a gearbox or other machinery. The main purpose of the clutch is to transfer the motion in a smooth and orderly manner, so that the gears and wheels (in the case of the motor car) will accelerate smoothly and not in a jerky manner.

The centrifugal clutch works on the principle that the rotating driving shaft will cause the centrifugal weights, shown in Figure 17.7, to move radially outwards with increasing speed of rotation of the driving shaft. These centrifugal weights will be restrained by the restraining springs shown, but when the speed of the driving shaft reaches the required value, the clutch material will engage with the driven shaft, through friction, and cause the driven shaft to rotate. The driven

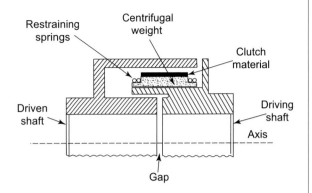

Figure 17.7

shaft will thus reach a high speed of rotation quite smoothly in the required time.

Centrifugal clutches are popular when it is required to exert a high starting torque quickly and smoothly.

A suitable clutch material is asbestos, but it is likely that asbestos will be replaced by more modern materials for health and safety reasons.

Now try the following Practise Exercises

1. The centrifugal force of a mass m moving at velocity v at a radius r is given by: ....

2. What is the potential energy at the top of a circle for the motion in a vertical circle?

3. What is the potential energy at the bottom of a circle for the motion in a vertical circle?

4. What is the potential energy at the 'middle' of a circle for the motion in a vertical circle?

(Answers on page 298)

1. To decrease the horizontal thrust on the outer rail of a train going round a bend, the outer rail should be:
   (a) lowered
   (b) raised
   (c) kept at the same level as the inner rail
   (d) made bigger

2. If the speed of rotation of a conical pendulum is increased, the height of the pendulum mass will:
    (a) fall
    (b) become zero
    (c) stay the same
    (d) rise

3. The minimum tension on the top of a vertical circle, for satisfactory motion in a circle is:
    (a) zero
    (b) mg
    (c) $\dfrac{mv^2}{r}$
    (d) negative

4. If v is the velocity at the 'middle' for the motion in a vertical circle, the tension is:
    (a) zero
    (b) $\dfrac{mv^2}{r}$
    (c) mg
    (d) negative

5. If the tension in the string is zero at the top of a circle for the motion in a vertical circle, the velocity at the bottom of the circle is:
    (a) zero
    (b) $\sqrt{5gr}$
    (c) $\sqrt{gr}$
    (d) $\sqrt{3gr}$

Chapter 18

Simple harmonic motion

Simple harmonic motion is of importance in a number of branches of engineering and physics, including structural and machine vibrations, alternating electrical currents, sound waves, light waves, tidal motion, and so on. This chapter explains simple harmonic motion, determines natural frequencies for spring-mass systems, calculates periodic times and explains simple and compound pendulums.

At the end of this chapter you should be able to:

- understand simple harmonic motion
- determine natural frequencies for simple spring-mass systems
- calculate periodic times
- understand the motion of a simple pendulum
- understand the motion of a compound pendulum

18.1 Introduction to simple harmonic motion (SHM)

A particle is said to be under SHM if its acceleration along a line is directly proportional to its displacement along that line, from a fixed point on that line.

Consider the motion of a particle A, rotating in a circle with a constant angular velocity ω, as shown in Figure 18.1.

Consider now the vertical displacement of A from xx, as shown by the distance y_C. If P is rotating at a constant angular velocity ω then the periodic time T to travel an angular distance of 2π, is given by:

$$T = \frac{2\pi}{\omega} \tag{18.1}$$

Let f = frequency of motion C (in Hertz), where

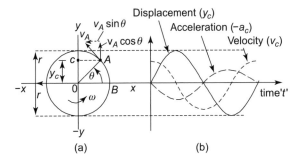

Figure 18.1

$$f = \frac{1}{T} = \frac{\omega}{2\pi} \tag{18.2}$$

To determine whether or not SHM is taking place, we will consider motion of A in the direction yy. Now $y_C = OA \sin \omega t$,

i.e. $\qquad y_C = r \sin \omega t \tag{18.3}$

where t = time in seconds.

Plotting of equation (18.3) against t results in the sinusoidal variation for displacement, as shown in Figure 18.1(b).

From Chapter 12, $v_A = \omega r$, which is the tangential velocity of the particle A.

From the velocity vector diagram, at the point A on the circle of Figure 18.1(a),

$$v_C = v_A \cos \theta = v_A \cos \omega t \tag{18.4}$$

Plotting of equation (18.4) against t results in the sinusoidal variation for the velocity v_C, as shown in Figure 18(b).

The centripetal acceleration of A

$$= a_A = \omega^2 r$$

Now $\qquad a_C = -a_A \sin \theta$

Mechanical Engineering Principles, Bird and Ross, ISBN 9780415517850

Therefore, $\qquad a_C = -\omega^2 r \sin \omega t \qquad$ (18.5)

Plotting of equation (18.5) against t results in the sinusoidal variation for the acceleration at C, a_C, as shown in Figure 18.1(b).

Substituting equation (18.3) into equation (18.5) gives:

$$a_C = -\omega^2 y_C \qquad (18.6)$$

Equation (18.6) shows that the acceleration along the line yy is directly proportional to the displacement along this line, therefore the point C is moving with SHM.

Now $\quad T = \dfrac{2\pi}{\omega}$, but from equation (18.6),

$$a_C = \omega^2 y, \text{ i.e. } \omega^2 = \frac{a}{y}$$

Therefore, $\qquad T = \dfrac{2\pi}{\sqrt{\dfrac{a}{y}}} \quad \text{or} \quad T = 2\pi \sqrt{\dfrac{y}{a}}$

i.e. $\qquad T = 2\pi \sqrt{\dfrac{\text{dispacement}}{\text{acceleration}}}$

In general, from equation (18.6),

$$a + \omega^2 y = 0 \qquad (18.7)$$

18.2 The spring-mass system

(a) Vibrating horizontally
Consider a mass m resting on a smooth surface and attached to a spring of stiffness k, as shown in Figure 18.2.

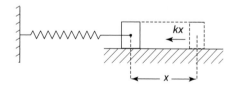

Figure 18.2

If the mass is given a small displacement x, the spring will exert a resisting force of kx, i.e. $F = -kx$

But, $\qquad\qquad F = ma$

hence, $\qquad\qquad ma = -kx$

or $\qquad\qquad ma + kx = 0$

or $\qquad\qquad a + \dfrac{k}{m} x = 0 \qquad (18.8)$

Equation (18.8) shows that this mass is oscillating (or vibrating) in SHM, or according to equation (18.7). Comparing equation (18.7) with equation (18.8) we see that

$$\omega^2 = \frac{k}{m}$$

from which, $\qquad \omega = \sqrt{\dfrac{k}{m}}$

Now $\qquad T = \dfrac{2\pi}{\omega} = 2\pi \sqrt{\dfrac{m}{k}}$

and f = frequency of oscillation or vibration

i.e. $\qquad f = \dfrac{\omega}{2\pi} = \dfrac{1}{2\pi} \sqrt{\dfrac{k}{m}} \qquad (18.9)$

(b) Vibrating vertically
Consider a mass m, supported by a vertical spring of stiffness k, as shown in Figure 18.3. In this equilibrium position, the mass has an initial downward static deflection of y_o. If the mass is given an additional downward displacement of y and then released, it will vibrate vertically.

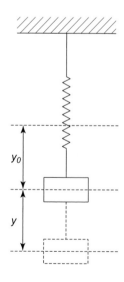

Figure 18.3

The force exerted by the spring $= -k(y_o + y)$

Therefore, F = accelerating force − resisting force
$\qquad\qquad = ma$

or $\qquad F = mg - k(y_o + y) = ma$

i.e. $F = mg - ky_o - ky = ma$

But, $ky_o = mg$,

hence $F = mg - mg - ky = ma$

Thus, $ma + ky = 0$

or $a + \dfrac{k}{m} y = 0$

i.e. SHM takes place, then $\omega = \sqrt{\dfrac{k}{m}}$ and $T = \dfrac{2\pi}{\omega}$

i.e. periodic time, $T = 2\pi\sqrt{\dfrac{m}{k}}$ (18.10)

and frequency, $f = \dfrac{\omega}{2\pi} = \dfrac{1}{2\pi}\sqrt{\dfrac{k}{m}}$ (18.11)

as before (from equation (18.9)).

Comparing equations (18.9) and (18.11), it can be seen that there is no difference in whether the spring is horizontal or vertical.

> **Problem 1.**   A mass of 1.5 kg is attached to a vertical spring, as shown in Figure 18.4. When the mass is displaced downwards a distance of 55 mm from its position of rest, it is observed to oscillate 60 times in 72 seconds. Determine (a) periodic time (b) the stiffness of the spring (c) the time taken to travel upwards a distance of 25 mm for the first time (d) the velocity at this point.

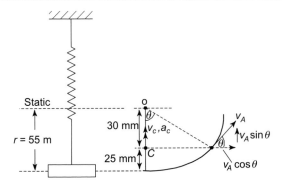

Figure 18.4

(a)   **Periodic time, $T = \dfrac{72\,\text{seconds}}{60\,\text{oscillations}} = \textbf{1.2 seconds}$**

(b)   From equation (18.10), $T = 2\pi\sqrt{\dfrac{m}{k}}$

   i.e. $1.2 = 2\pi\sqrt{\dfrac{1.5}{k}}$

   Hence, $1.2^2 = (2\pi)^2 \times \dfrac{1.5}{k}$

from which, $k = (2\pi)^2 \times \dfrac{1.5}{1.2^2}$

i.e. **stiffness of spring, k = 41.1 N/m**

(c)   From Figure 18.4, $\cos\theta = \dfrac{(r-25)}{r}$

   or $\cos\theta = \dfrac{(55-25)}{55} = 0.545$

   from which, $\theta = \cos^{-1} 0.545 = 56.94°$

   Now, $\omega = \dfrac{2\pi}{T} = \dfrac{2\pi}{1.2} = 5.236$ rad/s

   But $\theta = \omega t$, hence, time t taken to travel upwards a distance of 25 mm, is given by:

   $t = \dfrac{\theta}{\omega} = \dfrac{56.94°}{5.236\,\dfrac{\text{rad}}{\text{s}}} \times \dfrac{2\pi\,\text{rad}}{360°} = \textbf{0.19 s}$

(d)   Velocity at C in Figure 18.4,

   $v_C = v_A \sin\theta$

   $= \omega r \sin\theta$

   $= 5.236\,\dfrac{\text{rad}}{\text{s}} \times \dfrac{55}{1000}\,\text{m} \times \sin 56.94°$

   $= 0.288 \times 0.838$ m/s

   i.e. $v_C = \textbf{0.241 m/s}$ after 25 mm of travel

Now try the following Practise Exercise

> **Practise Exercise 95**   **Further problems on simple harmonic motion**
>
> 1.   A particle oscillates 50 times in 22 s. Determine the periodic time and frequency.
> $[T = 0.44\text{ s}, f = 2.27\text{ Hz}]$
>
> 2.   A yacht floats at a depth of 2.2 m. On a particular day, at a time of 09.30 h, the depth at low tide is 1.8 m and at a time of 17.30 h, the depth of water at high tide is 3.4 m. Determine the earliest time of day that the yacht is refloated.   [12 h, 10 min, 1 s]
>
> 3.   A mass of 2 kg is attached to a vertical spring. The initial state displacement of this mass is 74 mm. The mass is displaced downwards and then released. Determine (a) the stiffness of the spring, and (b) the frequency of oscillation of the mass.
>   [(a) 265.1 N/m (b) 1.83 Hz]

4. A particle of mass 4 kg rests on a smooth horizontal surface and is attached to a horizontal spring. The mass is then displaced horizontally outwards from the spring a distance of 26 mm and then released to vibrate. If the periodic time is 0.75 s, determine (a) the frequency f (b) the force required to give the mass the displacement of 26 mm, (c) the time taken to move horizontally inwards for the first 12 mm.

[(a) 1.33 Hz (b) 7.30 N (c) 0.12 s]

5. A mass of 3 kg rests on a smooth horizontal surface, as shown in Figure 18.5. If the stiffness of each spring is 1 kN/m, determine the frequency of vibration of the mass. It may be assumed that initially, the springs are un-stretched.

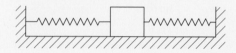

Figure 18.5

[4.11 Hz]

6. A helical spring has a mass of 10 kg attached to its top. If the mass vibrates vertically with a frequency of 1.5 Hz, determine the stiffness of the spring. [888.3 N/m]

18.3 The simple pendulum

A simple pendulum consists of a particle of mass m attached to a mass-less string of length L, as shown in Figure 18.6.

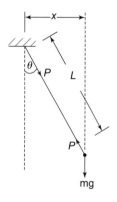

Figure 18.6

From Section 14.4, page 167,

$$T = I_o \alpha = -\text{restoring couple}$$
$$= -mg(L \sin \theta)$$

But, $I_o = mL^2$ = mass moment of inertia about the point of rotation

hence, $mL^2 \alpha + mgL \sin \theta = 0$

For small deflections, $\sin \theta = \theta$

Hence, $L^2 \alpha + gL\theta = 0$

or $\alpha + \dfrac{g\theta}{L} = 0$

But $\alpha + \omega^2\theta = 0$ (see Section 18.6)

Therefore, $\omega^2 = \dfrac{g}{L}$

and $\omega = \sqrt{\dfrac{g}{L}}$ (18.12)

Now $T = \dfrac{2\pi}{\omega} = 2\pi\sqrt{\dfrac{L}{g}}$ (18.13)

and $f = \dfrac{1}{T} = \dfrac{\sqrt{\dfrac{g}{L}}}{2\pi}$ (18.14)

Problem 2. If the simple pendulum of Figure 18.6 were of length 2 m, determine its frequency of vibration. Take $g = 9.81$ m/s².

From equation (18.14), **frequency,** $f = \dfrac{\sqrt{\dfrac{g}{L}}}{2\pi} = \dfrac{\sqrt{\dfrac{9.81}{2}}}{2\pi}$

$= \mathbf{0.352\ Hz}$

Problem 3. In order to determine the value of g at a certain point on the Earth's surface, a simple pendulum is used. If the pendulum is of length 3 m and its frequency of oscillation is 0.2875 Hz, determine the value of g.

From equation (18.14), frequency, $f = \dfrac{\sqrt{\dfrac{g}{L}}}{2\pi}$

i.e. $0.2875 = \dfrac{\sqrt{\dfrac{g}{3}}}{2\pi}$

and

$$(0.2875)^2 \times (2\pi)^2 = \frac{g}{3}$$

$$3.263 = \frac{g}{3}$$

from which, acceleration due to gravity, $g = 3 \times 3.263$

$$= \textbf{9.789 m/s}$$

18.4   The compound pendulum

Consider the compound pendulum of Figure 18.7, which oscillates about the point O. The point G in Figure 18.7 is the position of the pendulum's centre of gravity.

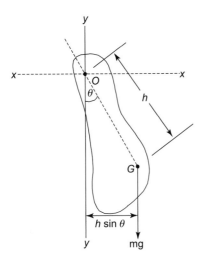

Figure 18.7

Let I_o = mass moment of inertia about O

Now $T = I_o \alpha = -$restoring couple

$$= -mgh \sin \theta$$

From the parallel axis theorem,

$$I_G = I_o - mh^2 = mk_G{}^2$$

or $I_o = mk_G{}^2 + mh^2$

where I_G = mass moment of inertia about G,

$k_G{}^2$ = radius of gyration about G

Now $T = I_o \alpha$

or $- mgh \sin \theta = I_o \alpha$

but $I_o = mk_G{}^2 + mh^2$

Therefore, $\left(mk_G{}^2 + mh^2\right)\alpha = -mgh \sin \theta$

but for small displacements, $\sin \theta = \theta$

Hence, $m\left(k_G{}^2 + h^2\right)\alpha = -mgh\,\theta$

i.e. $\left(k_G{}^2 + h^2\right)\alpha + gh\theta = 0$

or $\alpha + \dfrac{gh}{\left(k_G{}^2 + h^2\right)}\theta = 0$

However, $\alpha + \omega^2\theta = 0$

but this motion is simple harmonic motion (see equation (18.7))

Therefore, $\omega^2 = \dfrac{gh}{\left(k_G{}^2 + h^2\right)}$

and $\omega = \sqrt{\dfrac{gh}{\left(k_G{}^2 + h^2\right)}}$ (18.15)

$$T = \frac{2\pi}{\omega} = 2\pi\sqrt{\frac{\left(k_G{}^2 + h^2\right)}{gh}} \quad (18.16)$$

and $f = \dfrac{1}{T} = \dfrac{1}{2\pi}\sqrt{\dfrac{gh}{\left(k_G{}^2 + h^2\right)}}$ (18.17)

Problem 4.   It is required to determine the mass moment of inertia about G of a metal ring, which has a complex cross-sectional area. To achieve this, the metal ring is oscillated about a knife edge, as shown in Figure 18.8, where the frequency of oscillation was found to be 1.26 Hz. If the mass of the ring is 10.5 kg, determine the mass moment of inertia about the centre of gravity, I_G. Take $g = 9.81$ m/s^2.

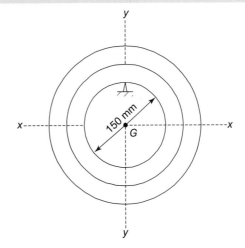

Figure 18.8

By inspection of Figure 18.8, $h = 75$ mm $= 0.075$ m.

Now frequency, $f = \dfrac{1}{2\pi}\sqrt{\dfrac{gh}{(k_G{}^2 + h^2)}}$

i.e. $\quad 1.26 = \dfrac{1}{2\pi}\sqrt{\dfrac{9.81 \times 0.075}{(k_G{}^2 + 0.075^2)}}$

i.e. $\quad (1.26)^2 = \dfrac{1}{(2\pi)^2} \times \dfrac{9.81 \times 0.075}{\left(k_G{}^2 + 0.075^2\right)}$

from which, $\left(k_G{}^2 + 0.005625\right) = \dfrac{0.73575}{1.5876 \times (2\pi)^2}$

$$= 0.011739$$

$$k_G{}^2 = 0.011739 - 0.005625$$

$$= 0.006114$$

from which, $\quad k_G = \sqrt{0.006114} = 0.0782$

The mass moment of inertia about the centre of gravity,

$$I_G = m\, k_G{}^2 = 10.5 \text{ kg} \times 0.006114 \text{ m}^2$$

i.e. $\quad \boldsymbol{I_G = 0.0642 \text{ kg m}^2}$

18.5 Torsional vibrations

From equation (18.7), it can be seen that for SHM in a linear direction,

$$a + \omega^2 y = 0$$

For SHM in a rotational direction,

$$\alpha r + \omega^2 y = 0$$

or $\quad \alpha + \omega^2\left(\dfrac{y}{r}\right) = 0$

or $\quad \alpha + \omega^2\theta = 0 \qquad$ (see equation (18.7))

i.e $\quad \ddot{\theta} + \omega^2\theta = 0 \qquad$ (18.18)

where $\theta = \dfrac{y}{r}$ = angular displacement, and

$\ddot{\theta} = \alpha$ = angular acceleration

Now try the following Practise Exercises

Practise Exercise 96   Further problems on pendulums

1. Determine the period of oscillation of a pendulum of length 2 m if $g = 9.81$ m/s^2.
    [0.3525 Hz]

2. What will be the period of oscillation if $g = 9.78$ m/s^2 for the pendulum of Problem 1?
    [0.3519 Hz]

3. What will be the period of oscillation if $g = 9.832$ m/^{2} for the pendulum of Problem 1?
    [0.3529 Hz]

4. What will be the value of the mass moment of inertia through the centre of gravity, I_G, for the compound pendulum of worked problem 4 on page 209, if the inner diameter of the disc of Figure 18.8 were 100 mm?
    [0.0559 kg m^2]

Practise Exercise 97   Short-answer questions on simple harmonic motion

1. State the relationship between the displacement (y) of a mass and its acceleration (a) for SHM to take place.

2. State the relationship between frequency f and periodic time T when SHM takes place.

3. State the formula for the frequency of oscillation for a simple pendulum.

4. State a simple method of increasing the period of oscillation of the pendulum of a 'grandfather' clock.

Practise Exercise 98   Multiple-choice questions on simple harmonic motion

(Answers on page 298)

1. Tidal motion is normally related to which mathematical function?
    (a) tangent   (b) sine
    (c) square root   (d) straight line

2. If the mass of a simple pendulum is doubled, its period of oscillation:
    (a) increases   (b) decreases
    (c) stays the same (d) doubles

3. A pendulum has a certain frequency of oscillation in London. Assuming that temperature remains the same, the frequency of

oscillation of the pendulum if it is measured on the equator:

(a) increases
(b) decreases
(c) remains the same
(d) doubles

4. The period of oscillation of a simple pendulum of length 9.81 m, given $g = 9.81$ m/s^2 is:

(a) 6.28 Hz
(b) 0.455 Hz
(c) 17.96 Hz
(d) 0.056 Hz

Chapter 19

Simple machines

This chapter commences by defining load, effort, mechanical advantage, velocity ratio and efficiency, where efficiency is then defined in terms of mechanical advantage and velocity ratio. These terms, together with other terms defined in earlier chapters, are applied to simple and quite complex pulley systems, screw-jacks, gear trains and levers. This chapter is fundamental to the study of the behaviour of machines.

At the end of this chapter you should be able to:

- define a simple machine
- define force ratio, movement ratio, efficiency and limiting efficiency
- understand and perform calculations with pulley systems
- understand and perform calculations with a simple screw-jack
- understand and perform calculations with gear trains
- understand and perform calculations with levers

19.1 Machines

A machine is a device that can change the magnitude or line of action, or both magnitude and line of action of a force. A simple machine usually amplifies an input force, called the **effort**, to give a larger output force, called the **load**. Some typical examples of simple machines include pulley systems, svcrew-jacks, gear systems and lever systems.

19.2 Force ratio, movement ratio and efficiency

The **force ratio** or **mechanical advantage** is defined as the ratio of load to effort, i.e.

$$\textbf{Force ratio} = \frac{\textbf{load}}{\textbf{effort}} = \textbf{mechanical advantage} \quad (19.1)$$

Since both load and effort are measured in newtons, force ratio is a ratio of the same units and thus is a dimension-less quantity.

The **movement ratio** or **velocity ratio** is defined as the ratio of the distance moved by the effort to the distance moved by the load, i.e.

$$\textbf{Movement ratio} = \frac{\textbf{distance moved by the effort}}{\textbf{distance moved by the load}}$$
$$= \textbf{velocity ratio} \quad (19.2)$$

Since the numerator and denominator are both measured in metres, movement ratio is a ratio of the same units and thus is a dimension-less quantity.

The **efficiency of a simple machine** is defined as the ratio of the force ratio to the movement ratio, i.e.

$$\text{Efficiency} = \frac{\text{force ratio}}{\text{movement ratio}}$$
$$= \frac{\text{mechanical advantage}}{\text{velocity ratio}}$$

Since the numerator and denominator are both dimension-less quantities, efficiency is a dimension-less quantity. It is usually expressed as a percentage, thus:

$$\textbf{Efficiency} = \frac{\textbf{force ratio}}{\textbf{movement ratio}} \times \textbf{100\%} \quad (19.3)$$

Mechanical Engineering Principles, Bird and Ross, ISBN 9780415517850

Due to the effects of friction and inertia associated with the movement of any object, some of the input energy to a machine is converted into heat and losses occur. Since losses occur, the energy output of a machine is less than the energy input, thus the mechanical efficiency of any machine cannot reach 100%

For simple machines, the relationship between effort and load is of the form: $F_e = aF_1 + b$, where F_e is the effort, F_1 is the load and a and b are constants. From equation (19.1),

$$\text{force ratio} = \frac{\text{load}}{\text{effort}} = \frac{F_l}{F_e} = \frac{F_l}{aF_l + b}$$

Dividing both numerator and denominator by F_1 gives:

$$\frac{F_l}{aF_l + b} = \frac{1}{a + \dfrac{b}{F_l}}$$

When the load is large, F_1 is large and $\dfrac{b}{F_l}$ is small compared with a. The force ratio then becomes approximately equal to $\dfrac{1}{a}$ and is called the **limiting force ratio**, i.e.

$$\textbf{limiting ratio} = \frac{1}{a}$$

The limiting efficiency of a simple machine is defined as the ratio of the limiting force ratio to the movement ratio, i.e.

$$\textbf{Limiting efficiency} = \frac{1}{a \times \textbf{movement ratio}} \times \textbf{100\%}$$

where a is the constant for **the law of the machine**:

$$F_e = aF_1 + b$$

Due to friction and inertia, the limiting efficiency of simple machines is usually well below 100%.

Problem 1. A simple machine raises a load of 160 kg through a distance of 1.6 m. The effort applied to the machine is 200 N and moves through a distance of 16 m. Taking g as 9.8 m/s², determine the force ratio, movement ratio and efficiency of the machine.

From equation (19.1),

$$\textbf{force ratio} = \frac{\text{load}}{\text{effort}} = \frac{160\,\text{kg}}{200\,\text{N}} = \frac{160 \times 9.81\,\text{N}}{200\,\text{N}}$$

$$= \textbf{7.84}$$

From equation (19.2),

$$\textbf{movement ratio} = \frac{\text{distance moved by the effort}}{\text{distance moved by the load}}$$

$$= \frac{16\,\text{m}}{1.6\,\text{m}} = \textbf{10}$$

From equation (19.3),

$$\textbf{efficiency} = \frac{\text{force ratio}}{\text{movement ratio}} \times 100\%$$

$$= \frac{7.84}{10} \times 100$$

$$= \textbf{78.4\%}$$

Problem 2. For the simple machine of Problem 1, determine: (a) the distance moved by the effort to move the load through a distance of 0.9 m (b) the effort which would be required to raise a load of 200 kg, assuming the same efficiency (c) the efficiency if, due to lubrication, the effort to raise the 160 kg load is reduced to 180 N.

(a) Since the movement ratio is 10, then from equation (19.2),

  distance moved by the effort

$$= 10 \times \text{distance moved by the load}$$
$$= 10 \times 0.9 = \textbf{9 m}$$

(b) Since the force ratio is 7.84, then from equation (19.1),

$$\textbf{effort} = \frac{\text{load}}{7.84} = \frac{200 \times 9.8}{7.84}$$

$$= \textbf{250 N}$$

(c) The new force ratio is given by

$$\frac{\text{load}}{\text{effort}} = \frac{160 \times 9.8}{180} = 8.711$$

Hence, **the new efficiency after lubrication**

$$= \frac{8.711}{10} \times 100 = \textbf{87.11\%}$$

Problem 3. In a test on a simple machine, the effort/load graph was a straight line of the form $F_e = aF_1 + b$. Two values lying on the graph were at $F_e = 10$ N, $F_1 = 30$ N, and at $F_e = 74$ N, $F_1 = 350$ N. The movement ratio of the machine was 17. Determine: (a) the limiting force ratio (b) the limiting efficiency of the machine.

(a) The equation $F_e = aF_1 + b$ is of the form $y = mx + c$, where m is the gradient of the graph. The slope of the line passing through points (x_1, y_1) and (x_2, y_2) of the graph $y = mx + c$ is given by:

$$m = \frac{y_2 - y_1}{x_2 - x_1}$$

Thus for $F_e = aF_1 + b$, the slope a is given by:

$$a = \frac{74 - 10}{350 - 30}$$

$$= \frac{64}{320} = 0.2$$

The **limiting force ratio** is $\dfrac{1}{a} = \dfrac{1}{0.2} = 5$

(b) The **limiting efficiency**

$$= \frac{1}{a \times \text{movement ratio}} \times 100$$

$$= \frac{1}{0.2 \times 17} \times 100 = \textbf{29.4\%}$$

Now try the following Practise Exercise

Practise Exercise 99 **Further problems on force ratio, movement ratio and efficiency**

1. A simple machine raises a load of 825 N through a distance of 0.3 m. The effort is 250 N and moves through a distance of 3.3 m. Determine: (a) the force ratio (b) the movement ratio (c) the efficiency of the machine at this load. [(a) 3.3 (b) 11 (c) 30%]

2. The efficiency of a simple machine is 50%. If a load of 1.2 kN is raised by an effort of 300 N, determine the movement ratio. [8]

3. An effort of 10 N applied to a simple machine moves a load of 40 N through a distance of 100 mm, the efficiency at this load being 80%. Calculate: (a) the movement ratio (b) the distance moved by the effort. [(a) 5 (b) 500 mm]

4. The effort required to raise a load using a simple machine, for various values of load is as shown:

Load F_1 (N)	2050	4120	7410	8240	10300
Effort F_e (N)	252	340	465	505	580

If the movement ratio for the machine is 30, determine (a) the law of the machine (b) the limiting force ratio (c) the limiting efficiency.
[(a) $F_e = 0.04 F_1 + 170$ (b) 25 (c) 83.3%]

5. For the data given in Question 4, determine the values of force ratio and efficiency for each value of the load. Hence plot graphs of effort, force ratio and efficiency to a base of load. From the graphs, determine the effort required to raise a load of 6 kN and the efficiency at this load. [410 N, 48%]

19.3 Pulleys

A **pulley system** is a simple machine. A single-pulley system, shown in Figure 19.1(a), changes the line of action of the effort, but does not change the magnitude of the force. A two-pulley system, shown in Figure 19.1(b), changes both the line of action and the magnitude of the force.

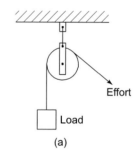

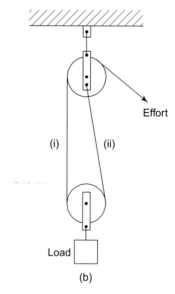

Figure 19.1

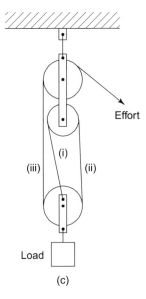

Figure 19.1 *(Continued)*

Theoretically, each of the ropes marked (i) and (ii) share the load equally, thus the theoretical effort is only half of the load, i.e. the theoretical force ratio is 2. In practice the actual force ratio is less than 2 due to losses. A three-pulley system is shown in Figure 19.1(c). Each of the ropes marked (i), (ii) and (iii) carry one-third of the load, thus the theoretical force ratio is 3. In general, for a multiple pulley system having a total of n pulleys, the theoretical force ratio is n. Since the theoretical efficiency of a pulley system (neglecting losses) is 100 and since from equation (18.3),

$$\text{efficiency} = \frac{\text{force ratio}}{\text{movement ratio}} \times 100\%$$

it follows that when the force ratio is n,

$$100 = \frac{n}{\text{movement ratio}} \times 100\%$$

that is, the movement ratio is also n.

Problem 4. A load of 80 kg is lifted by a three-pulley system similar to that shown in Figure 19.1(c) and the applied effort is 392 N. Calculate (a) the force ratio (b) the movement ratio (c) the efficiency of the system. Take g to be 9.8 m/s^2.

(a) From equation (19.1), the force ratio is given by
$$\frac{\text{load}}{\text{effort}}$$

The load is 80 kg, i.e. (80×9.8) N, hence,
$$\textbf{force ratio} = \frac{80 \times 9.8}{392} = \textbf{2}$$

(b) From above, for a system having n pulleys, the movement ratio is n. Thus, for a three-pulley system, the **movement ratio is 3**

(c) From equation (19.3),

$$\textbf{efficiency} = \frac{\text{force ratio}}{\text{movement ratio}} \times 100\%$$

$$= \frac{2}{3} \times 100 = \textbf{66.67\%}$$

Problem 5. A pulley system consists of two blocks, each containing three pulleys and connected as shown in Figure 19.2. An effort of 400 N is required to raise a load of 1500 N. Determine (a) the force ratio (b) the movement ratio (c) the efficiency of the pulley system.

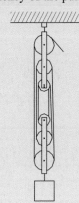

Figure 19.2

(a) From equation (19.1),

$$\textbf{force ratio} = \frac{\text{load}}{\text{effort}} = \frac{1500}{400} = \textbf{3.75}$$

(b) An n-pulley system has a movement ratio of n, hence this 6-pulley system has a **movement ratio of 6**

(c) From equation (19.3),

$$\textbf{efficiency} = \frac{\text{force ratio}}{\text{movement ratio}} \times 100\%$$

$$= \frac{3.75}{6} \times 100 = \textbf{62.5\%}$$

Now try the following Practise Exercise

Practise Exercise 100   Further problems on pulleys

1. A pulley system consists of four pulleys in an upper block and three pulleys in a lower

block. Make a sketch of this arrangement showing how a movement ratio of 7 may be obtained. If the force ratio is 4.2, what is the efficiency of the pulley?   [60%]

2. A three-pulley lifting system is used to raise a load of 4.5 kN. Determine the effort required to raise this load when losses are neglected. If the actual effort required is 1.6 kN, determine the efficiency of the pulley system at this load.
    [1.5 kN, 93.75%]

19.4   The screw-jack

A **simple screw-jack** is shown in Figure 19.3 and is a simple machine since it changes both the magnitude and the line of action of a force (see also Section 16.8).

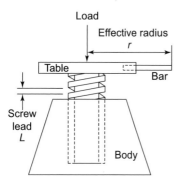

Figure 19.3

The screw of the table of the jack is located in a fixed nut in the body of the jack. As the table is rotated by means of a bar, it raises or lowers a load placed on the table. For a single-start thread, as shown, for one complete revolution of the table, the effort moves through a distance $2\pi r$, and the load moves through a distance equal to the lead of the screw, say, L.

$$\textbf{Movement ratio} = \frac{2\pi r}{L} \qquad (19.4)$$

For the efficiency of a screw-jack, see Section 16.8, page 192.

Problem 6.   A screw-jack is being used to support the axle of a car, the load on it being 2.4 kN. The screw jack has an effort of effective radius 200 mm and a single-start square thread, having a lead of 5 mm. Determine the efficiency of the jack if an effort of 60 N is required to raise the car axle.

From equation (19.3),

$$\text{efficiency} = \frac{\text{force ratio}}{\text{movement ratio}} \times 100\%$$

where force ratio $= \dfrac{\text{load}}{\text{effort}} = \dfrac{2400\,\text{N}}{60\,\text{N}} = 40$

From equation (19.4),

$$\text{movement ratio} = \frac{2\pi r}{L} = \frac{2\pi(200)\,\text{mm}}{5\,\text{mm}} = 251.3$$

Hence, **efficiency** $= \dfrac{\text{force ratio}}{\text{movement ratio}} \times 100\%$

$$= \frac{40}{251.3} \times 100 = \textbf{15.9\%}$$

Now try the following Practise Exercise

19.5   Gear trains

A **simple gear train** is used to transmit rotary motion and can change both the magnitude and the line of action of a force, hence it is a simple machine. The gear train shown in Figure 19.4 consists of **spur gears** and has an effort applied to one gear, called the driver, and a load applied to the other gear, called the **follower**.

In such a system, the teeth on the wheels are so spaced that they exactly fill the circumference with a whole number of identical teeth, and the teeth on the driver and follower mesh without interference. Under these conditions, the number of teeth on the driver and

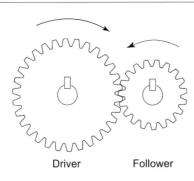

Figure 19.4

follower are in direct proportion to the circumference of these wheels, i.e.

$$\frac{\textbf{number of teeth on driver}}{\textbf{number of teeth on follower}} =$$

$$\frac{\textbf{circumference of driver}}{\textbf{circumference of follower}} \quad (19.5)$$

If there are, say, 40 teeth on the driver and 20 teeth on the follower then the follower makes two revolutions for each revolution of the driver. In general:

$$\frac{\text{number of revolutions made by driver}}{\text{number of revolutions made by the follower}} =$$

$$\frac{\text{number of teeth on follower}}{\text{number of teeth on driver}} \quad (19.6)$$

It follows from equation (19.6) that the speeds of the wheels in a gear train are inversely proportional to the number of teeth. The ratio of the speed of the driver wheel to that of the follower is the movement ratio, i.e.

$$\textbf{Movement ratio} = \frac{\textbf{speed of driver}}{\textbf{speed of follower}} =$$

$$\frac{\textbf{teeth on follower}}{\textbf{teeth on driver}} \quad (19.7)$$

When the same direction of rotation is required on both the driver and the follower an **idler wheel** is used as shown in Figure 19.5.

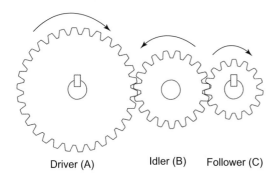

Driver (A)   Idler (B)   Follower (C)

Figure 19.5

Let the driver, idler, and follower be A, B and C, respectively, and let N be the speed of rotation and T be the number of teeth. Then from equation (19.7),

$$\frac{N_B}{N_A} = \frac{T_A}{T_B} \quad \text{or} \quad N_A = N_B \frac{T_B}{T_A} \quad \text{and}$$

$$\frac{N_C}{N_B} = \frac{T_B}{T_C} \quad \text{or} \quad N_C = N_B \frac{T_B}{T_C}$$

Thus,

$$\frac{\text{speed of } A}{\text{speed of } C} = \frac{N_A}{N_C} = \frac{N_B \dfrac{T_B}{T_A}}{N_B \dfrac{T_B}{T_C}}$$

$$= \frac{T_B}{T_A} \times \frac{T_C}{T_B} = \frac{T_C}{T_A}$$

This shows that the movement ratio is independent of the idler, only the direction of the follower being altered.

A **compound gear train** is shown in Figure 19.6, in which gear wheels B and C are fixed to the same shaft and hence $N_B = N_C$

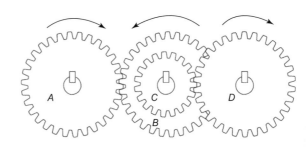

Figure 19.6

From equation (19.7), $\dfrac{N_A}{N_B} = \dfrac{T_B}{T_A}$ i.e. $N_B = N_A \times \dfrac{T_A}{T_B}$

Also, $\dfrac{N_D}{N_C} = \dfrac{T_C}{T_D}$ i.e. $N_D = N_C \times \dfrac{T_C}{T_D}$

But $N_B = N_C$ and $\quad N_D = N_B \times \dfrac{T_C}{T_D}$

therefore, $\quad N_D = N_A \times \dfrac{T_A}{T_B} \times \dfrac{T_C}{T_D} \quad (19.8)$

For compound gear trains having, say, P gear wheels,

$$N_P = N_A \times \frac{T_A}{T_B} \times \frac{T_C}{T_D} \times \frac{T_E}{T_F} \ldots\ldots \times \frac{T_O}{T_P}$$

from which,

$$\textbf{movement ratio} = \frac{N_A}{N_P} = \frac{T_B}{T_A} \times \frac{T_D}{T_C} \ldots\ldots \times \frac{T_P}{T_O}$$

Problem 7. A driver gear on a shaft of a motor has 35 teeth and meshes with a follower having 98 teeth. If the speed of the motor is 1400 revolutions per minute, find the speed of rotation of the follower.

From equation (19.7),

$$\frac{\text{speed of driver}}{\text{speed of follower}} = \frac{\text{teeth on follower}}{\text{teeth on driver}}$$

i.e. $\dfrac{1400}{\text{speed of follower}} = \dfrac{98}{35}$

Hence, **speed of follower** $= \dfrac{1400 \times 35}{98} = $ **500 rev/min**

Problem 8. A compound gear train similar to that shown in Figure 19.6 consists of a driver gear A, having 40 teeth, engaging with gear B, having 160 teeth. Attached to the same shaft as B, gear C has 48 teeth and meshes with gear D on the output shaft, having 96 teeth. Determine (a) the movement ratio of this gear system and (b) the efficiency when the force ratio is 6.

(a) From equation (19.8), the speed of D

$$= \text{speed of } A \times \frac{T_A}{T_B} \times \frac{T_C}{T_D}$$

From equation (19.7), **movement ratio**

$$= \frac{\text{speed of } A}{\text{speed of } D} = \frac{T_B}{T_A} \times \frac{T_D}{T_C} = \frac{160}{40} \times \frac{96}{48} = \mathbf{8}$$

(b) The efficiency of any simple machine

$$= \frac{\text{force ratio}}{\text{movement ratio}} \times 100\%$$

Thus, **efficiency** $= \dfrac{6}{8} \times 100 = $ **75%**

Now try the following Practise Exercise

Practise Exercise 102   Further problems on gear trains

1. The driver gear of a gear system has 28 teeth and meshes with a follower gear having 168 teeth. Determine the movement ratio and the speed of the follower when the driver gear rotates at 60 revolutions per second.
    [6, 10 rev/s]

2. A compound gear train has a 30-tooth driver gear A, meshing with a 90-tooth follower gear B. Mounted on the same shaft as B and attached to it is a gear C with 60 teeth, meshing with a gear D on the output shaft having 120 teeth. Calculate the movement and force ratios if the overall efficiency of the gears is 72%. [6, 4.32]

3. A compound gear train is as shown in Figure 19.6. The movement ratio is 6 and the numbers of teeth on gears A, C and D are 25, 100 and 60, respectively. Determine the number of teeth on gear B and the force ratio when the efficiency is 60%. [250, 3.6]

19.6   Levers

A **lever** can alter both the magnitude and the line of action of a force and is thus classed as a simple machine. There are three types or orders of levers, as shown in Figure 19.7.

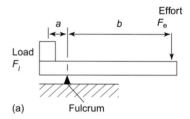
(a)

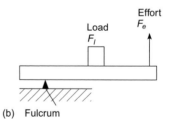
(b)

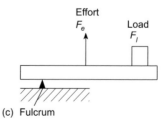
(c)

Figure 19.7

A lever of the first order has the fulcrum placed between the effort and the load, as shown in Figure 19.7(a).

A lever of the second order has the load placed between the effort and the fulcrum, as shown in Figure 19.7(b).

A lever of the third order has the effort applied between the load and the fulcrum, as shown in Figure 19.7(c).

Problems on levers can largely be solved by applying the principle of moments (see Chapter 5). Thus for the lever shown in Figure 19.7(a), when the lever is in equilibrium,

anticlockwise moment = clockwise moment

i.e. $\qquad a \times F_1 = b \times F_e$

Thus, **force ratio** $= \dfrac{F_l}{F_e} = \dfrac{b}{a}$

$$= \dfrac{\textbf{distance of effort from fulcrum}}{\textbf{distance of load from fulcrum}}$$

> **Problem 9.** The load on a first-order lever, similar to that shown in Figure 19.7(a), is 1.2 kN. Determine the effort, the force ratio and the movement ratio when the distance between the fulcrum and the load is 0.5 m and the distance between the fulcrum and effort is 1.5 m. Assume the lever is 100% efficient.

Applying the principle of moments, for equilibrium:

anticlockwise moment = clockwise moment

i.e. $\qquad 1200 \text{ N} \times 0.5 \text{ m} = \text{effort} \times 1.5 \text{ m}$

Hence, $\qquad$ **effort** $= \dfrac{1200 \times 0.5}{1.5} = \textbf{400 N}$

$$\textbf{force ratio} = \dfrac{F_l}{F_e} = \dfrac{1200}{400} = \textbf{3}$$

Alternatively, **force ratio** $= \dfrac{b}{a} = \dfrac{1.5}{0.5} = \textbf{3}$

This result shows that to lift a load of, say, 300 N, an effort of 100 N is required.

Since, from equation (19.3),

$$\text{efficiency} = \dfrac{\text{force ratio}}{\text{movement ratio}} \times 100\%$$

then, **movement ratio** $= \dfrac{\text{force ratio}}{\text{efficiency}} \times 100\%$

$$= \dfrac{3}{100} \times 100 = \textbf{3}$$

This result shows that to raise the load by, say, 100 mm, the effort has to move 300 mm.

> **Problem 10.** A second-order lever, AB, is in a horizontal position. The fulcrum is at point C. An effort of 60 N applied at B just moves a load at point D, when BD is 0.5 m and BC is 1.25 m. Calculate the load and the force ratio of the lever.

A second-order lever system is shown in Figure 19.7(b). Taking moments about the fulcrum as the load is just moving, gives:

anticlockwise moment = clockwise moment

i.e. $\qquad 60 \text{ N} \times 1.25 \text{ m} = \text{load} \times 0.75 \text{ m}$

Thus, $\qquad$ **load** $= \dfrac{60 \times 1.25}{0.75} = \textbf{100 N}$

From equation (19.1),

$$\textbf{force ratio} = \dfrac{\text{load}}{\text{effort}} = \dfrac{100}{60} = \textbf{1.67}$$

Alternatively,

$$\text{force ratio} = \dfrac{\text{distance of effort from fulcrum}}{\text{distance of load from fulcrum}}$$

$$= \dfrac{1.25}{0.75} = \textbf{1.67}$$

Now try the following Practise Exercises

> **Practise Exercise 103 Further problems on levers**
>
> 1. In a second-order lever system, the force ratio is 2.5. If the load is at a distance of 0.5 m from the fulcrum, find the distance that the effort acts from the fulcrum if losses are negligible. [1.25 m]
>
> 2. A lever AB is 2 m long and the fulcrum is at a point 0.5 m from B. Find the effort to be applied at A to raise a load of 0.75 kN at B when losses are negligible. [250 N]
>
> 3. The load on a third-order lever system is at a distance of 750 mm from the fulcrum and the effort required to just move the load is 1 kN when applied at a distance of 250 mm from the fulcrum. Determine the value of the load and the force ratio if losses are negligible. [333.3 N, 1/3]

Practise Exercise 104 Short-answer questions on simple machines

1. State what is meant by a simple machine.

2. Define force ratio.

3. Define movement ratio.

4. Define the efficiency of a simple machine in terms of the force and movement ratios.

5. State briefly why the efficiency of a simple machine cannot reach 100%.

6. With reference to the law of a simple machine, state briefly what is meant by the term 'limiting force ratio'.

7. Define limiting efficiency.

8. Explain why a four-pulley system has a force ratio of 4, when losses are ignored.

9. Give the movement ratio for a screw-jack in terms of the effective radius of the effort and the screw lead.

10. Explain the action of an idler gear.

11. Define the movement ratio for a two-gear system in terms of the teeth on the wheels.

12. Show that the action of an idler wheel does not affect the movement ratio of a gear system.

13. State the relationship between the speed of the first gear and the speed of the last gear in a compound train of four gears, in terms of the teeth on the wheels.

14. Define the force ratio of a first-order lever system in terms of the distances of the load and effort from the fulcrum.

15. Use sketches to show what is meant by: (a) a first-order (b) a second-order (c) a third-order lever system. Give one practical use for each type of lever.

Practise Exercise 105 Multiple-choice questions on simple machines

(Answers on page 298)

A simple machine requires an effort of 250 N moving through 10 m to raise a load of 1000 N

through 2 m. Use this data to find the correct answers to questions 1 to 3, selecting these answers from:

(a) 0.25   (b) 4
(c) 80%   (d) 20%
(e) 100   (f) 5
(g) 100%   (h) 0.2
(i) 25%

1. Find the force ratio.

2. Find the movement ratio.

3. Find the efficiency.

The law of a machine is of the form $F_e = aF_1 + b$. An effort of 12 N is required to raise a load of 40 N and an effort of 6 N is required to raise a load of 16 N. The movement ratio of the machine is 5. Use this data to find the correct answers to questions 4 to 6, selecting these answers from:

(a) 80%   (b) 4
(c) 2.8   (d) 0.25
(e) $\dfrac{1}{2.8}$ (f) 25%
(g) 100%   (h) 2
(i) 25%

4. Determine the constant 'a'.

5. Find the limiting force ratio.

6. Find the limiting efficiency.

7. Which of the following statements is false?

(a) A single-pulley system changes the line of action of the force but does not change the magnitude of the force, when losses are neglected.

(b) In a two-pulley system, the force ratio is $\dfrac{1}{2}$ when losses are neglected.

(c) In a two-pulley system, the movement ratio is 2.

(d) The efficiency of a two-pulley system is 100% when losses are neglected.

8. Which of the following statements concerning a screw-jack is false?

(a) A screw-jack changes both the line of action and the magnitude of the force.

(b) For a single-start thread, the distance moved in 5 revolutions of the table is 5ℓ, where ℓ is the lead of the screw.

(c) The distance moved by the effort is $2\pi r$, where r is the effective radius of the effort.

(d) The movement ratio is given by $\dfrac{2\pi r}{5\ell}$

9. In a simple gear train, a follower has 50 teeth and the driver has 30 teeth. The movement ratio is:

(a) 0.6   (b) 20

(c) 1.67   (d) 80

10. Which of the following statements is true?

(a) An idler wheel between a driver and a follower is used to make the direction of the follower opposite to that of the driver.

(b) An idler wheel is used to change the movement ratio.

(c) An idler wheel is used to change the force ratio.

(d) An idler wheel is used to make the direction of the follower the same as that of the driver.

11. Which of the following statements is false?

(a) In a first-order lever, the fulcrum is between the load and the effort.

(b) In a second-order lever, the load is between the effort and the fulcrum.

(c) In a third-order lever, the effort is applied between the load and the fulcrum.

(d) The force ratio for a first-order lever system is given by:

$$\frac{\text{distance of load from fulcrum}}{\text{distance of effort from fulcrum}}$$

12. In a second-order lever system, the load is 200 mm from the fulcrum and the effort is 500 mm from the fulcrum. If losses are neglected, an effort of 100 N will raise a load of:

(a) 100 N   (b) 250 N

(c) 400 N   (d) 40 N

This Revision Test covers the material contained in Chapters 16 to 19. *The marks for each question are shown in brackets at the end of each question.*

Assume, where necessary, that the acceleration due to gravity, $g = 9.81$ m/s^2

1. The material of a brake is being tested and it is found that the dynamic coefficient of friction between the material and steel is 0.90. Calculate the normal force when the frictional force is 0.630 kN. (4)

2. A mass of 10 kg rests on a plane, which is inclined at 30° to the horizontal. The coefficient of friction between the mass and the plane is 0.6. Determine the magnitude of a force, applied parallel to and up the plane, which will just move the mass up the plane. (10)

3. If in Problem 2, the force required to just move the mass up the plane, is applied horizontally, what will be the minimum value of this force? (10)

4. A train travels around a curve of radius 400 m. If the horizontal thrust on the outer rail is to be 1/30th the weight of the train, what is the velocity of the train (in km/h)? It may be assumed that the inner and outer rail rails are on the same level and that the inner rail takes no horizontal thrust. (6)

5. A conical pendulum of length 2.5 m rotates in a horizontal circle of diameter 0.6 m. Determine its angular velocity. (6)

6. Determine the time of oscillation for a simple pendulum of length 1.5 m. (6)

7. A simple machine raises a load of 120 kg through a distance of 1.2 m. The effort applied to the machine is 150 N and moves through a distance of 12 m. Taking g as 10 m/s^2, determine the force ratio, movement ratio and efficiency of the machine. (6)

8. A load of 30 kg is lifted by a three-pulley system and the applied effort is 140 N. Calculate, taking g to be 9.8 m/s^2, (a) the force ratio (b) the movement ratio (c) the efficiency of the system. (5)

9. A screw-jack is being used to support the axle of a lorry, the load on it being 5.6 kN. The screw jack has an effort of effective radius 318.3 mm and a single-start square thread, having a lead of 5 mm. Determine the efficiency of the jack if an effort of 70 N is required to raise the car axle. (6)

10. A driver gear on a shaft of a motor has 32 teeth and meshes with a follower having 96 teeth. If the speed of the motor is 1410 revolutions per minute, find the speed of rotation of the follower. (4)

11. The load on a first-order lever is 1.5 kN. Determine the effort, the force ratio and the movement ratio when the distance between the fulcrum and the load is 0.4 m and the distance between the fulcrum and effort is 1.6 m. Assume the lever is 100% efficient. (7)

Heat Transfer and Fluid Mechanics

Chapter 20

Heat energy and transfer

This chapter defines sensible and latent heat and provides the appropriate formulae to calculate the amount of energy required to convert a solid to a gas and vice-versa, and also for other combinations of solids, liquids and gases. This information is often required by engineers if they are required to design an artefact (say) to convert ice to steam via the state of liquid water. An example of a household requirement of when this type of calculation is required is that of the simple domestic kettle. When the designer is required to design a domestic electric kettle, it is important that the design is such that the powering arrangement is (just) enough to boil the required amount of water in a reasonable time. If the powering were too low, you may have great difficulty in boiling the water when the kettle is full. Similar calculations are required for large water containers, which are required to boil large quantities of water for other uses, including for kitchens in schools, to make tea/coffee, etc, and for large hotels, which have many uses for hot water. The chapter also describes the three main methods of heat transfer, namely conduction, convection and radiation, together with their uses.

At the end of this chapter you should be able to:

- distinguish between heat and temperature
- appreciate that temperature is measured on the Celsius or the thermodynamic scale
- convert temperatures from Celsius into Kelvin and vice versa
- recognise several temperature measuring devices
- define specific heat capacity, c and recognise typical values
- calculate the quantity of heat energy Q using $Q = mc(t_2 - t_1)$

- understand change of state from solid to liquid to gas, and vice versa
- distinguish between sensible and latent heat
- define specific latent heat of fusion
- define specific latent heat of vaporisation
- recognise typical values of latent heats of fusion and vaporisation
- calculate quantity of heat Q using $Q = mL$
- describe the principle of operation of a simple refrigerator

20.1 Introduction

Heat is a form of energy and is measured in joules. **Temperature** is the degree of hotness or coldness of a substance. Heat and temperature are thus not the same thing. For example, twice the heat energy is needed to boil a full container of water than half a container – that is, different amounts of heat energy are needed to cause an equal rise in the temperature of different amounts of the same substance.

Temperature is measured either (i) on the **Celsius (°C)** scale (formerly Centigrade), where the temperature at which ice melts, i.e. the freezing point of water, is taken as 0°C and the point at which water boils under normal atmospheric pressure is taken as 100°C, or (ii) on the **thermodynamic scale**, in which the unit of temperature is the **kelvin (K)**. The kelvin scale uses the same temperature interval as the Celsius scale but as its zero takes the 'absolute zero of temperature' which is at about – 273°C. Hence,

kelvin temperature = degree Celsius + 273

i.e.

K = (°C) + 273

Mechanical Engineering Principles, Bird and Ross, ISBN 9780415517850

Thus, for example, 0°C = 273 K, 25°C = 298 K and 100°C = 373 K

> **Problem 1.** Convert the following temperatures into the Kelvin scale:
> (a) 37°C   (b) −28°C

From above, Kelvin temperature = degree Celsius + 273

(a) 37°C corresponds to a Kelvin temperature of 37 + 273, i.e. **310 K**

(b) −28°C corresponds to a Kelvin temperature of −28 + 273, i.e. **245 K**

> **Problem 2.** Convert the following temperatures into the Celsius scale:
> (a) 365 K   (b) 213 K

From above, $K = (°C) + 273$

Hence, degree Celsius = Kelvin temperature −273

(a) 365 K corresponds to 365 − 273, i.e. **92°C**

(b) 213 K corresponds to 213 − 273, i.e. **− 60°C**

Now try the following Practise Exercise

> **Practise Exercise 106   Further problems on temperature scales**
>
> 1. Convert the following temperatures into the Kelvin scale:
>   (a) 51°C   (b) −78°C
>   (c) 183°C
>   [(a) 324 K   (b) 195 K   (c) 456 K]
>
> 2. Convert the following temperatures into the Celsius scale:
>   (a) 307 K   (b) 237 K
>   (c) 415 K
>   [(a) 34°C   (b) −36°C   (c) 142°C]

20.2   The measurement of temperature

A **thermometer** is an instrument that measures temperature. Any substance that possesses one or more properties that vary with temperature can be used to measure temperature. These properties include changes in length, area or volume, electrical resistance or in colour. Examples of temperature measuring devices include:

(i) **liquid-in-glass thermometer**, which uses the expansion of a liquid with increase in temperature as its principle of operation,

(ii) **thermocouples**, which use the e.m.f. set up, when the junction of two dissimilar metals is heated,

(iii) **resistance thermometer**, which uses the change in electrical resistance caused by temperature change, and

(iv) **pyrometers**, which are devices for measuring very high temperatures, using the principle that all substances emit radiant energy when hot, the rate of emission depending on their temperature.

Each of these temperature measuring devices, together with others, are described in Chapter 25, page 281 .

20.3   Specific heat capacity

The **specific heat capacity** of a substance is the quantity of heat energy required to raise the temperature of 1 kg of the substance by 1°C. The symbol used for specific heat capacity is c and the units are J/(kg °C) or J/(kg K). (Note that these units may also be written as J kg^{-1} °C^{-1} or J kg^{-1} K^{-1}).

Some typical values of specific heat capacity for the range of temperature 0°C to 100°C include:

Water	4190 J/(kg °C),
Ice	2100 J/(kg °C)
Aluminium	950 J/(kg °C),
Copper	390 J/(kg °C)
Iron	500 J/(kg °C),
Lead	130 J/(kg °C)

Hence to raise the temperature of 1 kg of iron by 1°C requires 500 J of energy, to raise the temperature of 5 kg of iron by 1°C requires (500 × 5) J of energy, and to raise the temperature of 5 kg of iron by 40°C requires (500 × 5 × 40) J of energy, i.e. 100 kJ.

In general, the quantity of heat energy, Q, required to raise a mass m kg of a substance with a specific heat capacity of c J/(kg °C), from temperature t_1 °C to t_2 °C is given by:

$$Q = mc(t_2 − t_1) \text{ joules}$$

> **Problem 3.** Calculate the quantity of heat required to raise the temperature of 5 kg of water from 0°C to 100°C. Assume the specific heat capacity of water is 4200 J/(kg °C).

Quantity of heat energy,

$$Q = mc(t_2 - t_1)$$
$$= 5 \text{ kg} \times 4200 \text{ J/(kg °C)} \times (100 - 0)°C$$
$$= 5 \times 4200 \times 100$$
$$= \textbf{2100000 J} \text{ or } \textbf{2100 kJ} \text{ or } \textbf{2.1 MJ}$$

Problem 4. A block of cast iron having a mass of 10 kg cools from a temperature of 150°C to 50°C. How much energy is lost by the cast iron? Assume the specific heat capacity of iron is 500 J/(kg °C).

Quantity of heat energy,

$$Q = mc(t_2 - t_1)$$
$$= 10 \text{ kg} \times 500 \text{ J/(kg °C)} \times (50 - 150)°C$$
$$= 10 \times 500 \times (-100)$$
$$= \textbf{-500000 J} \text{ or } \textbf{-500 kJ} \text{ or } \textbf{-0.5 MJ}$$

(Note that the minus sign indicates that heat is given out or lost).

Problem 5. Some lead having a specific heat capacity of 130 J/(kg °C) is heated from 27°C to its melting point at 327°C. If the quantity of heat required is 780 kJ, determine the mass of the lead.

Quantity of heat, $Q = mc(t_2 - t_1)$, hence,

$$780 \times 10^3 \text{ J} = m \times 130 \text{ J/(kg °C)} \times (327 - 27)°C$$

i.e. $\qquad 780000 = m \times 130 \times 300$

from which, **mass, m** $= \dfrac{780000}{130 \times 300}$ kg $= \textbf{20 kg}$

Problem 6. 273 kJ of heat energy are required to raise the temperature of 10 kg of copper from 15°C to 85°C. Determine the specific heat capacity of copper.

Quantity of heat, $Q = mc(t_2 - t_1)$, hence:

$$273 \times 10^3 \text{ J} = 10 \text{ kg} \times c \times (85 - 15)°C$$

where c is the specific heat capacity,

i.e. $\qquad 273000 = 10 \times c \times 70$

from which, **specific heat capacity,** $c = \dfrac{273000}{10 \times 70}$

$$= \textbf{390 J/(kg °C)}$$

Problem 7. 5.7 MJ of heat energy are supplied to 30 kg of aluminium that is initially at a temperature of 20°C. If the specific heat capacity of aluminium is 950 J(kg°C), determine its final temperature.

Quantity of heat,

$$Q = mc(t_2 - t_1), \text{ hence,}$$
$$5.7 \times 10^6 \text{ J} = 30 \text{ kg} \times 950 \text{ J/(kg °C)} \times (t_2 - 20)°C$$

from which, $(t_2 - 20) = \dfrac{5.7 \times 10^6}{30 \times 950} = 200$

Hence, the **final temperature, t_2** $= 200 + 20$

$$= \textbf{220°C}$$

Problem 8. A copper container of mass 500 g contains 1 litre of water at 293 K. Calculate the quantity of heat required to raise the temperature of the water and container to boiling point, assuming there are no heat losses. Assume that the specific heat capacity of copper is 390 J/(kg K), the specific heat capacity of water is 4.2 kJ(kg K) and 1 litre of water has a mass of 1 kg.

Heat is required to raise the temperature of the water, and also to raise the temperature of the copper container.

For the water: $m = 1$ kg, $t_1 = 293$ K,

$$t_2 = 373 \text{ K (i.e. boiling point) and}$$
$$c = 4.2 \text{ kJ/(kg K)}$$

Quantity of heat required for the water is given by:

$$Q_w = mc(t_2 - t_1) = (1 \text{ kg})\left(4.2 \dfrac{\text{kJ}}{\text{kg K}}\right)(373 - 293) \text{ K}$$
$$= 4.2 \times 80 \text{ kJ}$$

i.e. $\quad \boldsymbol{Q_w = 336 \text{ kJ}}$

For the copper container: $m = 500$ g $= 0.5$ kg,

$$t_1 = 293 \text{ K,}$$
$$t_2 = 373 \text{ K and}$$
$$c = 390 \text{ J/(kg K)}$$
$$= 0.39 \text{ kJ(kg K)}$$

Quantity of heat required for the copper container is given by:

$$Q_C = mc(t_2 - t_1) = (0.5 \text{ kg})(0.39 \text{ kJ/(kg K)})(80 \text{ K})$$

i.e. $\quad \boldsymbol{Q_C = 15.6 \text{ kJ}}$

Total quantity of heat required, $Q = Q_w + Q_C$

$$= 336 + 15.6$$
$$= \textbf{351.6 kJ}$$

Now try the following Practise Exercise

> **Practise Exercise 107** **Further problems on specific heat capacity**
>
> 1. Determine the quantity of heat energy (in megajoules) required to raise the temperature of 10 kg of water from 0°C to 50°C. Assume the specific heat capacity of water is 4200 J/(kg °C). [2.1 MJ]
>
> 2. Some copper, having a mass of 20 kg, cools from a temperature of 120°C to 70°C. If the specific heat capacity of copper is 390 J/(kg °C), how much heat energy is lost by the copper ? [390 kJ]
>
> 3. A block of aluminium having a specific heat capacity of 950 J/(kg °C) is heated from 60°C to its melting point at 660°C. If the quantity of heat required is 2.85 MJ, determine the mass of the aluminium block. [5 kg]
>
> 4. 20.8 kJ of heat energy is required to raise the temperature of 2 kg of lead from 16°C to 96°C. Determine the specific heat capacity of lead. [130 J/kg °C]
>
> 5. 250 kJ of heat energy is supplied to 10 kg of iron which is initially at a temperature of 15°C. If the specific heat capacity of iron is 500 J/(kg °C) determine its final temperature. [65°C]

20.4 Change of state

A material may exist in any one of three states – solid, liquid or gas. If heat is supplied at a constant rate to some ice initially at, say, –30°C, its temperature rises as shown in Figure 20.1. Initially the temperature increases from –30°C to 0°C as shown by the line *AB*. It then remains constant at 0°C for the time *BC* required for the ice to melt into water.

When melting commences the energy gained by continual heating is offset by the energy required for the change of state and the temperature remains constant even though heating is continued. When the ice is completely melted to water, continual heating raises the temperature to 100°C, as shown by *CD* in Figure 20.1. The water then begins to boil and the temperature

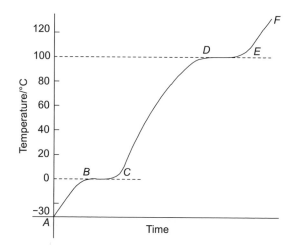

Figure 20.1

again remains constant at 100°C, shown as *DE*, until all the water has vaporised.

Continual heating raises the temperature of the steam as shown by *EF* in the region where the steam is termed superheated.

Changes of state from solid to liquid or liquid to gas occur without change of temperature and such changes are reversible processes. When heat energy flows to or from a substance and causes a change of temperature, such as between *A* and *B*, between *C* and *D* and between *E* and *F* in Figure 20.1, it is called **sensible heat** (since it can be 'sensed' by a thermometer).

Heat energy which flows to or from a substance while the temperature remains constant, such as between *B* and *C* and between *D* and *E* in Figure 20.1, is called **latent heat** (latent means concealed or hidden).

> **Problem 9.** Steam initially at a temperature of 130°C is cooled to a temperature of 20°C below the freezing point of water, the loss of heat energy being at a constant rate. Make a sketch, and briefly explain, the expected temperature/time graph representing this change.

A temperature/time graph representing the change is shown in Figure 20.2. Initially steam cools until it reaches the boiling point of water at 100°C. Temperature then remains constant, i.e. between *A* and *B*, even though it is still giving off heat (i.e. latent heat). When all the steam at 100°C has changed to water at 100°C it starts to cool again until it reaches the freezing point of water at 0°C. From *C* to *D* the temperature again remains constant (i.e. latent heat), until all the water is converted to ice. The temperature of the ice then decreases as shown.

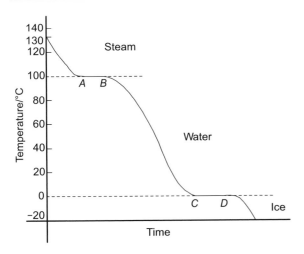

Figure 20.2

Now try the following Practise Exercise

20.5   Latent heats of fusion and vaporisation

The **specific latent heat of fusion** is the heat required to change 1 kg of a substance from the solid state to the liquid state (or vice versa) at constant temperature. The **specific latent heat of vaporisation** is the heat required to change 1 kg of a substance from a liquid to a gaseous state (or vice versa) at constant temperature.

The units of the specific latent heats of fusion and vaporisation are J/kg, or more often kJ/kg, and some typical values are shown in Table 20.1

The quantity of heat Q supplied or given out during a change of state is given by:

$$Q = mL$$

where m is the mass in kilograms and L is the specific latent heat.

Table 20.1

	Latent heat of fusion (kJ/kg)	Melting point (°C)
Mercury	11.8	– 39
Lead	22	327
Silver	100	957
Ice	335	0
Aluminium	387	660

	Latent heat of vaporisation (kJ/kg)	Boiling point (°C)
Oxygen	214	– 183
Mercury	286	357
Ethyl alcohol	857	79
Water	2257	100

Thus, for example, the heat required to convert 10 kg of ice at 0°C to water at 0°C is given by

$$10 \text{ kg} \times 335 \text{ kJ/kg} = 3350 \text{ kJ or } 3.35 \text{ MJ}$$

Besides changing temperature, the effects of supplying heat to a material can involve changes in dimensions, as well as in colour, state and electrical resistance. Most substances expand when heated and contract when cooled, and there are many practical applications and design implications of thermal movement (see Chapter 21).

Problem 10.   How much heat is needed to melt completely 12 kg of ice at 0°C ? Assume the latent heat of fusion of ice is 335 kJ/kg.

Quantity of heat required, $Q = mL = 12 \text{ kg} \times 335 \text{ kJ/kg}$

$$= \textbf{4020 kJ} \text{ or } \textbf{4.02 MJ}$$

Problem 11.   Calculate the heat required to convert 5 kg of water at 100°C to superheated steam at 100°C. Assume the latent heat of vaporisation of water is 2260 kJ/kg.

Quantity of heat required, $Q = mL = 5 \text{ kg} \times 2260 \text{ kJ/kg}$

$$= \textbf{11300 kJ} \text{ or } \textbf{11.3 MJ}$$

Problem 12.   Determine the heat energy needed to convert 5 kg of ice initially at – 20°C completely to water at 0°C. Assume the specific heat capacity of ice is 2100 J/(kg °C) and the specific latent heat of fusion of ice is 335 kJ/kg.

Quantity of heat energy needed, Q = sensible heat + latent heat.

The quantity of heat needed to raise the temperature of ice from –20°C to 0°C i.e. sensible heat,

$$Q_1 = mc(t_2 - t_1)$$
$$= 5 \text{ kg} \times 2100 \text{ J/(kg°C)} \times (0 - -20)°C$$
$$= (5 \times 2100 \times 20) \text{ J} = \textbf{210 kJ}$$

The quantity of heat needed to melt 5 kg of ice at 0°C,

i.e. the latent heat, $Q_2 = mL$
$$= 5 \text{ kg} \times 335 \text{ kJ/kg}$$
$$= \textbf{1675 kJ}$$

Total heat energy needed, $Q = Q_1 + Q_2$
$$= 210 + 1675 = \textbf{1885 kJ}$$

Problem 13. Calculate the heat energy required to convert completely 10 kg of water at 50°C into steam at 100°C, given that the specific heat capacity of water is 4200 J/(kg °C) and the specific latent heat of vaporisation of water is 2260 kJ/kg.

Quantity of heat required = sensible heat + latent heat.

Sensible heat,

$$Q_1 = mc(t_2 - t_1)$$
$$= 10 \text{ kg} \times 4200 \text{ J/(kg °C)} \times (100 - 50)°C$$
$$= \textbf{2100 kJ}$$

Latent heat, $Q_2 = mL = 10 \text{ kg} \times 2260 \text{ kJ/kg}$
$$= \textbf{22600 kJ}$$

Total heat energy required, $Q = Q_1 + Q_2$
$$= (2100 + 22600) \text{ kJ}$$
$$= \textbf{24700 kJ or 24.70 MJ}$$

Problem 14. Determine the amount of heat energy needed to change 400 g of ice, initially at –20°C, into steam at 120°C. Assume the following: latent heat of fusion of ice = 335 kJ/kg, latent heat of vaporisation of water = 2260 kJ/kg, specific heat capacity of ice = 2.14 kJ/(kg °C), specific heat capacity of water = 4.2 kJ(kg °C) and specific heat capacity of steam = 2.01 kJ/(kg °C).

The energy needed is determined in five stages:

(i) Heat energy needed to change the temperature of ice from –20°C to 0°C is given by:
$$Q_1 = mc(t_2 - t_1)$$
$$= 0.4 \text{ kg} \times 2.14 \text{ kJ/(kg °C)} \times (0 - -20)°C$$
$$= \textbf{17.12 kJ}$$

(ii) Latent heat needed to change ice at 0°C into water at 0°C is given by:
$$Q_2 = mL_f = 0.4 \text{ kg} \times 335 \text{ kJ/kg} = \textbf{134 kJ}$$

(iii) Heat energy needed to change the temperature of water from 0°C (i.e. melting point) to 100°C (i.e. boiling point) is given by:
$$Q_3 = mc(t_2 - t_1)$$
$$= 0.4 \text{ kg} \times 4.2 \text{ kJ/(kg °C)} \times 100°C$$
$$= \textbf{168 kJ}$$

(iv) Latent heat needed to change water at 100°C into steam at 100°C is given by:
$$Q_4 = mL_v = 0.4 \text{ kg} \times 2260 \text{ kJ/kg} = \textbf{904 kJ}$$

(v) Heat energy needed to change steam at 100°C into steam at 120°C is given by:
$$Q_5 = mc(t_1 - t_2)$$
$$= 0.4 \text{ kg} \times 2.01 \text{ kJ/(kg °C)} \times 20°C$$
$$= \textbf{16.08 kJ}$$

Total heat energy needed,
$$Q = Q_1 + Q_2 + Q_3 + Q_4 + Q_5$$
$$= 17.12 + 134 + 168 + 904 + 16.08$$
$$= \textbf{1239.2 kJ}$$

Now try the following Practise Exercise

Practise Exercise 109   Further problems on the latent heats of fusion and vaporisation

1. How much heat is needed to melt completely 25 kg of ice at 0°C. Assume the specific latent heat of fusion of ice is 335 kJ/kg.
[8.375 MJ]

2. Determine the heat energy required to change 8 kg of water at 100°C to superheated steam at 100°C. Assume the specific latent heat of vaporisation of water is 2260 kJ/kg.
[18.08 MJ]

3. Calculate the heat energy required to convert 10 kg of ice initially at $-30°C$ completely into water at $0°C$. Assume the specific heat capacity of ice is 2.1 kJ/(kg °C) and the specific latent heat of fusion of ice is 335 kJ/kg.
[3.98 MJ]

4. Determine the heat energy needed to convert completely 5 kg of water at $60°C$ to steam at $100°C$, given that the specific heat capacity of water is 4.2 kJ/(kg °C) and the specific latent heat of vaporisation of water is 2260 kJ/kg.
[12.14 MJ]

20.6   A simple refrigerator

The boiling point of most liquids may be lowered if the pressure is lowered. In a simple refrigerator a working fluid, such as ammonia or freon, has the pressure acting on it reduced. The resulting lowering of the boiling point causes the liquid to vaporise. In vaporising, the liquid takes in the necessary latent heat from its surroundings, i.e. the freezer, which thus becomes cooled. The vapour is immediately removed by a pump to a condenser that is outside of the cabinet, where it is compressed and changed back into a liquid, giving out latent heat. The cycle is repeated when the liquid is pumped back to the freezer to be vaporised.

20.7   Conduction, convection and radiation

Heat may be **transferred** from a hot body to a cooler body by one or more of three methods, these being: (a) by **conduction** (b) by **convection** or (c) by **radiation**.

Conduction

Conduction is the transfer of heat energy from one part of a body to another (or from one body to another) without the particles of the body moving.

Conduction is associated with solids. For example, if one end of a metal bar is heated, the other end will become hot by conduction. Metals and metallic alloys are good conductors of heat, whereas air, wood, plastic, cork, glass and gases are examples of poor conductors (i.e. they are heat insulators).

Practical applications of conduction include:

(i)   A domestic saucepan or dish conducts heat from the source to the contents. Also, since wood and plastic are poor conductors of heat they are used for saucepan handles.

(ii)   The metal of a radiator of a central heating system conducts heat from the hot water inside to the air outside.

Convection

Convection is the transfer of heat energy through a substance by the actual movement of the substance itself. Convection occurs in liquids and gases, but not in solids. When heated, a liquid or gas becomes less dense. It then rises and is replaced by a colder liquid or gas and the process repeats. For example, electric kettles and central heating radiators always heat up at the top first.

Examples of convection are:

(i)   Natural circulation hot water heating systems depend on the hot water rising by convection to the top of the house and then falling back to the bottom of the house as it cools, releasing the heat energy to warm the house as it does so.

(ii)   Convection currents cause air to move and therefore affect climate.

(iii)   When a radiator heats the air around it, the hot air rises by convection and cold air moves in to take its place.

(iv)   A cooling system in a car radiator relies on convection.

(v)   Large electrical transformers dissipate waste heat to an oil tank. The heated oil rises by convection to the top, then sinks through cooling fins, losing heat as it does so.

(vi)   In a refrigerator, the cooling unit is situated near the top. The air surrounding the cold pipes become heavier as it contracts and sinks towards the bottom. Warmer, less dense air is pushed upwards and in turn is cooled. A cold convection current is thus created.

Radiation

Radiation is the transfer of heat energy from a hot body to a cooler one by electromagnetic waves. Heat radiation is similar in character to light waves – it travels at the same speed and can pass through a vacuum – except that the frequency of the waves is different. Waves are emitted by a hot body, are transmitted through space (even a vacuum) and are not detected until they fall on to another body. Radiation is reflected from shining, polished surfaces but absorbed by dull, black surfaces.

Practical applications of radiation include:

(i)   heat from the sun reaching earth
(ii) heat felt by a flame
(iii) cooker grills
(iv) industrial furnaces
(v)   infra-red space heaters.

20.8 Vacuum flask

A cross-section of a typical vacuum flask is shown in Figure 20.3 and is seen to be a double-walled bottle with a vacuum space between them, the whole supported in a protective outer case.

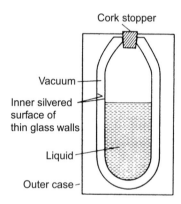

Figure 20.3

Very little heat can be transferred by conduction because of the vacuum space and the cork stopper (cork is a bad conductor of heat). Also, because of the vacuum space, no convection is possible. Radiation is minimised by silvering the two glass surfaces (radiation is reflected off shining surfaces).

Thus a vacuum flask is an example of prevention of all three types of heat transfer and is therefore able to keep hot liquids hot and cold liquids cold.

20.9 Use of insulation in conserving fuel

Fuel used for heating a building is becoming increasingly expensive. By the careful use of insulation, heat can be retained in a building for longer periods and the cost of heating thus minimised.

(i)   Since convection causes hot air to rise it is important to insulate the roof space, which is probably the greatest source of heat loss in the home. This can be achieved by laying fibre-glass between the wooden joists in the roof space.

(ii) Glass is a poor conductor of heat. However, large losses can occur through thin panes of glass and such losses can be reduced by using double-glazing. Two sheets of glass, separated by air, are used. Air is a very good insulator but the air space must not be too large otherwise convection currents can occur which would carry heat across the space.

(iii) Hot water tanks should be lagged to prevent conduction and convection of heat to the surrounding air.

(iv) Brick, concrete, plaster and wood are all poor conductors of heat. A house is made from two walls with an air gap between them. Air is a poor conductor and trapped air minimises losses through the wall. Heat losses through the walls can be prevented almost completely by using cavity wall insulation, i.e. plastic-foam.

Besides changing temperature, the effects of supplying heat to a material can involve changes in dimensions, as well as in colour, state and electrical resistance.

Most substances expand when heated and contract when cooled, and there are many practical applications and design implications of thermal movement as explained in Chapter 21 following.

Now try the following Practise Exercises

Practise Exercise 110   Short-answer questions on heat energy

1. Differentiate between temperature and heat.

2. Name two scales on which temperature is measured.

3. Name any four temperature measuring devices.

4. Define specific heat capacity and name its unit.

5. Differentiate between sensible and latent heat.

6. The quantity of heat, Q, required to raise a mass m kg from temperature $t_1°$ C to t_2 °C, the specific heat capacity being c, is given by $Q = $ ......

7. What is meant by the specific latent heat of fusion?

8. Define the specific latent heat of vaporisation.

9. Explain briefly the principle of operation of a simple refrigerator.

10. State three methods of heat transfer.

11. Define conduction and state two practical examples of heat transfer by this method.

12. Define convection and give three examples of heat transfer by this method.

13. What is meant by radiation? Give three uses.

14. How can insulation conserve fuel in a typical house?

Practise Exercise 111   Multiple-choice questions on heat energy

(Answers on page 298)

1. Heat energy is measured in:
    (a) kelvin   (b) watts
    (c) kilograms   (d) joules

2. A change of temperature of 20°C is equivalent to a change in thermodynamic temperature of:
    (a) 293 K   (b) 20 K
    (c) 80 K   (d) 120 K

3. A temperature of 20°C is equivalent to:
    (a) 293 K   (b) 20 K
    (c) 80 K   (d) 120 K

4. The unit of specific heat capacity is:
    (a) joules per kilogram
    (b) joules
    (c) joules per kilogram kelvin
    (d) cubic metres

5. The quantity of heat required to raise the temperature of 500 g of iron by 2°C,

given that the specific heat capacity is 500 J/(kg °C), is:
    (a) 500 kJ   (b) 0.5 kJ
    (c) 2 J   (d) 250 kJ

6. The heat energy required to change 1 kg of a substance from a liquid to a gaseous state at the same temperature is called:
    (a) specific heat capacity
    (b) specific latent heat of vaporisation
    (c) sensible heat
    (d) specific latent heat of fusion

7. The temperature of pure melting ice is:
    (a) 373 K   (b) 273 K
    (c) 100 K   (d) 0 K

8. 1.95 kJ of heat is required to raise the temperature of 500 g of lead from 15°C to its final temperature. Taking the specific heat capacity of lead to be 130 J/(kg °C), the final temperature is:
    (a) 45°C   (b) 37.5°C
    (c) 30°C   (d) 22.5°C

9. Which of the following temperature is absolute zero ?
    (a) 0°C   (b) – 173°C
    (c) – 273°C   (d) – 373°C

10. When two wires of different metals are twisted together and heat applied to the junction, an e.m.f. is produced. This effect is used in a thermocouple to measure:
    (a) e.m.f.   (b) temperature
    (c) expansion   (d) heat

11. Which of the following statements is false?
    (a) – 30°C is equivalent to 243 K
    (b) Convection only occurs in liquids and gases
    (c) Conduction and convection cannot occur in a vacuum
    (d) Radiation is absorbed by a silver surface

12. The transfer of heat through a substance by the actual movement of the particles of the substance is called:
    (a) conduction
    (b) radiation

(c) convection

(d) specific heat capacity

13. Which of the following statements is true?

(a) Heat is the degree of hotness or coldness of a body.

(b) Heat energy that flows to or from a substance while the temperature remains constant is called sensible heat.

(c) The unit of specific latent heat of fusion is J/(kg K).

(d) A cooker-grill is a practical application of radiation.

Chapter 21

Thermal expansion

Thermal expansion and contraction are very important features in engineering science. For example, if the metal railway lines of a railway track are heated or cooled due to weather conditions or time of day, their lengths can increase or decrease accordingly. If the metal lines are heated due to the weather effects, then the railway lines will attempt to expand, and depending on their construction, they can buckle, rendering the track useless for transporting trains. In countries with large temperature variations, this effect can be much worse, and the engineer may have to choose a superior metal to withstand these changes. The effect of metals expanding and contracting due to the rise and fall of temperatures, accordingly, can also be put to good use. A classic example of this is the simple humble domestic thermostat, which when the water gets too hot, will cause the metal thermostat to expand and switch off the electric heater; conversely, when the water becomes too cool, the metal thermostat shrinks, causing the electric heater to switch on again. All sorts of materials, besides metals, are affected by thermal expansion and contraction. The chapter also defines the coefficients of linear, superficial and cubic expansion.

At the end of this chapter you should be able to:

- appreciate that expansion and contraction occurs with change of temperature
- describe practical applications where expansion and contraction must be allowed for
- understand the expansion and contraction of water
- define the coefficient of linear expansion α
- recognise typical values for the coefficient of linear expansion
- calculate the new length L_2, after expansion or contraction, using $L_2 = L_1 [1 + \alpha(t_2 - t_1)]$

- define the coefficient of superficial expansion β
- calculate the new surface area A_2, after expansion or contraction, using $A_2 = A_1 [1 + \beta(t_2 - t_1)]$
- appreciate that $\beta \approx 2\alpha$
- define the coefficient of cubic expansion γ
- recognise typical values for the coefficient of cubic expansion
- appreciate that $\gamma \approx 3\alpha$
- calculate the new volume V_2, after expansion or contraction, using $V_2 = V_1 [1 + \gamma(t_2 - t_1)]$

21.1 Introduction

When heat is applied to most materials, **expansion** occurs in all directions (see Section 2.12, page 33). Conversely, if heat energy is removed from a material (i.e. the material is cooled) **contraction** occurs in all directions. The effects of expansion and contraction each depend on the **change of temperature** of the material.

21.2 Practical applications of thermal expansion

Some practical applications where expansion and contraction of solid materials must be allowed for include:

(i) Overhead electrical transmission lines are hung so that they are slack in summer, otherwise their contraction in winter may snap the conductors or bring down pylons.

(ii) Gaps need to be left in lengths of railway lines to prevent buckling in hot weather (except where these are continuously welded).

Mechanical Engineering Principles, Bird and Ross, ISBN 9780415517850

(iii) Ends of large bridges are often supported on rollers to allow them to expand and contract freely.

(iv) Fitting a metal collar to a shaft or a steel tyre to a wheel is often achieved by first heating the collar or tyre so that they expand, fitting them in position, and then cooling them so that the contraction holds them firmly in place; this is known as a 'shrink-fit'. By a similar method hot rivets are used for joining metal sheets.

(v) The amount of expansion varies with different materials. Figure 21.1(a) shows a bimetallic strip at room temperature (i.e. two different strips of metal riveted together – see Section 2.12, page 33). When heated, brass expands more than steel, and since the two metals are riveted together the bimetallic strip is forced into an *arc* as shown in Figure 21.1(b). Such a movement can be arranged to make or break an electric circuit and bimetallic strips are used, in particular, in thermostats (which are temperature-operated switches) used to control central heating systems, cookers, refrigerators, toasters, irons, hot-water and alarm systems.

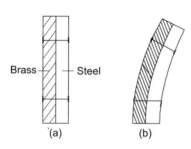

Brass — — Steel

(a)   (b)

Figure 21.1

(vi) Motor engines use the rapid expansion of heated gases to force a piston to move.

(vii) Designers must predict, and allow for, the expansion of steel pipes in a steam-raising plant so as to avoid damage and consequent danger to health.

21.3 Expansion and contraction of water

Water is a liquid that at low temperature displays an unusual effect. If cooled, contraction occurs until, at about 4°C, the volume is at a minimum. As the temperature is further decreased from 4°C to 0°C expansion occurs, i.e. the volume increases. (For cold, deep fresh water, the temperature at the bottom is more likely to be about 4°C, somewhat warmer than less deep water). When ice is formed, considerable expansion occurs and it is this expansion that often causes frozen water pipes to burst.

A practical application of the expansion of a liquid is with thermometers, where the expansion of a liquid, such as mercury or alcohol, is used to measure temperature.

21.4 Coefficient of linear expansion

The amount by which unit length of a material expands when the temperature is raised one degree is called the **coefficient of linear expansion** of the material and is represented by α (Greek alpha) – see Section 2.12, page 33.

The units of the coefficient of linear expansion are m/(mK), although it is usually quoted as just/K or K^{-1}. For example, copper has a coefficient of linear expansion value of 17×10^{-6} K^{-1}, which means that a 1 m long bar of copper expands by 0.000017 m if its temperature is increased by 1 K (or 1°C). If a 6 m long bar of copper is subjected to a temperature rise of 25 K then the bar will expand by $(6 \times 0.000017 \times 25)$ m, i.e. 0.00255 m or 2.55 mm. (Since the Kelvin scale uses the same temperature interval as the Celsius scale, a **change** of temperature of, say, 50°C, is the same as a change of temperature of 50 K).

If a material, initially of length L_1 and at a temperature of t_1 and having a coefficient of linear expansion α, has its temperature increased to t_2, then the new length L_2 of the material is given by:

New length = original length + expansion

i.e. $\qquad L_2 = L_1 + L_1\,\alpha(t_2 - t_1)$

i.e. $\qquad \boldsymbol{L_2 = L_1\,[1 + \alpha(t_2 - t_1)]}$ $\qquad$ (21.1)

Some typical values for the coefficient of linear expansion include:

Aluminium	$23 \times 10^{-6}\,K^{-1}$	Brass	$18 \times 10^{-6}\,K^{-1}$
Concrete	$12 \times 10^{-6}\,K^{-1}$	Copper	$17 \times 10^{-6}\,K^{-1}$
Gold	$14 \times 10^{-6}\,K^{-1}$	Invar (nickel-steel alloy)	$0.9 \times 10^{-6}\,K^{-1}$
Iron	$11 - 12 \times 10^{-6}\,K^{-1}$	Nylon	$100 \times 10^{-6}\,K^{-1}$

Steel	$15 - 16 \times 10^{-6}$ K^{-1}	Tungsten 4.5×10^{-6} K^{-1}
Zinc	31×10^{-6} K^{-1}	

Problem 1. The length of an iron steam pipe is 20.0 m at a temperature of 18°C. Determine the length of the pipe under working conditions when the temperature is 300°C. Assume the coefficient of linear expansion of iron is 12×10^{-6} K^{-1}.

Length $L_1 = 20.0$ m, temperature $t_1 = 18$°C,
$t_2 = 300$°C and $\alpha = 12 \times 10^{-6}$ K^{-1}

Length of pipe at 300°C is given by:

$$L_2 = L_1 [1 + \alpha(t_2 - t_1)]$$
$$= 20.0[1 + (12 \times 10^{-6})(300 - 18)]$$
$$= 20.0[1 + 0.003384]$$
$$= 20.0[1.003384] = \mathbf{20.06768 \ m}$$

i.e. an increase in length of 0.06768 m or 67.68 mm.

In practice, allowances are made for such expansions. U-shaped expansion joints are connected into pipelines carrying hot fluids to allow some 'give' to take up the expansion.

Problem 2. An electrical overhead transmission line has a length of 80.0 m between its supports at 15°C. Its length increases by 92 mm at 65°C. Determine the coefficient of linear expansion of the material of the line.

Length $L_1 = 80.0$ m, $L_2 = 80.0$ m + 92 mm
$= 80.092$ m,
temperature $t_1 = 15$°C and temperature $t_2 = 65$°C

Length $L_2 = L_1 [1 + \alpha(t_2 - t_1)]$

i.e. $80.092 = 80.0[1 + \alpha(65 - 15)]$

$80.092 = 80.0 + (80.0)(\alpha)(50)$

i.e. $80.092 - 80.0 = (80.0)(\alpha)(50)$

Hence, the coefficient of linear expansion,

$$\alpha = \frac{0.092}{(80.0)(50)} = 0.000023$$

i.e. $\alpha = \mathbf{23 \times 10^{-6} \ K^{-1}}$ (which is aluminium – see above)

Problem 3. A measuring tape made of copper measures 5.0 m at a temperature of 288 K.

Calculate the percentage error in measurement when the temperature has increased to 313 K. Take the coefficient of linear expansion of copper as 17×10^{-6} K^{-1}.

Length $L_1 = 5.0$ m, temperature $t_1 = 288$ K,
$t_2 = 313$ K and $\alpha = 17 \times 10^{-6}$ K^{-1}

Length at 313 K is given by:

Length $L_2 = L_1 [1 + \alpha(t_2 - t_1)]$
$$= 5.0[1 + (17 \times 10^{-6})(313 - 288)]$$
$$= 5.0[1 + (17 \times 10^{-6})(25)]$$
$$= 5.0[1 + 0.000425]$$
$$= 5.0[1.000425] = 5.002125 \text{ m}$$

i.e. the length of the tape has increased by 0.002125 m

Percentage error in measurement at 313 K

$$= \frac{\text{increase in length}}{\text{original length}} \times 100\%$$

$$= \frac{0.002125}{5.0} \times 100 = \mathbf{0.0425\%}$$

Problem 4. The copper tubes in a boiler are 4.20 m long at a temperature of 20°C. Determine the length of the tubes (a) when surrounded only by feed water at 10°C (b) when the boiler is operating and the mean temperature of the tubes is 320°C. Assume the coefficient of linear expansion of copper to be 17×10^{-6} K^{-1}.

(a) Initial length, $L_1 = 4.20$ m, initial temperature, $t_1 = 20$°C, final temperature, $t_2 = 10$°C and $\alpha = 17 \times 10^{-6}$ K^{-1}

Final length at 10°C is given by:

$$L_2 = L_1 [1 + \alpha(t_2 - t_1)]$$
$$= 4.20[1 + (17 \times 10^{-6})(10 - 20)]$$
$$= 4.20[1 - 0.00017] = \mathbf{4.1993 \ m}$$

i.e. **the tube contracts by 0.7 mm when the temperature decreases from 20°C to 10°C**

(b) Length, $L_1 = 4.20$ m, $t_1 = 20$°C, $t_2 = 320$°C and $\alpha = 17 \times 10^{-6}$ K^{-1}

Final length at 320°C is given by:

$$L_2 = L_1 [1 + \alpha(t_2 - t_1)]$$
$$= 4.20[1 + (17 \times 10^{-6})(320 - 20)]$$
$$= 4.20[1 + 0.0051] = \mathbf{4.2214 \ m}$$

i.e. **the tube extends by 21.4 mm when the temperature rises from 20°C to 320°C**

Now try the following Practise Exercise

> **Practise Exercise 112** **Further problems on the coefficient of linear expansion**
>
> 1. A length of lead piping is 50.0 m long at a temperature of 16°C. When hot water flows through it the temperature of the pipe rises to 80°C. Determine the length of the hot pipe if the coefficient of linear expansion of lead is $29 \times 10^{-6} \text{ K}^{-1}$. [50.0928 m]
>
> 2. A rod of metal is measured at 285 K and is 3.521 m long. At 373 K the rod is 3.523 m long. Determine the value of the coefficient of linear expansion for the metal. [$6.45 \times 10^{-6} \text{ K}^{-1}$]
>
> 3. A copper overhead transmission line has a length of 40.0 m between its supports at 20°C. Determine the increase in length at 50°C if the coefficient of linear expansion of copper is $17 \times 10^{-6} \text{ K}^{-1}$. [20.4 mm]
>
> 4. A brass measuring tape measures 2.10 m at a temperature of 15°C. Determine
>   (a) the increase in length when the temperature has increased to 40°C
>   (b) the percentage error in measurement at 40°C. Assume the coefficient of linear expansion of brass to be $18 \times 10^{-6} \text{ K}^{-1}$.
>   [(a) 0.945 mm (b) 0.045%]
>
> 5. A pendulum of a 'grandfather' clock is 2.0 m long and made of steel. Determine the change in length of the pendulum if the temperature rises by 15 K. Assume the coefficient of linear expansion of steel to be $15 \times 10^{-6} \text{ K}^{-1}$. [0.45 mm]
>
> 6. A temperature control system is operated by the expansion of a zinc rod which is 200 mm long at 15°C. If the system is set so that the source of heat supply is cut off when the rod has expanded by 0.20 mm, determine the temperature to which the system is limited. Assume the coefficient of linear expansion of zinc to be $31 \times 10^{-6} \text{ K}^{-1}$. [47.26°C]
>
> 7. A length of steel railway line is 30.0 m long when the temperature is 288 K. Determine the increase in length of the line when the temperature is raised to 303 K. Assume the coefficient of linear expansion of steel to be $15 \times 10^{-6} \text{ K}^{-1}$. [6.75 mm]
>
> 8. A brass shaft is 15.02 mm in diameter and has to be inserted in a hole of diameter 15.0 mm. Determine by how much the shaft must be cooled to make this possible, without using force. Take the coefficient of linear expansion of brass as $18 \times 10^{-6} \text{ K}^{-1}$. [74 K]

21.5 Coefficient of superficial expansion

The amount by which unit area of a material increases when the temperature is raised by one degree is called the **coefficient of superficial (i.e. area) expansion** and is represented by β (Greek beta).

If a material having an initial surface area A_1 at temperature t_1 and having a coefficient of superficial expansion β, has its temperature increased to t_2, then the new surface area A_2 of the material is given by:

New surface area = original surface area + increase in area

i.e.
$$A_2 = A_1 + A_1 \beta(t_2 - t_1)$$

i.e.
$$\boldsymbol{A_2 = A_1 [1 + \beta(t_2 - t_1)]} \qquad (21.2)$$

It is shown in Problem 5 below that the coefficient of superficial expansion is twice the coefficient of linear expansion, i.e. $\beta = 2\alpha$, to a very close approximation.

> **Problem 5.** Show that for a rectangular area of material having dimensions L by b, the coefficient of superficial expansion $\beta \approx 2\alpha$, where α is the coefficient of linear expansion.

Initial area, $A_1 = Lb$. For a temperature rise of 1 K, side L will expand to $(L + L\alpha)$ and side b will expand to $(b + b\alpha)$. Hence the new area of the rectangle, A_2, is given by:

$$A_2 = (L + L\alpha)(b + b\alpha)$$
$$= L(1 + \alpha)b(1 + \alpha) = Lb(1 + \alpha)^2$$
$$= Lb(1 + 2\alpha + \alpha^2) \approx Lb(1 + 2\alpha)$$

since α^2 is very small (see typical values in Section 21.4)

Hence, $A_2 \approx A_1 (1 + 2\alpha)$

For a temperature rise of $(t_2 - t_1)$ K

$$A_2 \approx A_1 [1 + 2\alpha(t_2 - t_1)]$$

Thus, from equation (21.2), $\boldsymbol{\beta \approx 2\alpha}$

21.6 Coefficient of cubic expansion

The amount by which unit volume of a material increases for a one degree rise of temperature is called the **coefficient of cubic (or volumetric) expansion** and is represented by γ (Greek gamma).

If a material having an initial volume V_1 at temperature t_1 and having a coefficient of cubic expansion γ, has its temperature raised to t_2, then the new volume V_2 of the material is given by:

New volume = initial volume + increase in volume

i.e. $V_2 = V_1 + V_1 \gamma(t_2 - t_1)$

i.e. $\boldsymbol{V_2 = V_1 [1 + \gamma(t_2 - t_1)]}$ (21.3)

It is shown in Problem 6 below that the coefficient of cubic expansion is three times the coefficient of linear expansion, i.e. $\gamma = 3\alpha$, to a very close approximation. A liquid has no definite shape and only its cubic or volumetric expansion need be considered. Thus with expansions in liquids, equation (21.3) is used.

> **Problem 6.**   Show that for a rectangular block of material having dimensions L, b and h, the coefficient of cubic expansion $\gamma \approx 3\alpha$, where α is the coefficient of linear expansion.

Initial volume, $V_1 = Lbh$. For a temperature rise of 1 K, side L expands to $(L + L\alpha)$, side b expands to $(b + b\alpha)$ and side h expands to $(h + h\alpha)$
Hence the new volume of the block V_2 is given by:

$V_2 = (L + L\alpha)(b + b\alpha)(h + h\alpha)$

$\quad = L(1 + \alpha)b(1 + \alpha)h(1 + \alpha)$

$\quad = Lbh(1 + \alpha)^3 = Lbh(1 + 3\alpha + 3\alpha^2 + \alpha^3)$

$\quad \approx Lbh(1 + 3\alpha)$

since terms in α^2 and α^3 are very small

Hence, $V_2 \approx V_1 (1 + 3\alpha)$

For a temperature rise of $(t_2 - t_1)$ K,

$$V_2 \approx V_1 [1 + 3\alpha(t_2 - t_1)]$$

Thus, from equation (21.3), $\gamma \approx 3\alpha$

Some **typical values** for the coefficient of cubic expansion measured at 20°C (i.e. 293 K) include:

Ethyl alcohol	1.1×10^{-3} K^{-1}	Mercury	1.82×10^{-4} K^{-1}
Paraffin oil	9×10^{-2} K^{-1}	Water	2.1×10^{-4} K^{-1}

The coefficient of cubic expansion γ is only constant over a limited range of temperature.

> **Problem 7.**   A brass sphere has a diameter of 50 mm at a temperature of 289 K. If the temperature of the sphere is raised to 789 K, determine the increase in (a) the diameter (b) the surface area (c) the volume of the sphere. Assume the coefficient of linear expansion for brass is 18×10^{-6} K^{-1}.

(a) Initial diameter, $L_1 = 50$ mm, initial temperature, $t_1 = 289$ K, final temperature, $t_2 = 789$ K and $\alpha = 18 \times 10^{-6}$ K^{-1}.

New diameter at 789 K is given by:

$L_2 = L_1 [1 + \alpha(t_2 - t_1)]$ from equation (21.1)

i.e. $L_2 = 50[1 + (18 \times 10^{-6})(789 - 289)]$

$\quad = 50[1 + 0.009] = 50.45$ mm

Hence the increase in the diameter is 0.45 mm

(b) Initial surface area of sphere,

$$A_1 = 4\pi r^2 = 4\pi \left(\frac{50}{2}\right)^2 = 2500\pi \text{ mm}^2$$

New surface area at 789 K is given by:

$A_2 = A_1 [1 + \beta(t_2 - t_1)]$ from equation (21.2)

i.e. $A_2 = A_1 [1 + 2\alpha(t_2 - t_1)]$

since $\beta = 2\alpha$, to a very close approximation

Thus $A_2 = 2500\pi[1 + 2(18 \times 10^{-6})(500)]$

$\quad = 2500\pi[1 + 0.018]$

$\quad = 2500\pi + 2500\pi(0.018)$

Hence increase in surface area = $2500\pi(0.018)$
$\qquad\qquad\qquad\qquad\qquad = \textbf{141.4 mm}^2$

(c) Initial volume of sphere, $V_1 = \dfrac{4}{3}\pi r^3$

$$= \frac{4}{3}\pi \left(\frac{50}{2}\right)^3 \text{ mm}^3$$

New volume at 789 K is given by:

$V_2 = V_1 [1 + \gamma(t_2 - t_1)]$ from equation (21.3)

i.e. $V_2 = V_1 [1 + 3\alpha(t_2 - t_1)]$

since $\gamma = 3\alpha$, to a very close approximation

Thus $V_2 = \dfrac{4}{3}\pi(25)^3 [1 + 3(18 \times 10^{-6})(500)]$

$= \dfrac{4}{3}\pi(25)^3 [1 + 0.027]$

$= \dfrac{4}{3}\pi(25)^3 + \dfrac{4}{3}\pi(25)^3 (0.027)$

Hence, the increase in volume $= \dfrac{4}{3}\pi(25)^3 0.027)$

$= \textbf{1767 mm}^3$

Problem 8. Mercury contained in a thermometer has a volume of 476 mm³ at 15°C. Determine the temperature at which the volume of mercury is 478 mm³, assuming the coefficient of cubic expansion for mercury to be 1.8×10^{-4} K⁻¹.

Initial volume, $V_1 = 476$ mm³, final volume, $V_2 = 478$ mm³, initial temperature, $t_1 = 15$°C and $\gamma = 1.8 \times 10^{-4}$ K⁻¹

Final volume, $\quad V_2 = V_1 [1 + \gamma(t_2 - t_1)]$

from equation (21.3)

i.e. $\quad V_2 = V_1 + V_1 \gamma(t_2 - t_1)$

from which, $\quad (t_2 - t_1) = \dfrac{V_2 - V_1}{V_1 \gamma}$

$= \dfrac{478 - 476}{(476)(1.8 \times 10^{-4})}$

$= 23.34$°C

Hence, $\quad t_2 = 23.34 + t_1 = 23.34 + 15 = 38.34$°C

Hence, the temperature at which the volume of mercury is 478 mm³ is 38.34°C

Problem 9. A rectangular glass block has a length of 100 mm, width 50 mm and depth 20 mm at 293 K. When heated to 353 K its length increases by 0.054 mm. What is the coefficient of linear expansion of the glass? Find also (a) the increase in surface area (b) the change in volume resulting from the change of length.

Final length, $L_2 = L_1 [1 + \alpha(t_2 - t_1)]$

from equation (21.1),

hence increase in length is given by:

$L_2 - L_1 = L_1 \alpha(t_2 - t_1)$

Hence $\quad 0.054 = (100)(\alpha)(353 - 293)$

from which, the **coefficient of linear expansion** is given by:

$$\alpha = \frac{0.054}{(100)(60)} = \textbf{9} \times \textbf{10}^{-6} \textbf{ K}^{-1}$$

(a) Initial surface area of glass,

$A_1 = (2 \times 100 \times 50) + (2 \times 50 \times 20)$

$+ (2 \times 100 \times 20)$

$= 10000 + 2000 + 4000 = 16000$ mm²

Final surface area of glass (from equation (21.2)),

$A_2 = A_1 [1 + \beta(t_2 - t_1)] = A_1 [1 + 2\alpha(t_2 - t_1)]$

since $\beta = 2\alpha$ to a very close approximation

Hence, **increase in surface area**

$= A_1 (2\alpha)(t_2 - t_1) = (16000)(2 \times 9 \times 10^{-6})(60)$

$= \textbf{17.28 mm}^2$

(b) Initial volume of glass, $V_1 = 100 \times 50 \times 20$

$= 100000$ mm³

Final volume of glass (from equation (21.3)),

$V_2 = V_1 [1 + \gamma(t_2 - t_1)]$

$= V_1 [1 + 3\alpha(t_2 - t_1)]$

since $\gamma = 3\alpha$ to a very close approximation

Hence, **increase in volume of glass**

$= V_1 (3\alpha)(t_2 - t_1)$

$= (100000)(3 \times 9 \times 10^{-6})(60) = \textbf{162 mm}^3$

Now try the following Practise Exercises

Practise Exercise 113 **Further problems on the coefficients of superficial and cubic expansion**

1. A silver plate has an area of 800 mm² at 15°C. Determine the increase in the area of the plate when the temperature is raised to 100°C. Assume the coefficient of linear expansion of silver to be 19×10^{-6} K⁻¹.
   [2.584 mm²]

2. At 283 K a thermometer contains 440 mm³ of alcohol. Determine the temperature at which the volume is 480 mm³ assuming that the coefficient of cubic expansion of the alcohol is 12×10^{-4} K⁻¹. [358.8 K]

3. A zinc sphere has a radius of 30.0 mm at a temperature of 20°C. If the temperature of the sphere is raised to 420°C, determine the

increase in: (a) the radius, (b) the surface area, (c) the volume of the sphere. Assume the coefficient of linear expansion for zinc to be 31×10^{-6} K^{-1}.
[(a) 0.372 mm (b) 280.5 mm^2 (c) 4207 mm^3]

4. A block of cast iron has dimensions of 50 mm by 30 mm by 10 mm at 15°C. Determine the increase in volume when the temperature of the block is raised to 75°C. Assume the coefficient of linear expansion of cast iron to be 11×10^{-6} K^{-1}. [29.7 mm^3]

5. Two litres of water, initially at 20°C, is heated to 40°C. Determine the volume of water at 40°C if the coefficient of volumetric expansion of water within this range is 30×10^{-5} K^{-1}. [2.012 litres]

6. Determine the increase in volume, in litres, of 3 m^3 of water when heated from 293 K to boiling point if the coefficient of cubic expansion is 2.1×10^{-4} K^{-1} (1 litre $\approx 10^{-3}$ m^3). [50.4 litres]

7. Determine the reduction in volume when the temperature of 0.5 litre of ethyl alcohol is reduced from 40°C to –15°C. Take the coefficient of cubic expansion for ethyl alcohol as 1.1×10^{-3} K^{-1}. [0.03025 litres]

Practise Exercise 114   Short-answer questions on thermal expansion

1. When heat is applied to most solids and liquids .... occurs.

2. When solids and liquids are cooled they usually ....

3. State three practical applications where the expansion of metals must be allowed for.

4. State a practical disadvantage where the expansion of metals occurs.

5. State one practical advantage of the expansion of liquids.

6. What is meant by the 'coefficient of expansion'.

7. State the symbol and the unit used for the coefficient of linear expansion.

8. Define the 'coefficient of superficial expansion' and state its symbol.

9. Describe how water displays an unexpected effect between 0°C and 4°C.

10. Define the 'coefficient of cubic expansion' and state its symbol.

Practise Exercise 115   Multiple-choice questions on thermal expansion

(Answers on page 298)

1. When the temperature of a rod of copper is increased, its length:
   (a) stays the same   (b) increases
   (c) decreases

2. The amount by which unit length of a material increases when the temperature is raised one degree is called the coefficient of:
   (a) cubic expansion
   (b) superficial expansion
   (c) linear expansion

3. The symbol used for volumetric expansion is:
   (a) γ (b) β
   (c) L (d) α

4. A material of length L_1 at temperature θ_1 K is subjected to a temperature rise of θ K. The coefficient of linear expansion of the material is α K^{-1}. The material expands by:
   (a) $L_2 (1 + \alpha\theta)$
   (b) $L_1 \alpha(\theta - \theta_1)$
   (c) $L_1 [1 + \alpha(\theta - \theta_1)]$
   (d) $L_1 \alpha\theta$

5. Some iron has a coefficient of linear expansion of 12×10^{-6} K^{-1}. A 100 mm length of iron piping is heated through 20 K. The pipe extends by:
   (a) 0.24 mm   (b) 0.024 mm
   (c) 2.4 mm   (d) 0.0024 mm

6. If the coefficient of linear expansion is A, the coefficient of superficial expansion is B and the coefficient of cubic expansion is C, which of the following is false?

(a) $C = 3A$ (b) $A = B/2$

(c) $B = \dfrac{3}{2}C$ (d) $A = C/3$

7. The length of a 100 mm bar of metal increases by 0.3 mm when subjected to a temperature rise of 100 K. The coefficient of linear expansion of the metal is:

   (a) $3 \times 10^{-3}\ \mathrm{K}^{-1}$ (b) $3 \times 10^{-4}\ \mathrm{K}^{-1}$
   (c) $3 \times 10^{-5}\ \mathrm{K}^{-1}$ (d) $3 \times 10^{-6}\ \mathrm{K}^{-1}$

8. A liquid has a volume V_1 at temperature θ_1. The temperature is increased to θ_2. If γ is the coefficient of cubic expansion, the increase in volume is given by:

   (a) $V_1\, \gamma(\theta_2 - \theta_1)$
   (b) $V_1\, \gamma\, \theta_2$
   (c) $V_1 + V_1\, \gamma\, \theta_2$
   (d) $V_1\, [1 + \gamma(\theta_2 - \theta_1)]$

9. Which of the following statements is false?

   (a) Gaps need to be left in lengths of railway lines to prevent buckling in hot weather.

   (b) Bimetallic strips are used in thermostats, a thermostat being a temperature-operated switch.

   (c) As the temperature of water is decreased from 4°C to 0°C contraction occurs.

   (d) A change of temperature of 15°C is equivalent to a change of temperature of 15 K

10. The volume of a rectangular block of iron at a temperature t_1 is V_1. The temperature is raised to t_2 and the volume increases to V_2. If the coefficient of linear expansion of iron is α, then volume V_1 is given by:

   (a) $V_2\, [1 + \alpha(t_2 - t_1)]$

   (b) $\dfrac{V_2}{1 + 3\alpha\,(t_2 - t_1)}$

   (c) $3V_2\, \alpha(t_2 - t_1)$

   (d) $\dfrac{1 + \alpha\,(t_2 - t_1)}{V_2}$

This Revision Test covers the material contained in Chapters 20 and 21. *The marks for each question are shown in brackets at the end of each question.*

1. A block of aluminium having a mass of 20 kg cools from a temperature of 250°C to 80°C. How much energy is lost by the aluminium? Assume the specific heat capacity of aluminium is 950 J/(kg °C).   (5)

2. Calculate the heat energy required to convert completely 12 kg of water at 30°C to superheated steam at 100°C. Assume that the specific heat capacity of water is 4200 J/(kg °C), and the specific latent heat of vaporisation of water is 2260 kJ/(kg °C).   (7)

3. A copper overhead transmission line has a length of 60 m between its supports at 15°C. Calculate its length at 40°C, if the coefficient of linear expansion of copper is 17×10^{-6} K^{-1}.   (6)

4. A gold sphere has a diameter of 40 mm at a temperature of 285 K. If the temperature of the sphere is raised to 785 K, determine the increase in (a) the diameter (b) the surface area (c) the volume of the sphere. Assume the coefficient of linear expansion for gold is 14×10^{-6} K^{-1}.   (12)

Chapter 22

Hydrostatics

This chapter describes fluid pressure, together with buoyancy and hydrostatic stability. The chapter also defines Archimedes' principle, which is used to determine the buoyancy of boats, yachts, ships, etc. The chapter also describes metacentric height, which is used to determine the hydrostatic stability of the aforementioned vessels, and the explanation of this topic is aided with a number of simple worked examples. The chapters also describe gauges used in fluid mechanics, such as barometers, manometers, and the Bourdon pressure and vacuum gauges. These gauges are used to determine the properties and behaviour of fluids when they are met in practice. Calculations are given of simple floating structures and reference is made to the mid-ordinate rule, described earlier in Chapter 15, which can be used for determining the areas and volumes of complex shapes, such as those often met in naval architecture and civil engineering and many other branches of engineering.

At the end of this chapter you should be able to:

- define pressure and state its unit
- understand pressure in fluids
- distinguish between atmospheric, absolute and gauge pressures
- state and apply Archimedes' principle
- describe the construction and principle of operation of different types of barometer
- describe the construction and principle of operation of different types of manometer
- describe the construction and principle of operation of the Bourdon pressure gauge
- describe the construction and principle of operation of different types of vacuum gauge
- calculate hydrostatic pressure on submerged surfaces

- understand hydrostatic thrust on curved surfaces
- define buoyancy
- appreciate and perform calculations on the stability of floating bodies

22.1 Pressure

The pressure acting on a surface is defined as the perpendicular force per unit area of surface. The unit of pressure is the **pascal, Pa**, where 1 pascal is equal to 1 newton per square metre. Thus pressure,

$$p = \frac{F}{A} \text{ pascals}$$

where F is the force in newtons acting at right angles to a surface of area A square metres.

When a force of 20 N acts uniformly over, and perpendicular to, an area of 4 m^2, then the pressure on the area, p, is given by:

$$p = \frac{20 \, \text{N}}{4 \, \text{m}^2} = 5 \, \text{Pa}$$

It should be noted that for **irregular shaped flat surfaces**, such as the water planes of ships, their areas can be calculated using the mid-ordinate rule, described in Chapter 15.

Problem 1. A table loaded with books has a force of 250 N acting in each of its legs. If the contact area between each leg and the floor is 50 mm^2, find the pressure each leg exerts on the floor.

From above, pressure $p = \dfrac{\text{force}}{\text{area}}$

Mechanical Engineering Principles, Bird and Ross, ISBN 9780415517850

Hence, $$p = \frac{250 \text{ N}}{50 \text{ mm}^2} = \frac{250 \text{ N}}{50 \times 10^{-6}\text{m}^2}$$
$$= 5 \times 10^6 \text{ N/m}^2$$
$$= \textbf{5 MPa}$$

That is, **the pressure exerted by each leg on the floor is 5 MPa**

Problem 2. Calculate the force exerted by the atmosphere on a pool of water that is 30 m long by 10 m wide, when the atmospheric pressure is 100 kPa.

From above, pressure $= \dfrac{\text{force}}{\text{area}}$
hence, force = pressure × area.

The area of the pool is 30 m × 10 m = 300 m²

Thus, force on pool, F = pressure × area
$$= 100 \text{ kPa} \times 300 \text{ m}^2 \text{ and}$$

since 1 Pa = 1 N/m²,
$$F = (100 \times 10^3)\ \frac{\text{N}}{\text{m}^2} \times 300 \text{ m}^2$$
$$= 3 \times 10^7 \text{ N} = 30 \times 10^6 \text{ N}$$
$$= 30 \text{ MN}$$

That is, **the force on the pool of water is 30 MN**

Problem 3. A circular piston exerts a pressure of 80 kPa on a fluid, when the force applied to the piston is 0.2 kN. Find the diameter of the piston.

From above, pressure $= \dfrac{\text{force}}{\text{area}}$

hence, area $= \dfrac{\text{force}}{\text{pressure}}$

Force in newtons = 0.2 kN
$$= 0.2 \times 10^3 \text{ N} = 200 \text{ N, and}$$

pressure in pascals = 80 kPa = 80000 Pa
$$= 80000 \text{ N/m}^2.$$

Hence, area $= \dfrac{\text{force}}{\text{pressure}} = \dfrac{200 \text{ N}}{80000 \text{ N/m}^2}$
$$= 0.0025 \text{ m}^2$$

Since the piston is circular, its area is given by $\pi d^2/4$, where d is the diameter of the piston.

Hence, area $= \dfrac{\pi d^2}{4} = 0.0025$

from which, $d^2 = 0.0025 \times \dfrac{4}{\pi} = 0.003183$

i.e. $d = \sqrt{0.003183} = 0.0564$ m
$$= 56.4 \text{ mm}$$

Hence, **the diameter of the piston is 56.4 mm**

Now try the following Practise Exercise

Practise Exercise 116   Further problems on pressure

1. A force of 280 N is applied to a piston of a hydraulic system of cross-sectional area 0.010 m². Determine the pressure produced by the piston in the hydraulic fluid.
   [28 kPa]

2. Find the force on the piston of Question 1 to produce a pressure of 450 kPa.   [4.5 kN]

3. If the area of the piston in Question 1 is halved and the force applied is 280 N, determine the new pressure in the hydraulic fluid.
   [56 kPa]

22.2   Fluid pressure

A fluid is either a liquid or a gas and there are four basic factors governing the pressure within fluids.

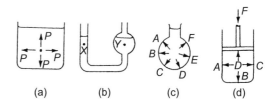

Figure 22.1

(a) The pressure at a given depth in a fluid is equal in all directions; see Figure 22.1(a).

(b) The pressure at a given depth in a fluid is independent of the shape of the container in which the fluid is held. In Figure 22.1(b), the pressure at X is the same as the pressure at Y.

(c) Pressure acts at right angles to the surface containing the fluid. In Figure 22.1(c), the pressures at points A to F all act at right angles to the container.

(d) When a pressure is applied to a fluid, this pressure is transmitted equally in all directions. In Figure 22.1(d), if the mass of the fluid is neglected, the pressures at points A to D are all the same.

The pressure, p, at any point in a fluid depends on three factors:

(a) the density of the fluid, ρ, in kg/m^3

(b) the gravitational acceleration, g, taken as approximately 9.8 m/s^2 (or the gravitational field force in N/kg), and

(c) the height of fluid vertically above the point, h metres.

The relationship connecting these quantities is:

$$p = \rho g h \text{ pascals}$$

When the container shown in Figure 22.2 is filled with water of density 1000 kg/m^3, the pressure due to the water at a depth of 0.03 m below the surface is given by:

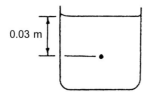

Figure 22.2

$$p = \rho g h = (1000 \times 9.8 \times 0.03)\text{Pa} = 294 \text{ Pa}$$

In the case of the **Mariana Trench**, which is situated in the Pacific ocean, near Guam, the hydrostatic pressure is about 115.2 MPa or 1152 bar, where 1 bar = 10^5 Pa and the density of sea water being 1020 kg/m^3.

Problem 4. A tank contains water to a depth of 600 mm. Calculate the water pressure (a) at a depth of 350 mm, and (b) at the base of the tank. Take the density of water as 1000 kg/m^3 and the gravitational acceleration as 9.8 m/s^2.

From above, pressure p at any point in a fluid is given by $p = \rho g h$ pascals, where ρ is the density in kg/m^3, g is the gravitational acceleration in m/s^2 and h is the height of fluid vertically above the point in metres.

(a) At a depth of 350 mm = 0.35 m,

$$p = \rho g h = 1000 \times 9.8 \times 0.35$$

$$= 3430 \text{ Pa} = 3.43 \text{ kPa}$$

(b) At the base of the tank, the vertical height of the water is 600 mm = 0.6 m.

Hence, $p = 1000 \times 9.8 \times 0.6$

$$= 5880 \text{ Pa} = 5.88 \text{ kPa}$$

Problem 5. A storage tank contains petrol to a height of 4.7 m. If the pressure at the base of the tank is 32.2 kPa, determine the density of the petrol. Take the gravitational field force as 9.8 m/s^2.

From above, pressure $p = \rho g h$ pascals, where ρ is the density in kg/m^3, g is the gravitational acceleration in m/s^2 and h is the vertical height of the petrol in metres.

Transposing gives: $\rho = \dfrac{p}{g h}$

Pressure p is 32.2 kPa = 32200 Pa

hence, density, $\rho = \dfrac{32200}{9.8 \times 4.7} = 699 \text{ kg/m}^3$

That is, **the density of the petrol is 699 kg/m^3**.

Problem 6. A vertical tube is partly filled with mercury of density 13600 kg/m^3. Find the height, in millimetres, of the column of mercury, when the pressure at the base of the tube is 101 kPa. Take the gravitational field force as 9.8 m/s^2.

From above, pressure $p = \rho g h$, hence vertical height h is given by:

$$h = \dfrac{p}{\rho g}$$

Pressure p = 101 kPa = 101000 Pa,

thus, $h = \dfrac{101000}{13600 \times 9.8} = 0.758 \text{ m}$

That is, **the height of the column of mercury is 758 mm**

Now try the following Practise Exercise

Practise Exercise 117   Further problems on fluid pressure

(Take the gravitational acceleration as 9.8 m/s^2)

1. Determine the pressure acting at the base of a dam, when the surface of the water is 35 m above base level. Take the density of water as 1000 kg/m^3.   [343 kPa]

2. An uncorked bottle is full of sea water of density 1030 kg/m^3. Calculate, correct to 3 significant figures, the pressures on the side wall of the bottle at depths of (a) 30 mm, and (b) 70 mm below the top of the bottle.

[(a) 303 Pa (b) 707 Pa]

3. A U-tube manometer is used to determine the pressure at a depth of 500 mm below the free surface of a fluid. If the pressure at this depth is 6.86 kPa, calculate the density of the liquid used in the manometer. [1400 kg/m^3]

4. A submarine pressure hull in the form of a circular cylinder is of external diameter 10 m and length 200 m. It dives to the bottom of the Mariana Tench which is 11.52 km deep. What will be the mass of water acting on the submarine's circular surface in terms of the number of London double-decker buses, given that the mass of a London double-decker bus is 7 tonnes. Assume that the density of water, $\rho = 1020$ kg/m^3 and gravitational acceleration, $g = 9.81$ m/s^2.   [10.55 million]

22.3   Atmospheric pressure

The air above the Earth's surface is a fluid, having a density, ρ, which varies from approximately 1.225 kg/m^3 at sea level to zero in outer space. Since $p = \rho g h$, where height h is several thousands of metres, the air exerts a pressure on all points on the earth's surface. This pressure, called **atmospheric pressure**, has a value of approximately 100 kilopascals (or 1 bar). Two terms are commonly used when measuring pressures:

(a) **absolute pressure**, meaning the pressure above that of an absolute vacuum (i.e. zero pressure), and

(b) **gauge pressure**, meaning the pressure above that normally present due to the atmosphere.

Thus, **absolute pressure = atmospheric pressure + gauge pressure.**

Thus, a gauge pressure of 50 kPa is equivalent to an absolute pressure of (100 + 50) kPa, i.e. 150 kPa, since the atmospheric pressure is approximately 100 kPa.

Problem 7.   Calculate the absolute pressure at a point on a submarine, at a depth of 30 m below the surface of the sea, when the atmospheric pressure is 101 kPa. Take the density of sea water as 1030 kg/m^3 and the gravitational acceleration as 9.8 m/s^2.

From Section 22.2, the pressure due to the sea, that is, the gauge pressure (p_g) is given by:

$$p_g = \rho g h \text{ pascals}$$

i.e. $p_g = 1030 \times 9.8 \times 30 = 302820$ Pa $= 302.82$ kPa

From above, absolute pressure

= atmospheric pressure + gauge pressure

= (101 + 302.82) kPa = 403.82 kPa

That is, **the absolute pressure at a depth of 30 m is 403.82 kPa**

Now try the following Practise Exercise

Practise Exercise 118   Further problems on atmospheric pressure

Take the gravitational acceleration as 9.8 m/s^2, the density of water as 1000 kg/m^3, and the density of mercury as 13600 kg/m^3.

1. The height of a column of mercury in a barometer is 750 mm. Determine the atmospheric pressure, correct to 3 significant figures.   [100 kPa]

2. A U-tube manometer containing mercury gives a height reading of 250 mm of mercury when connected to a gas cylinder. If the barometer reading at the same time is 756 mm of mercury, calculate the absolute pressure of the gas in the cylinder, correct to 3 significant figures.   [134 kPa]

3. A water manometer connected to a condenser shows that the pressure in the condenser is 350 mm below atmospheric pressure. If the barometer is reading 760 mm of mercury, determine the absolute pressure in the condenser, correct to 3 significant figures.   [97.9 kPa]

4. A Bourdon pressure gauge shows a pressure of 1.151 MPa. If the absolute pressure is 1.25 MPa, find the atmospheric pressure in millimetres of mercury.   [743 mm]

22.4 Archimedes' principle

Archimedes' principle states that:

If a solid body floats, or is submerged, in a liquid, the liquid exerts an upthrust on the body equal to the gravitational force on the liquid displaced by the body.

In other words, if a solid body is immersed in a liquid, the apparent loss of weight is equal to the weight of liquid displaced.

If V is the volume of the body below the surface of the liquid, then the apparent loss of weight W is given by:

$$W = V\omega = V\rho g$$

where ω is the specific weight (i.e. weight per unit volume) and ρ is the density.

If a body floats on the surface of a liquid all of its weight appears to have been lost. The weight of liquid displaced is equal to the weight of the floating body.

Problem 8. A body weighs 2.760 N in air and 1.925 N when completely immersed in water of density 1000 kg/m³. Calculate (a) the volume of the body (b) the density of the body and (c) the relative density of the body. Take the gravitational acceleration as 9.81 m/s².

(a) The apparent loss of weight is 2.760 N – 1.925 N = 0.835 N. This is the weight of water displaced, i.e. $V\rho g$, where V is the volume of the body and ρ is the density of water,

i.e. $0.835 \text{ N} = V \times 1000 \text{ kg/m}^3 \times 9.81 \text{ m/s}^2$

$$= V \times 9.81 \text{ kN/m}^3$$

Hence, $V = \dfrac{0.835}{9.81 \times 10^3} \text{ m}^3$

$$= 8.512 \times 10^{-5} \text{ m}^3$$

$$= \mathbf{8.512 \times 10^4 \ mm^3}$$

(b) The density of the body

$$= \frac{\text{mass}}{\text{volume}} = \frac{\text{weight}}{g \times V}$$

$$= \frac{2.760 \text{ N}}{9.81 \text{ m/s}^2 \times 8.512 \times 10^{-5} \text{ m}^3}$$

$$= \frac{\dfrac{2.760}{9.81} \text{ kg} \times 10^5}{8.512 \text{ m}^3} = \mathbf{3305 \ kg/m^3}$$

$$= \mathbf{3.305 \ tonne/m^3}$$

(c) Relative density $= \dfrac{\text{density}}{\text{density of water}}$

Hence, **the relative density of the body**

$$= \frac{3305 \text{ kg/m}^3}{1000 \text{ kg/m}^3} = \mathbf{3.305}$$

Problem 9. A rectangular watertight box is 560 mm long, 420 mm wide and 210 mm deep. It weighs 223 N. (a) If it floats with its sides and ends vertical in water of density 1030 kg/m³, what depth of the box will be submerged? (b) If the box is held completely submerged in water of density 1030 kg/m³, by a vertical chain attached to the underside of the box, what is the force in the chain?

(a) The apparent weight of a floating body is zero. That is, the weight of the body is equal to the weight of liquid displaced. This is given by:

$$V\rho g$$

where V is the volume of liquid displaced, and ρ is the density of the liquid.

Here, $223 \text{ N} = V \times 1030 \text{ kg/m}^3 \times 9.81 \text{ m/s}^2$

$$= V \times 10.104 \text{ kN/m}^3$$

Hence, $V = \dfrac{223 \text{ N}}{10.104 \text{ kN/m}^3}$

$$= 22.07 \times 10^{-3} \text{ m}^3$$

This volume is also given by Lbd, where L = length of box, b = breadth of box, and d = depth of box submerged,

i.e. $22.07 \times 10^{-3} \text{ m}^3 = L \times b \times d$

$$= 0.56 \text{ m} \times 0.42 \text{ m} \times d$$

Hence, **depth submerged, $d = \dfrac{22.07 \times 10^{-3}}{0.56 \times 0.42}$**

$$= 0.09384 \text{ m} = \mathbf{93.84 \ mm}$$

(b) The volume of water displaced is the total volume of the box. The upthrust or buoyancy of the water, i.e. the 'apparent loss of weight', is greater than the weight of the box. The force in the chain accounts for the difference.

Volume of water displaced,

$$V = 0.56 \text{ m} \times 0.42 \text{ m} \times 0.21 \text{ m}$$

$$= 4.9392 \times 10^{-2} \text{ m}^3$$

Weight of water displaced
$$= V\rho g = 4.9392 \times 10^{-2} \text{ m}^3 \times 1030 \text{ kg/m}^3 \times 9.81 \text{ m/s}^2$$

$$= 499.1 \text{ N}$$

The **force in the chain**

= weight of water displaced – weight of box

= 499.1 N – 223 N = **276.1 N**

Now try the following **Practise Exercise**

Practise Exercise 119 Further problems on Archimedes' principle

Take the gravitational acceleration as 9.8 m/s^2, the density of water as 1000 kg/m^3 and the density of mercury as 13600 kg/m^3.

1. A body of volume 0.124 m^3 is completely immersed in water of density 1000 kg/m^3. What is the apparent loss of weight of the body? [1.215 kN]

2. A body of weight 27.4 N and volume 1240 cm^2 is completely immersed in water of specific weight 9.81 kN/m^3. What is its apparent weight? [15.25 N]

3. A body weighs 512.6 N in air and 256.8 N when completely immersed in oil of density 810 kg/m^3. What is the volume of the body? [0.03222 m^3]

4. A body weighs 243 N in air and 125 N when completely immersed in water. What will it weigh when completely immersed in oil of relative density 0.8? [148.6 N]

5. A watertight rectangular box, 1.2 m long and 0.75 m wide, floats with its sides and ends vertical in water of density 1000 kg/m^3. If the depth of the box in the water is 280 mm, what is its weight? [2.47 kN]

6. A body weighs 18 N in air and 13.7 N when completely immersed in water of density 1000 kg/m^3. What is the density and relative density of the body? [4.186 tonne/m^3, 4.186]

7. A watertight rectangular box is 660 mm long and 320 mm wide. Its weight is 336 N. If it floats with its sides and ends vertical in water of density 1020 kg/m^3, what will be its depth in the water? [159 mm]

8. A watertight drum has a volume of 0.165 m^3 and a weight of 115 N. It is completely submerged in water of density 1030 kg/m^3, held in position by a single vertical chain attached to the underside of the drum. What is the force in the chain? [1.551 kN]

22.5 Measurement of pressure

As stated earlier, pressure is the force exerted by a fluid per unit area. A fluid (i.e. liquid, vapour or gas) has a negligible resistance to a shear force, so that the force it exerts always acts at right angles to its containing surface.

The *SI* unit of pressure is the **pascal**, *Pa*, which is unit force per unit area, i.e. **1 Pa = 1 N/m^2**

The pascal is a very small unit and a commonly used larger unit is the bar, where **1 bar = 10^{5}Pa**

Atmospheric pressure is due to the mass of the air above the Earth's surface being attracted by Earth's gravity. Atmospheric pressure changes continuously. A standard value of atmospheric pressure, called 'standard atmospheric pressure', is often used, having a value of 101325 Pa or 1.01325 bars or 1013.25 millibars. This latter unit, the millibar, is usually used in the measurement of meteorological pressures. (Note that when atmospheric pressure varies from 101325 Pa it is no longer standard.)

Pressure indicating instruments are made in a wide variety of forms because of their many different applications. Apart from the obvious criteria such as pressure range, accuracy and response, many measurements also require special attention to material, sealing and temperature effects. The fluid whose pressure is being measured may be corrosive or may be at high temperatures. Pressure indicating devices used in science and industry include:

(i) barometers (see Section 22.6),
(ii) manometers (see Section 22.8),
(iii) Bourdon pressure gauge (see Section 22.9), and
(iv) McLeod and Pirani gauges (see Section 22.10).

22.6 Barometers

Introduction

A barometer is an instrument for measuring atmospheric pressure. It is affected by seasonal changes of temperature. Barometers are therefore also used for the measurement of altitude and also as one of the aids in weather forecasting. The value of atmospheric pressure will thus vary with climatic conditions, although not usually by more than about 10% of standard atmospheric pressure.

Construction and principle of operation

A simple barometer consists of a glass tube, just less than 1 m in length, sealed at one end, filled with mercury and then inverted into a trough containing more mercury. Care must be taken to ensure that no air enters the tube during this latter process. Such a barometer is shown in Figure 22.3(a) and it is seen that the level of the mercury column falls, leaving an empty space, called a vacuum. Atmospheric pressure acts on the surface of the mercury in the trough as shown and this pressure is equal to the pressure at the base of the column of mercury in the inverted tube, i.e. the pressure of the atmosphere is supporting the column of mercury. If the atmospheric pressure falls the barometer height h decreases. Similarly, if the atmospheric pressure

rises, then h increases. Thus atmospheric pressure can be measured in terms of the height of the mercury column. It may be shown that for mercury the height h is 760 mm at standard atmospheric pressure, i.e. a vertical column of mercury 760 mm high exerts a pressure equal to the standard value of atmospheric pressure.

There are thus several ways in which atmospheric pressure can be expressed:

Standard atmospheric pressure

$$= 101325 \text{ Pa} \text{ or } 101.325 \text{ kPa}$$
$$= 101325 \text{ N/m}^2 \text{ or } 101.325 \text{ kN/m}^2$$
$$= 1.01325 \text{ bars} \text{ or } 1013.25 \text{ mbars}$$
$$= 760 \text{ mm of mercury}$$

Another arrangement of a typical barometer is shown in Figure 22.3(b) where a U-tube is used instead of an inverted tube and trough, the principle being similar.

If, instead of mercury, water was used as the liquid in a barometer, then the barometric height h at standard atmospheric pressure would be 13.6 times more than for mercury, i.e. about 10.4 m high, which is not very practicable. This is because the relative density of mercury is 13.6.

Types of barometer

The **Fortin barometer** is an example of a mercury barometer that enables barometric heights to be measured to a high degree of accuracy (in the order of one-tenth of a millimetre or less). Its construction is merely a more sophisticated arrangement of the inverted tube and trough shown in Figure 22.3(a), with the addition of a vernier scale to measure the barometric height with great accuracy. A disadvantage of this type of barometer is that it is not portable.

A Fortin barometer is shown in Figure 22.4. Mercury is contained in a leather bag at the base of the mercury reservoir, and height, H, of the mercury in the reservoir can be adjusted using the screw at the base of the barometer to depress or release the leather bag. To measure the atmospheric pressure the screw is adjusted until the pointer at H is just touching the surface of the mercury and the height of the mercury column is then read using the main and vernier scales. The measurement of atmospheric pressure using a Fortin barometer is achieved much more accurately than by using a simple barometer.

A portable type often used is the **aneroid barometer**. Such a barometer consists basically of a circular, hollow,

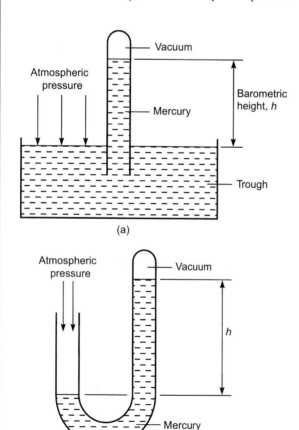

(a)

(b)

Figure 22.3

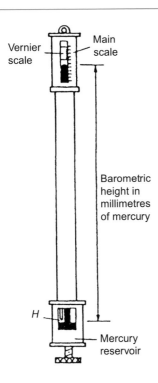

Figure 22.4

22.7 Absolute and gauge pressure

A barometer measures the true or absolute pressure of the atmosphere. The term absolute pressure means the pressure above that of an absolute vacuum (which is zero pressure), as stated earlier. In Figure 22.6 a pressure scale is shown with the line *AB* representing absolute zero pressure (i.e. a vacuum) and line *CD* representing atmospheric pressure. With most practical pressure-measuring instruments the part of the instrument that is subjected to the pressure being measured is also subjected to atmospheric pressure. Thus practical instruments actually determine the difference between the pressure being measured and atmospheric pressure. The pressure that the instrument is measuring is then termed the gauge pressure. In Figure 22.6, the line *EF* represents an absolute pressure which has a value greater than atmospheric pressure, i.e. the 'gauge' pressure is positive.

sealed vessel, *S*, usually made from thin flexible metal. The air pressure in the vessel is reduced to nearly zero before sealing, so that a change in atmospheric pressure will cause the shape of the vessel to expand or contract. These small changes can be magnified by means of a lever and be made to move a pointer over a calibrated scale. Figure 22.5 shows a typical arrangement of an aneroid barometer. The scale is usually circular and calibrated in millimetres of mercury. These instruments require frequent calibration.

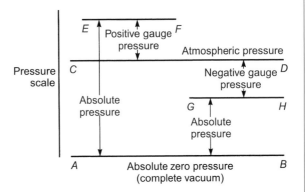

Figure 22.6

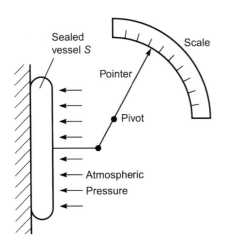

Figure 22.5

Thus, **absolute pressure = gauge pressure + atmospheric pressure.**
Hence a gauge pressure of, say, 60 *kPa* recorded on an indicating instrument when the atmospheric pressure is 101 *kPa* is equivalent to an absolute pressure of 60 *kPa* + 101 *kPa*, or 161 *kPa*.

Pressure-measuring indicating instruments are referred to generally as **pressure gauges** (which acts as a reminder that they measure 'gauge' pressure).

It is possible, of course, for the pressure indicated on a pressure gauge to be below atmospheric pressure, i.e. the gauge pressure is negative. Such a gauge pressure is often referred to as a vacuum, even though it does not necessarily represent a complete vacuum at absolute zero pressure. Such a pressure is shown by the line *GH* in Figure 22.6. An indicating instrument used for measuring such pressures is called a **vacuum gauge**.

A vacuum gauge indication of, say, 0.4 bar, means that the pressure is 0.4 bar less than atmospheric pressure. If atmospheric pressure is 1 bar, then the absolute pressure is 1 − 0.4 or 0.6 bar.

22.8 The manometer

A manometer is a device for measuring or comparing fluid pressures, and is the simplest method of indicating such pressures.

U-tube manometer

A U-tube manometer consists of a glass tube bent into a U shape and containing a liquid such as mercury. A U-tube manometer is shown in Figure 22.7(a). If limb A is connected to a container of gas whose pressure is above atmospheric, then the pressure of the gas will cause the levels of mercury to move as shown in Figure 22.7(b), such that the difference in height is h_1. The measuring scale can be calibrated to give the gauge pressure of the gas as h_1 mm of mercury.

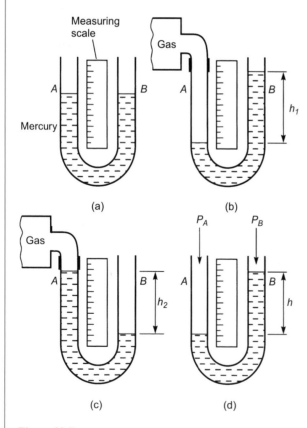

(a)   (b)

(c)   (d)

Figure 22.7

If limb A is connected to a container of gas whose pressure is below atmospheric then the levels of mercury will move as shown in Figure 22.7(c), such that their pressure difference is h_2 mm of mercury.

It is also possible merely to compare two pressures, say, P_A and P_B, using a U-tube manometer. Figure 22.7(d) shows such an arrangement with $(P_B − P_A)$ equivalent to h mm of mercury. One application of this differential pressure-measuring device is in determining the velocity of fluid flow in pipes (see Chapter 23).

For the measurement of lower pressures, water or paraffin may be used instead of mercury in the U-tube to give larger values of h and thus greater sensitivity.

Inclined manometers

For the measurement of very low pressures, greater sensitivity is achieved by using an inclined manometer, a typical arrangement of which is shown in Figure 22.8. With the inclined manometer the liquid used is water and the scale attached to the inclined tube is calibrated in terms of the vertical height h. Thus when a vessel containing gas under pressure is connected to the reservoir, movement of the liquid levels of the manometer occurs. Since small-bore tubing is used the movement of the liquid in the reservoir is very small compared with the movement in the inclined tube and is thus neglected. Hence the scale on the manometer is usually used in the range 0.2 mbar to 2 mbar.

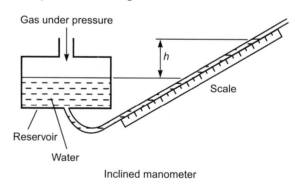

Inclined manometer

Figure 22.8

The pressure of a gas that a manometer is capable of measuring is naturally limited by the length of tube used. Most manometer tubes are less than 2 m in length and this restricts measurement to a maximum pressure of about 2.5 bar (or 250 kPa) when mercury is used.

22.9 The Bourdon pressure gauge

Pressures many times greater than atmospheric can be measured by the Bourdon pressure gauge, which is the most extensively used of all pressure-indicating instruments. It is a robust instrument. Its main component is a piece of metal tube (called the Bourdon tube), usually made of phosphor bronze or alloy steel, of oval or elliptical cross-section, sealed at one end and bent into an arc. In some forms the tube is bent into a spiral for greater sensitivity. A typical arrangement is shown in Figure 22.9(a). One end, E, of the Bourdon tube is fixed and the fluid whose pressure is to be measured is connected to this end. The pressure acts at right angles to the metal tube wall as shown in the cross-section of the tube in Figure 22.9(b). Because of its elliptical shape it is clear that the sum of the pressure components, i.e. the total force acting on the sides A and C, exceeds the sum of the pressure

components acting on ends B and D. The result is that sides A and C tend to move outwards and B and D inwards tending to form a circular cross-section. As the pressure in the tube is increased the tube tends to uncurl, or if the pressure is reduced the tube curls up further. The movement of the free end of the tube is, for practical purposes, proportional to the pressure applied to the tube, this pressure, of course, being the gauge pressure (i.e. the difference between atmospheric pressure acting on the outside of the tube and the applied pressure acting on the inside of the tube). By using a link, a pivot and a toothed segment as shown in Figure 22.9(a), the movement can be converted into the rotation of a pointer over a graduated calibrated scale.

The Bourdon tube pressure gauge is capable of measuring high pressures up to 10^4 bar (i.e. 7600 m of mercury) with the addition of special safety features.

A pressure gauge must be calibrated, and this is done either by a manometer, for low pressures, or by a piece of equipment called a 'dead weight tester'. This tester consists of a piston operating in an oil-filled cylinder of known bore, and carrying accurately known weights as shown in Figure 22.10. The gauge under test is attached to the tester and a screwed piston or ram applies the required pressure, until the weights are just lifted. While the gauge is being read, the weights are turned to reduce friction effects.

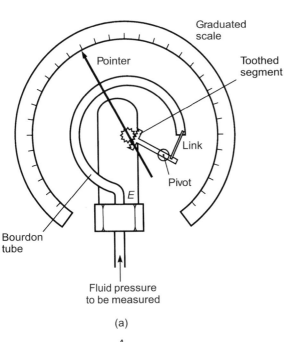

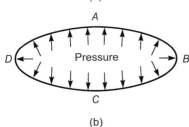

(a)

(b)

Figure 22.9

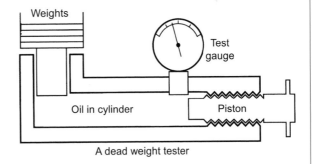

A dead weight tester

Figure 22.10

22.10 Vacuum gauges

Vacuum gauges are instruments for giving a visual indication, by means of a pointer, of the amount by which the pressure of a fluid applied to the gauge is less than the pressure of the surrounding atmosphere. Two examples of vacuum gauges are the McLeod gauge and the Pirani gauge.

Part Four

McLeod gauge

The McLeod gauge is normally regarded as a standard and is used to calibrate other forms of vacuum gauges. The basic principle of this gauge is that it takes a known volume of gas at a pressure so low that it cannot be measured, then compresses the gas in a known ratio until the pressure becomes large enough to be measured by an ordinary manometer. This device is used to measure low pressures, often in the range 10^{-6} to 1.0 mm of mercury. A disadvantage of the McLeod gauge is that it does not give a continuous reading of pressure and is not suitable for registering rapid variations in pressure.

Pirani gauge

The Pirani gauge measures the resistance and thus the temperature of a wire through which current is flowing. The thermal conductivity decreases with the pressure in the range 10^{-1} to 10^{-4} mm of mercury so that the increase in resistance can be used to measure pressure in this region. The Pirani gauge is calibrated by comparison with a McLeod gauge.

22.11 Hydrostatic pressure on submerged surfaces

From Section 22.2, it can be seen that hydrostatic pressure increases with depth according to the formula:

$$p = \rho g h$$

Problem 10. The deepest part of the oceans is the Marianas Trench, where its depth is approximately 11.52 km (7.16 miles). What is the gauge pressure at this depth, assuming that $\rho = 1020$ kg/m^3 and $g = 9.81$ m/s^2?

Gauge pressure, $p = \rho g h$

$$= 1020 \frac{\text{kg}}{\text{m}^3} \times 9.81 \frac{\text{m}}{\text{s}^2} \times 11.52 \times 10^3 \text{m}$$

$$= 11.527 \times 10^7 \text{ N/m}^2 \times \frac{1 \text{ bar}}{10^5 \text{ N/m}^2}$$

i.e.   pressure, $p = \mathbf{1152.7\ bar}$

Note that from the above calculation, it can be seen that a gauge pressure of 1 bar is approximately equivalent to a depth of 10 m.

Problem 11. Determine an expression for the thrust acting on a submerged plane surface, which is inclined to the horizontal by an angle θ, as shown in Figure 22.11.

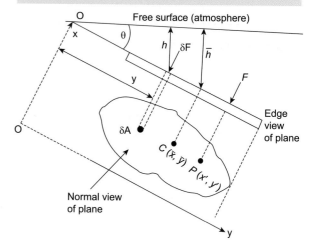

Figure 22.11

From Figure 22.11, δF = elemental thrust on dA

$$= \rho g h \times dA$$

But $h = y \sin \theta$

Hence, $\delta F = \rho g y \sin \theta\, dA$

Total thrust on plane surface $= F = \int dF = \int \rho g y \sin \theta\, dA$

or $\mathbf{F = \rho g \sin \theta \int y\, dA}$

However, $\int y\, dA = A\bar{h}$ where A = area of the surface, and $\bar{h}$ = distance of the centroid of the plane from the free surface.

Problem 12. Determine an expression for the position of the centre of pressure of the plane surface P(x′, y′) of Figure 22.11; this is also the position of the centre of thrust.

Taking moments about O gives:

$$F y' = \int \rho g\, y \sin \theta\, dA \times y$$

However, $F = \rho g \sin \theta \int y\, dA$

Hence, $y' = \dfrac{\int \rho g y^2 \sin \theta\, dA}{\rho g \sin \theta \int y\, dA} = \dfrac{\rho g \sin \theta \int y^2 dA}{\rho g \sin \theta \int y\, dA}$

$$= \dfrac{\left(A k^2\right)_{Ox}}{A \bar{y}}$$

where

$$\left(Ak^2\right)_{Ox}$$

= the second moment of area about Ox

k = the radius of gyration from O.

Now try the following Practise Exercise

Practise Exercise 120 Further problems on hydrostatic pressure on submerged surfaces

(Take $g = 9.81$ m/s^2)

1. Determine the gauge pressure acting on the surface of a submarine that dives to a depth of 500 m. Take water density as 1020 kg/m^3.
   [50.03 bar]

2. Solve Problem 1, when the submarine dives to a depth of 780 m. [78.05 bar]

3. If the gauge pressure measured on the surface of the submarine of Problem 1 were 92 bar, at what depth has the submarine dived to? [919.4 m]

4. A tank has a flat rectangular end, which is of size 4 m depth by 3 m width. If the tank is filled with water to its brim and the flat end is vertical, determine the thrust on this end and the position of its centre of pressure. Take water density as 1000 kg/m^3.
   [0.235 MN; 2.667 m]

5. If another vertical flat rectangular end of the tank of Problem 4 is of size 6 m depth by 4 m width, determine the thrust on this end and position of the centre of pressure. The depth of water at this end may be assumed to be 6 m. [0.706 MN; 4 m]

6. A tank has a flat rectangular end, which is inclined to the horizontal surface, so that $\theta = 30°$, where θ is as defined in Figure 22.11, page 254. If this end is of size 6 m height and 4 m width, determine the thrust on this end and the position of the centre of pressure from the top. The tank may be assumed to be just full.
   [0.235 MN; 2 m]

22.12 Hydrostatic thrust on curved surfaces

As hydrostatic pressure acts perpendicularly to a surface, the integration of δF over the surface can be complicated. One method of determining the thrust on a curved surface is to project its area on flat vertical and horizontal surfaces, as shown by AB and DE, respectively, in Figure 22.12.

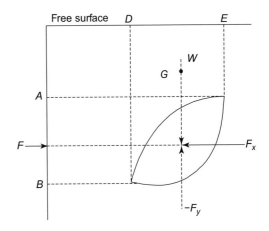

Figure 22.12

From equilibrium considerations, $F = F_x$ and $W = F_y$ and these thrusts must act through the centre of pressures of the respective vertical and horizontal planes. The resultant thrust can be obtained by adding F_x and F_y vectorially, where W = weight of the fluid enclosed by the curved surface and the vertical projection lines to the free surface, and G = centre of gravity of W.

22.13 Buoyancy

The upward force exerted by the fluid on a body that is wholly or partially immersed in it is called the buoyancy of the body.

22.14 The stability of floating bodies

For most ships and boats the centre of buoyancy (B) of the vessel is usually below the vessels' centre of gravity (G), as shown in Figure 22.13(a). When this vessel is subjected to a small angle of keel (θ), as shown in Figure 22.13(b), the centre of buoyancy moves to the position B', where BM = the centre of curvature of the centre of buoyancy = $\dfrac{I}{V}$, (given without proof)

GM = the metacentric height,

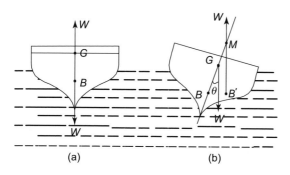

Figure 22.13

M = the position of the metacentre,

I = the second moment of area of the water plane about its centreline, and

V = displaced volume of the vessel.

The metacentric height GM can be found by a simple inclining experiment, where a weight P is moved transversely a distance x, as shown in Figure 22.14.

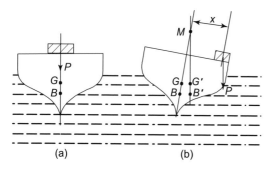

Figure 22.14

From rotational equilibrium considerations,

$$W(GM)\tan\theta = Px$$

Therefore, $\qquad GM = \dfrac{Px}{W}\cot\theta \qquad$ (22.1)

where W = the weight of the vessel,

and $\cot\theta = \dfrac{1}{\tan\theta}$

Problem 13. A naval architect has carried out hydrostatic calculations on a yacht, where he has found the following:

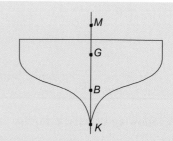

Figure 22.15

M = mass of yacht = 100 tonnes,

KB = vertical distance of the centre of buoyancy (B) above the keel (K) = 1.2 m (see Figure 22.15),

BM = distance of the metacentre (M) above the centre of buoyancy = 2.4 m.

He then carries out an inclining experiment, where he moves a mass of 50 kg through a transverse distance of 10 m across the yacht's deck. In doing this, he finds that the resulting angle of keel, $\theta = 1°$. What is the metacentric height (GM) and the position of the centre of gravity of the yacht above the keel, namely KG? Assume $g = 9.81$ m/s^2.

$$P = 50 \text{ kg} \times 9.81 = 490.5 \text{ N},$$

$$W = 100 \text{ tonnes} \times 1000\,\frac{\text{kg}}{\text{tonne}} \times 9.81\,\frac{\text{m}}{\text{s}^2} = 981 \text{ kN},$$

$$x = 10 \text{ m},$$

$$\theta = 1° \text{ from which,}$$

$$\tan\theta = 0.017455 \text{ and } \cot\theta = \frac{1}{\tan\theta} = 57.29$$

From equation (22.1), $GM = \dfrac{Px}{W}\cot\theta$

$$= \frac{490.5\,\text{N} \times 10\,\text{m} \times 57.29}{981 \times 10^3\,\text{N}}$$

i.e. **metacentric height, GM = 0.286 m**

Now $\quad KM = KB + BM = 1.2$ m + 2.4 m = 3.6 m

$$KG = KM - GM = 3.6 - 0.286 = 3.314 \text{ m}$$

i.e. centre of gravity above the keel, **KG = 3.314 m**, (where 'K' is a point on the keel).

Problem 14. A barge of length 30 m and width 8 m floats on an even keel at a depth of 3 m. What is the value of its buoyancy? Take density of water, ρ, as 1000 kg/m^3 and g as 9.81 m/s^2.

The displaced volume of the barge,
$V = 30 \text{ m} \times 8 \text{ m} \times 3 \text{ m} = 720 \text{ m}^3$.

From Section 22.4,

buoyancy $= V\rho g = 720 \text{ m}^3 \times 1000 \dfrac{\text{kg}}{\text{m}^3} \times 9.81 \dfrac{\text{m}}{\text{s}^2}$

$\qquad\qquad = \textbf{7.063 MN}$

Problem 15.   If the vertical centre of gravity of the barge in Problem 14 is 2 m above the keel, (i.e. $KG = 2$ m), what is the metacentric height of the barge?

Now $KB =$ the distance of the centre of buoyancy of the barge from the keel $= \dfrac{3 \text{ m}}{2}$ i.e. $KB = 1.5$ m.

From page 256, $BM = \dfrac{I}{V}$ and for a rectangle, $I = \dfrac{Lb^3}{12}$ from Table 8.1, page 108, where $L =$ length of the waterplane $= 30$ m, and
$b =$ width of the waterplane $= 8$ m.

Hence, moment of inertia, $I = \dfrac{30 \times 8^3}{12} = 1280 \text{ m}^4$

From Problem 14, volume, $V = 720 \text{ m}^3$,

hence, $BM = \dfrac{I}{V} = \dfrac{1280}{720} = 1.778$ m

Now, $KM = KB + BM = 1.5 \text{ m} + 1.778 \text{ m} = 3.278 \text{ m}$

i.e. the metacentre above the keel, $KM = 3.278$ m.

Since $KG = 2$ m (given), then
$\qquad GM = KM - KG = 3.278 - 2 = 1.278$ m,

i.e. **the metacentric height of the barge, $GM = 1.278$ m**

Problem 16.   A circular cylindrical steel buoy, made from 10 mm thick steel plate, is of a hollow box-like disc shape, as shown in Figure 22.16. It is sealed off at its top and bottom by circular plates so that it is watertight. (a) If the external diameter of the buoy, D, is 1 m and its height, h, is 0.5 m, determine its weight, W, given that the density of steel, $\rho_S = 7860 \text{ kg/m}^3$. (b) At what depth, H, will the buoy float if the density of water, $\rho_W = 1020 \text{ kg/m}^3$? (c) What is its GM? Take g to be 9.81 m/s^2

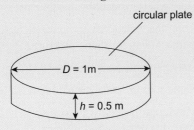

circular plate

Figure 22.16 Circular cylindrical buoy

(a)   Weight of two end plates,

$W_1 = \pi R^2 \times t \times \rho_S \times g \times 2$ where radius $R = D/2$

$\qquad = \pi (0.5)^2 \times (10 \times 10^{-3}) \times 7860 \times 9.81 \times 2$

$\qquad = 1211.2 \text{ N}$

Weight of circular cylinder,

$W_2 = 2\pi R_{\text{mean}}\, t \times h \times \rho_S \times g$

$\qquad = 2\pi \times \left(0.5 - 5 \times 10^{-3}\right) \times (10 \times 10^{-3})$

$\qquad\qquad\qquad \times 0.5 \times 7860 \times 9.81 = 1199 \text{ N}$

Total weight of buoy, $W = W_1 + W_2$

$\qquad\qquad\qquad = 1211 + 1199 = \textbf{2410 N}$

(b)   Buoyancy, $B = W = 2410 \text{ N} =$ weight of water displaced   (1)

Let $H =$ draught of water of the buoy, so that:

$B = \pi R^2 H \times \rho_W \times 9.81$

$\qquad = \pi (0.5)^2 \times H \times 1020 \times 9.81 = 7858.9\,H$ (2)

Equating (1) and (2) gives: $7858.9\,H = 2410$

from which, **depth, $H = \dfrac{2410}{7858.9} = \textbf{0.307 m}$**

(c) $KB = \dfrac{H}{2} = \dfrac{0.307}{2} = 0.1533$ m

$BM = \dfrac{I}{V} = \dfrac{\pi R^4}{4} \times \dfrac{1}{\pi R^2 H}$

$\qquad = \dfrac{R^2}{4H} = \dfrac{0.5^2}{4 \times 0.307} = 0.204$ m

$KM = KB + BM = 0.153 + 0.204$

$\qquad\qquad = 0.357$ m

$GM = KM - KG = 0.357 - \dfrac{0.5}{2}$

Hence, $\textbf{GM = 0.107 m}$

Problem 17.   A submarine pressure hull is in the form of a watertight circular cylindrical shell, of length 80 m, blocked off by flat ends, and of external diameter 10 m, and it descends to the bottom of the Mariana Trench, which is 11.52 km deep. What will be the hydrostatic pressure acting on it? If a double-decker London bus is of mass 7 tonnes what will be the equivalent number of double-decker London buses acting on this hull? Density of water, $\rho_W = 1020 \text{ kg/m}^3$. Take g to be 9.81 m/s^2.

Hydrostatic pressure,

$$p = \rho g h = 1020 \times 9.81 \times 11520$$

$$= \textbf{115.27 MPa} = \frac{115.27 \times 10^6}{10^5 \, \text{Pa/bar}}$$

$$= \textbf{1152.7 bar} \equiv \textbf{1152.7 atmosphere}$$

(since 10^5 pascals = 1 bar $\equiv$ 14.5 psi)

Let A = area of the external surface of the pressure hull assuming flat ends = $\pi R^2 \times 2 + 2\pi RL$

where R = external cylinder radius = $\dfrac{10}{2}$ = 5 m and L = length between its ends = 80 m

Hence, **area** $A = \pi(5)^2 \times 2 + 2\pi \times 5 \times 80$

$$= 157.1 + 2513.3 = \textbf{2670.4 m}^2$$

Total hydrostatic head on the submarine hull
$= p \times A = 115.27 \times 10^6 \times 2670.4 = 307817$ MN

Hence, $W = 307817$ MN $\times \dfrac{10^6 \text{N}}{\text{MN}} \times \dfrac{\text{kg}}{9.81 \, \text{N}}$

$$= 3.1378 \times 10^{10} \text{kg} \times \frac{\text{tonne}}{1000 \, \text{kg}}$$

$$= 31.378 \times 10^6 \text{ tonnes}$$

Number of double-decker London buses,

$$N = \frac{31.378 \times 10^6}{7} = 4.48 \times 10^6$$

Thus, N = **4.48 million equivalent weight of double-decker London buses**

Because of the huge heads suffered by submarine pressure hulls, they are one of the most difficult structures to design. (see [1], page 260)

Now try the following Practise Exercises

Practise Exercise 121   Further problems on hydrostatics

(In the following problems, where necessary, take $g = 9.81$ m/s^2 and density of water $\rho = 1020$ kg/m^3)

1. A ship is of mass 10000 kg. If the ship floats in the water, what is the value of its buoyancy?   [98.1 kN]

2. A submarine may be assumed to be in the form of a circular cylinder of 10 m external diameter and of length 100 m. If the submarine floats just below the surface of the water, what is the value of its buoyancy?   [78.59 MN]

3. A barge of length 20 m and of width 5 m floats on an even keel at a depth of 2 m. What is the value of its buoyancy?   [2 MN]

4. An inclining experiment is carried out on the barge of Problem 3 where a mass of 20 kg is moved transversely across the deck by a distance of 2.2 m. The resulting angle of keel is 0.8°. Determine the metacentric height, GM.   [0.0155 m]

5. Determine the value of the radius of curvature of the centre of buoyancy, namely, BM, for the barge of Problems 3 and 4, and hence the position of the centre of gravity above the keel, KG.   [2.026 m]

6. If the submarine of Problem 2 floats so that its top is 2 m above the water, determine the centre of curvature of the centre of buoyancy, BM.   [0.633 m]

Practise Exercise 122   Short-answer questions on hydrostatics

1. Define pressure.

2. State the unit of pressure.

3. Define a fluid.

4. State the four basic factors governing the pressure in fluids.

5. Write down a formula for determining the pressure at any point in a fluid in symbols, defining each of the symbols and giving their units.

6. What is meant by atmospheric pressure?

7. State the approximate value of atmospheric pressure.

8. State what is meant by gauge pressure.

9. State what is meant by absolute pressure.

10. State the relationship between absolute, gauge and atmospheric pressures.

11. State Archimedes' principle.

12. Name four pressure measuring devices.

13. Standard atmospheric pressure is 101325 Pa. State this pressure in millibars.

14. Briefly describe how a barometer operates.

15. State the advantage of a Fortin barometer over a simple barometer.

16. What is the main disadvantage of a Fortin barometer?

17. Briefly describe an aneroid barometer.

18. What is a vacuum gauge?

19. Briefly describe the principle of operation of a *U*-tube manometer.

20. When would an inclined manometer be used in preference to a U-tube manometer?

21. Briefly describe the principle of operation of a Bourdon pressure gauge.

22. What is a 'dead weight tester'?

23. What is a Pirani gauge?

24. What is a McLeod gauge used for?

25. What is buoyancy?

26. What does the abbreviation *BM* mean?

27. What does the abbreviation *GM* mean?

28. Define *BM* in terms of the second moment of area I of the water plane, and the displaced volume *V* of a vessel.

29. What is the primary purpose of a ship's inclining experiment?

Practise Exercise 123   Multiple-choice questions on hydrostatics

(Answers on page 298)

1. A force of 50 N acts uniformly over and at right angles to a surface. When the area of the surface is 5 m^2, the pressure on the area is:
   (a) 250 Pa   (b) 10 Pa
   (c) 45 Pa   (d) 55 Pa

2. Which of the following statements is false? The pressure at a given depth in a fluid
   (a) is equal in all directions
   (b) is independent of the shape of the container

(c) acts at right angles to the surface containing the fluid
(d) depends on the area of the surface

3. A container holds water of density 1000 kg/m^3. Taking the gravitational acceleration as 10 m/s^2, the pressure at a depth of 100 mm is:
   (a) 1 kPa   (b) 1 MPa
   (c) 100 Pa   (d) 1 Pa

4. If the water in Question 3 is now replaced by a fluid having a density of 2000 kg/m^3, the pressure at a depth of 100 mm is:
   (a) 2 kPa   (b) 500 kPa
   (c) 200 Pa   (d) 0.5 Pa

5. The gauge pressure of fluid in a pipe is 70 kPa and the atmospheric pressure is 100 kPa. The absolute pressure of the fluid in the pipe is:
   (a) 7 MPa   (b) 30 kPa
   (c) 170 kPa   (d) 10/7 kPa

6. A *U*-tube manometer contains mercury of density 13600 kg/m^3. When the difference in the height of the mercury levels is 100 mm and taking the gravitational acceleration as 10 m/s^2, the gauge pressure is:
   (a) 13.6 Pa   (b) 13.6 MPa
   (c) 13 710 Pa   (d) 13.6 kPa

7. The mercury in the U-tube of Question 6 is to be replaced by water of density 1000 kg/m^3. The height of the tube to contain the water for the same gauge pressure is:
   (a) (1/13.6) of the original height
   (b) 13.6 times the original height
   (c) 13.6 m more than the original height
   (d) 13.6 m less than the original height

8. Which of the following devices does not measure pressure?
   (a) barometer   (b) McLeod gauge
   (c) thermocouple (d) manometer

9. A pressure of 10 kPa is equivalent to:
   (a) 10 millibars (b) 1 bar
   (c) 0.1 bar   (d) 0.1 millibars

10. A pressure of 1000 mbars is equivalent to:
    (a) 0.1 kN/m^2 (b) 10 kPa
    (c) 1000 Pa   (d) 100 kN/m^2

11. Which of the following statements is false?

    (a) Barometers may be used for the measurement of altitude.

    (b) Standard atmospheric pressure is the pressure due to the mass of the air above the ground.

    (c) The maximum pressure that a mercury manometer, using a 1 m length of glass tubing, is capable of measuring is in the order of 130 kPa.

    (d) An inclined manometer is designed to measure higher values of pressure than the U-tube manometer.

In Questions 12 and 13 assume that atmospheric pressure is 1 bar.

12. A Bourdon pressure gauge indicates a pressure of 3 bars. The absolute pressure of the system being measured is:

    (a) 1 bar   (b) 2 bars

    (c) 3 bars   (d) 4 bars

13. In question 12, the gauge pressure is:

    (a) 1 bar   (b) 2 bars

    (c) 3 bars   (d) 4 bars

In Questions 14 to 18 select the most suitable pressure-indicating device from the following list:

(a) Mercury filled *U*-tube manometer

(b) Bourdon gauge

(c) McLeod gauge

(d) aneroid barometer

(e) Pirani gauge

(f) Fortin barometer

(g) water-filled inclined barometer

14. A robust device to measure high pressures in the range 0 – 30 MPa.

15. Calibration of a Pirani gauge.

16. Measurement of gas pressures comparable with atmospheric pressure.

17. To measure pressures of the order of 1 MPa.

18. Measurement of atmospheric pressure to a high degree of accuracy.

19. Figure 22.7(b), on page 252, shows a *U*-tube manometer connected to a gas under pressure. If atmospheric pressure is 76 cm of mercury and h_1 is measured in centimetres then the gauge pressure (in cm of mercury) of the gas is:

    (a) h_1 (b) $h_1 + 76$

    (c) $h_1 - 76$ (d) $76 - h_1$

20. In question 19 the absolute pressure of the gas (in cm of mercury) is:

    (a) h_1 (b) $h_1 + 76$

    (c) $h_1 - 76$ (d) $76 - h_1$

21. Which of the following statements is true?

    (a) Atmospheric pressure of 101.325 kN/m^2 is equivalent to 101.325 millibars.

    (b) An aneroid barometer is used as a standard for calibration purposes.

    (c) In engineering, 'pressure' is the force per unit area exerted by fluids.

    (d) Water is normally used in a barometer to measure atmospheric pressure.

22. Which of the following statements is true for a ship floating in equilibrium?

    (a) The weight is larger than the buoyancy.

    (b) The weight is smaller than the buoyancy.

    (c) The weight is equal to the buoyancy.

    (d) The weight is independent of the buoyancy.

23. For a ship to be initially stable, the metacentric height must be:

    (a) positive

    (b) negative

    (c) zero

    (d) equal to the buoyancy

24. For a ship to be stable, it is helpful if KG is:

    (a) negative   (b) large

    (c) small   (d) equal to KM

References

[1] ROSS, C.T.F, *Pressure Vessels; External Pressure Technology* – 2nd Edition, Woodhead Publishers, Cambridge, UK.

[2] Pressure – www.routledge.com/cw/bird

A video reference on YouTube on various types of pressure, including atmospheric pressure, hydrostatic pressure and wind pressure, which in the case of the last type, was experienced by Hurricane Katrina & the tornadoes experienced in Alabama and elsewhere.

[3] Hydrostatic Stability – www.routledge.com/cw/bird

A video reference on YouTube on the damage stability of ro-ro 'car' ferries.

Chapter 23

Fluid flow

The measurement of fluid flow is of great importance in many industrial processes, some examples including air flow in the ventilating ducts of a coal mine, the flow rate of water in a condenser at a power station, the flow rate of liquids in chemical processes, the control and monitoring of the fuel, lubricating and cooling fluids of ships and aircraft engines, and so on. Fluid flow is one of the most difficult of industrial measurements to carry out, since flow behaviour depends on a great many variables concerning the physical properties of a fluid. There are available a large number of fluid flow measuring instruments generally called flowmeters, which can measure the flow rate of liquids (in m^3/s) or the mass flow rate of gaseous fluids (in kg/s). The two main categories of flowmeters are differential pressure flowmeters and mechanical flowmeters. This chapter also contains calculations on Bernoulli's equation and the impact of a jet on a stationary plate.

At the end of this chapter you should be able to:

- appreciate the importance of measurement of fluid flow
- describe the construction, principle of operation, advantages and disadvantages, and practical applications of orifice plates, Venturi tubes, flow nozzles and Pitot-static tube and describe the principle of operation of deflecting vane and turbine type flowmeters
- describe the construction, principle of operation, advantages and disadvantages, and practical applications of float and tapered tube flowmeters, electromagnetic flowmeters, and hot-wire anemometer
- select the most appropriate flowmeter for a particular application

- state the continuity equation (i.e. the principle of conservation of mass)
- state and perform calculations on Bernoulli's equation
- state and perform calculations on the impact of a jet on a stationary plate

23.1 Differential pressure flowmeters

When certain flowmeters are installed in pipelines they often cause an obstruction to the fluid flowing in the pipe by reducing the cross-sectional area of the pipeline. This causes a change in the velocity of the fluid, with a related change in pressure. Figure 23.1 shows a section through a pipeline into which a flowmeter has been inserted. The flow rate of the fluid may be determined from a measurement of the difference between the pressures on the walls of the pipe at specified distances upstream and downstream of the flowmeter. Such devices are known as **differential pressure flowmeters**.

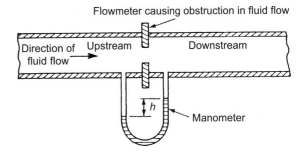

Figure 23.1

The pressure difference in Figure 23.1 is measured using a manometer connected to appropriate pressure

Mechanical Engineering Principles, Bird and Ross, ISBN 9780415517850

tapping points. The pressure is seen to be greater upstream of the flowmeter than downstream, the pressure difference being shown as *h*.

Calibration of the manometer depends on the shape of the obstruction, the positions of the pressure tapping points and the physical properties of the fluid.

In industrial applications the pressure difference is detected by a differential pressure cell, the output from which is either an amplified pressure signal or an electrical signal.

Examples of differential pressure flowmeters commonly used include:

(a)　Orifice plate (see Section 23.2)
(b)　Venturi tube (see Section 23.3)
(c)　Flow nozzles (see Section 23.4)
(d)　Pitot-static tube (see Section 23.5)

British Standard reference BS 1042: Part 1: 1964 and Part 2A: 1973 'Methods for the measurement of fluid flow in pipes' gives specifications for measurement, manufacture, tolerances, accuracy, sizes, choice, and so on, of differential flowmeters.

23.2　Orifice plate

Construction

An orifice plate consists of a circular, thin, flat plate with a hole (or orifice) machined through its centre to fine limits of accuracy. The orifice has a diameter less than the pipeline into which the plate is installed and a typical section of an installation is shown in Figure 23.2(a). Orifice plates are manufactured in stainless steel, monel metal, polyester glass fibre, and for large pipes, such as sewers or hot gas mains, in brick and concrete.

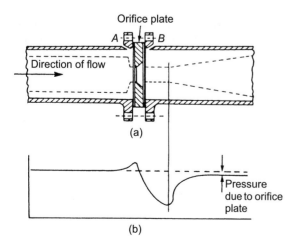

Figure 23.2

Principles of operation

When a fluid moves through a restriction in a pipe, the fluid accelerates and a reduction in pressure occurs, the magnitude of which is related to the flow rate of the fluid. The variation of pressure near an orifice plate is shown in Figure 23.2(b). The position of minimum pressure is located downstream from the orifice plate where the flow stream is narrowest. This point of minimum cross-sectional area of the jet is called the 'vena contracta'. Beyond this point the pressure rises but does not return to the original upstream value and there is a permanent pressure loss. This loss depends on the size and type of orifice plate, the positions of the upstream and downstream pressure tappings and the change in fluid velocity between the pressure tappings that depends on the flow rate and the dimensions of the orifice plate.

In Figure 23.2(a) corner pressure tappings are shown at *A* and *B*. Alternatively, with an orifice plate inserted into a pipeline of diameter *d*, pressure tappings are often located at distances of *d* and *d*/2 from the plate respectively upstream and downstream. At distance *d* upstream the flow pattern is not influenced by the presence of the orifice plate, and distance *d*/2 coincides with the vena contracta.

Advantages of orifice plates
(i)　They are relatively inexpensive.
(ii)　They are usually thin enough to fit between an existing pair of pipe flanges.

Disadvantages of orifice plates
(i)　The sharpness of the edge of the orifice can become worn with use, causing calibration errors.
(ii)　The possible build-up of matter against the plate.
(iii)　A considerable loss in the pumping efficiency due to the pressure loss downstream of the plate.

Applications
Orifice plates are usually used in medium and large pipes and are best suited to the indication and control of essentially constant flow rates. Several applications are found in the general process industries.

23.3　Venturi tube

Construction

The Venturi tube or venturimeter is an instrument for measuring with accuracy the flow rate of fluids in pipes. A typical arrangement of a section through such a device is shown in Figure 23.3, and consists of a short converging conical tube called the inlet or upstream cone, leading to a cylindrical portion called the throat. A diverging section called the outlet or recovery cone follows this. The entrance and exit diameter is the same as that of the pipeline into which it is installed. Angle β is usually a maximum of 21°, giving a taper of $\beta/2$ of 10.5°. The length of the throat is made equal to the diameter of the throat. Angle α is about 5° to 7° to ensure a minimum loss of energy but where this is unimportant α can be as large as 14° to 15°.

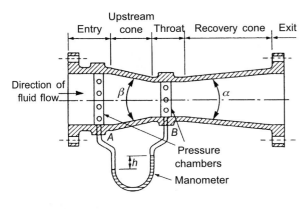

Figure 23.3

Pressure tappings are made at the entry (at A) and at the throat (at B) and the pressure difference h which is measured using a manometer, a differential pressure cell or similar gauge, is dependent on the flow rate through the meter. Usually pressure chambers are fitted around the entrance pipe and the throat circumference with a series of tapping holes made in the chamber to which the manometer is connected. This ensures that an average pressure is recorded. The loss of energy due to turbulence that occurs just downstream with an orifice plate is largely avoided in the venturimeter due to the gradual divergence beyond the throat.

Venturimeters are usually made a permanent installation in a pipeline and are manufactured usually from stainless steel, cast iron, monel metal or polyester glass fibre.

Advantages of venturimeters
 (i) High accuracy results are possible.
 (ii) There is a low pressure loss in the tube (typically only 2% to 3% in a well proportioned tube).
 (iii) Venturimeters are unlikely to trap any matter from the fluid being metered.

Disadvantages of venturimeters
 (i) High manufacturing costs.
 (ii) The installation tends to be rather long (typically 120 mm for a pipe of internal diameter 50 mm).

23.4 Flow nozzle

The flow nozzle lies between an orifice plate and the venturimeter both in performance and cost. A typical section through a flow nozzle is shown in Figure 23.4, where pressure tappings are located immediately adjacent to the upstream and downstream faces of the nozzle (i.e. at points A and B). The fluid flow does not contract any further as it leaves the nozzle and the pressure loss created is considerably less than that occurring with orifice plates. Flow nozzles are suitable for use with high velocity flows for they do not suffer the wear that occurs in orifice plate edges during such flows.

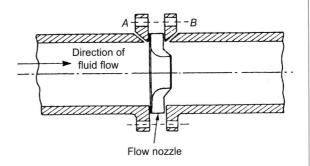

Figure 23.4

23.5 Pitot-static tube

A Pitot-static tube is a device for measuring the velocity of moving fluids or of the velocity of bodies moving through fluids. It consists of one tube, called the Pitot tube, with an open end facing the direction of the fluid motion, shown as pipe R in Figure 23.5, and a second tube, called the piezometer tube, with the opening at 90° to the fluid flow, shown as T in Figure 23.5. Pressure recorded by a pressure gauge moving with

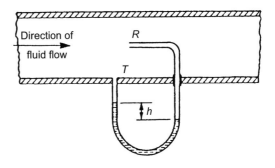

Figure 23.5

the flow, i.e. static or stationary relative to the fluid, is called free stream pressure and connecting a pressure gauge to a small hole in the wall of a pipe, such as point T in Figure 23.5, is the easiest method of recording this pressure. The difference in pressure $(p_R - p_T)$, shown as h in the manometer of Figure 23.5, is an indication of the speed of the fluid in the pipe.

Figure 23.6 shows a practical Pitot-static tube consisting of a pair of concentric tubes. The centre tube is the impact probe that has an open end which faces 'head-on' into the flow. The outer tube has a series of holes around its circumference located at right angles to the flow, as shown by A and B in Figure 23.6. The manometer, showing a pressure difference of h, may be calibrated to indicate the velocity of flow directly.

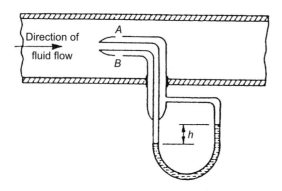

Figure 23.6

Applications
A Pitot-static tube may be used for both turbulent and non-turbulent flow. The tubes can be made very small compared with the size of the pipeline and the monitoring of flow velocity at particular points in the cross-section of a duct can be achieved. The device is generally unsuitable for routine measurements and in industry is often used for making preliminary tests of flow rate in order to specify permanent flow measuring equipment for a pipeline. The main use of Pitot tubes is to measure the velocity of solid bodies moving through fluids, such as the velocity of ships. In these cases, the tube is connected to a Bourdon pressure gauge that can be calibrated to read velocity directly. A development of the Pitot tube, a **pitometer**, tests the flow of water in water mains and detects leakages.

Advantages of Pitot-static tubes
 (i) They are inexpensive devices.
 (ii) They are easy to install.
 (iii) They produce only a small pressure loss in the tube.
 (iv) They do not interrupt the flow.

Disadvantages of Pitot-static tubes
 (i) Due to the small pressure difference, they are only suitable for high velocity fluids.
 (ii) They can measure the flow rate only at a particular position in the cross-section of the pipe.
 (iii) They easily become blocked when used with fluids carrying particles.

23.6 Mechanical flowmeters

With mechanical flowmeters, a sensing element situated in a pipeline is displaced by the fluid flowing past it. Examples of mechanical flowmeters commonly used include:

(a) Deflecting vane flowmeter (see Section 23.7)
(b) Turbine type meters (see Section 23.8)

23.7 Deflecting vane flowmeter

The deflecting vane flowmeter consists basically of a pivoted vane suspended in the fluid flow stream as shown in Figure 23.7.

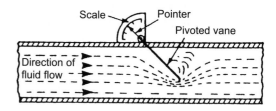

Figure 23.7

When a jet of fluid impinges on the vane it deflects from its normal position by an amount proportional to the flow rate. The movement of the vane is indicated on a scale that may be calibrated in flow units. This type of meter is normally used for measuring liquid flow rates in open channels or for measuring the velocity of air in ventilation ducts. The main disadvantages of this device are that it restricts the flow rate and it needs to be recalibrated for fluids of differing densities.

23.8 Turbine type meters

Turbine type flowmeters are those that use some form of multi-vane rotor and are driven by the fluid being investigated. Three such devices are the cup anemometer, the rotary vane positive displacement meter and the turbine flowmeter.

(a) **Cup anemometer**. An anemometer is an instrument that measures the velocity of moving gases and is most often used for the measurement of wind speed. The cup anemometer has three or four cups of hemispherical shape mounted at the end of arms radiating horizontally from a fixed point. The cup system spins round the vertical axis with a speed approximately proportional to the velocity of the wind. With the aid of a mechanical and/or electrical counter the wind speed can be determined and the device is easily adapted for automatic recording.

(b) **Rotary vane positive displacement meters** measure the flow rate by indicating the quantity of liquid flowing through the meter in a given time. A typical such device is shown in section in Figure 23.8 and consists of a cylindrical chamber into which is placed a rotor containing a number of vanes (six in this case). Liquid entering the chamber turns the rotor and a known amount of liquid is trapped and carried round to the outlet. If x is the volume displaced by one blade then for each revolution of the rotor in Figure 23.8 the total volume displaced is $6x$. The rotor shaft may be coupled to a mechanical counter and electrical devices which may be calibrated to give flow volume. This type of meter in its various forms is used widely for the measurement of domestic and industrial water consumption, for the accurate measurement of petrol in petrol pumps and for the consumption and batch control measurements in the general process and food industries for measuring flows as varied as solvents, tar and molasses (i.e. thickish treacle).

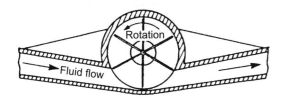

Figure 23.8

(c) A **turbine flowmeter** contains in its construction a rotor to which blades are attached which spin at a velocity proportional to the velocity of the fluid which flows through the meter. A typical section through such a meter is shown in Figure 23.9. The number of revolutions made by the turbine blades may be determined by a mechanical or electrical device enabling the flow rate or total flow to be determined. Advantages of turbine flowmeters include a compact durable form, high accuracy, wide temperature and pressure capability and good response characteristics. Applications include the volumetric measurement of both crude and refined petroleum products in pipelines up to 600 mm bore, and in the water, power, aerospace, process and food industries, and with modification may be used for natural, industrial and liquid gas measurements. Turbine flowmeters require periodic inspection and cleaning of the working parts.

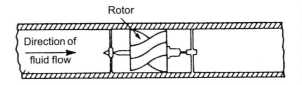

Figure 23.9

23.9   Float and tapered-tube meter

Principle of operation

With orifice plates and venturimeters the area of the opening in the obstruction is fixed and any change in the flow rate produces a corresponding change in pressure. With the float and tapered-tube meter the area of the restriction may be varied so as to maintain a steady pressure differential. A typical meter of this type is shown diagrammatically in Figure 23.10 where a vertical tapered tube contains a 'float' that has a density greater than the fluid.

The float in the tapered tube produces a restriction to the fluid flow. The fluid can only pass in the annular area between the float and the walls of the tube. This reduction in area produces an increase in velocity and hence a pressure difference, which causes the float to rise. The greater the flow rate, the greater is the rise in the float position, and vice versa. The position of the float is a measure of the flow rate of the fluid and this is shown on a vertical scale engraved on a transparent tube of plastic or glass. For air, a small sphere is used for the float but for liquids there is a tendency to instability and the float is then designed with vanes that cause it to spin and thus stabilize itself as the liquid flows past. Such meters are often called '**rotameters**'. Calibration of float and tapered tube flowmeters can be achieved using a Pitot-static tube or,

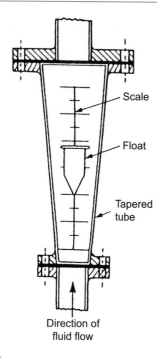

Scale

Float

Tapered tube

Direction of fluid flow

Figure 23.10

more often, by using a weighing meter in an instrument repair workshop.

Advantages of float and tapered-tube flowmeters
(i) They have a very simple design.
(ii) They can be made direct reading.
(iii) They can measure very low flow rates.

Disadvantages of float and tapered-tube flowmeters
(i) They are prone to errors, such as those caused by temperature fluctuations.
(ii) They can only be installed vertically in a pipeline.
(iii) They cannot be used with liquids containing large amounts of solids in suspension.
(iv) They need to be recalibrated for fluids of different densities.

Practical applications of float and tapered-tube meters are found in the medical field, in instrument purging, in mechanical engineering test rigs and in simple process applications, in particular for very low flow rates. Many corrosive fluids can be handled with this device without complications.

23.10   Electromagnetic flowmeter

The flow rate of fluids that conduct electricity, such as water or molten metal, can be measured using an electromagnetic flowmeter whose principle of operation is based on the laws of electromagnetic induction. When a conductor of length L moves at right angles to a magnetic field of density B at a velocity v, an induced e.m.f. e is generated, given by: $e = BLv$.

With the electromagnetic flowmeter arrangement shown in Figure 23.11, the fluid is the conductor and the e.m.f. is detected by two electrodes placed across the diameter of the non-magnetic tube.

Rearranging $e = BLv$ gives:

$$\text{velocity, } v = \frac{e}{BL}$$

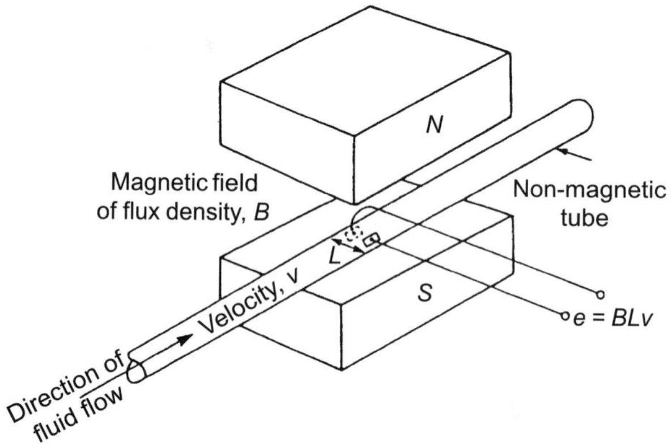

Magnetic field of flux density, B

Non-magnetic tube

N

S

L

Velocity, v

Direction of fluid flow

$e = BLv$

Figure 23.11

Thus with B and L known, when e is measured, the velocity of the fluid can be calculated.

Main advantages of electromagnetic flowmeters
(i) Unlike other methods, there is nothing directly to impede the fluid flow.
(ii) There is a linear relationship between the fluid flow and the induced e.m.f.
(iii) Flow can be metered in either direction by using a centre-zero measuring instrument.

Applications of electromagnetic flowmeters are found in the measurement of speeds of slurries, pastes and viscous liquids, and they are also widely used in the water production, supply and treatment industry.

23.11   Hot-wire anemometer

A simple hot-wire anemometer consists of a small piece of wire which is heated by an electric current and positioned in the air or gas stream whose velocity is to be measured. The stream passing the wire cools it, the rate of cooling being dependent on the flow velocity. In practice there are various ways in which this is achieved:

(i)   If a constant current is passed through the wire, variation in flow results in a change of temperature of the wire and hence a change in resistance which may be measured by a Wheatstone bridge arrangement. The change in resistance may be related to fluid flow.
(ii) If the wire's resistance, and hence temperature, is kept constant, a change in fluid flow results in a corresponding change in current which can be calibrated as an indication of the flow rate.
(iii) A thermocouple may be incorporated in the assembly, monitoring the hot wire and recording the temperature which is an indication of the air or gas velocity.

Advantages of the hot-wire anemometer
(a)   Its size is small.
(b)   It has great sensitivity.

23.12   Choice of flowmeter

Problem 1.   Choose the most appropriate fluid flow measuring device for the following circumstances:
(a)   The most accurate, permanent installation for measuring liquid flow rate.
(b)   To determine the velocity of low-speed aircraft and ships.
(c)   Accurate continuous volumetric measurement of crude petroleum products in a duct of 500 mm bore.
(d)   To give a reasonable indication of the mean flow velocity, while maintaining a steady pressure difference on a hydraulic test rig.
(e)   For an essentially constant flow rate with reasonable accuracy in a large pipe bore, with a cheap and simple installation.

(a)   **Venturimeter**
(b)   **Pitot-static tube**
(c)   **Turbine flowmeter**
(d)   **Float and tapered-tube flowmeter**
(e)   **Orifice plate.**

Now try the following Practise Exercise

Practise Exercise 124   Further problems on the measurement of fluid flow

For the flow measurement devices listed 1 to 5, (a) describe briefly their construction (b) state their principle of operation (c) state their characteristics and limitations (d) state typical practical applications (e) discuss their advantages and disadvantages.

1.   Orifice plate
2.   Venturimeter
3.   Pitot-static tube
4.   Float and tapered-tube meter
5.   Turbine flowmeter.

23.13   Equation of continuity

The calibrations of many of the flowmeters described earlier are based on the equation of continuity and Bernoulli's equation.

The **equation of continuity** states that for the steady flow of a fluid through a pipe of varying cross-section the rate of mass entering the pipe must be equal to the rate of mass leaving the pipe; this is really a statement of the **principle of conservation of mass.** Thus, for an incompressible fluid:

$$a_1 v_1 = a_2 v_2$$

where a_1 = cross-sectional area at section 1, a_2 = cross-sectional area at section 2, v_1 = velocity of fluid at section 1, and v_2 = velocity of fluid at section 2

23.14   Bernoulli's equation

Bernoulli's equation states that for a fluid flowing through a pipe from section 1 to section 2:

$$\frac{P_1}{\rho} + \frac{v_1^2}{2} + gz_1 = \frac{P_2}{\rho} + \frac{v_2^2}{2} + g(z_2 + h_f)$$

where ρ = density of the fluid,
    P_1 = pressure at section 1,
    P_2 = pressure at section 2,

v_1 = velocity at section 1,

v_2 = velocity at section 2,

z_1 = 'height' of pipe at section 1,

z_2 = 'height' of pipe at section 2,

h_f = friction losses (in m) due to the fluid flowing from section 1 to section 2,

and $\quad g = 9.81$ m/s^2 (assumed).

Problem 2. A storage tank contains oil whose free surface is 5 m above an outlet pipe, as shown in Figure 23.12. Determine the mass rate of flow at the exit of the outlet pipe, assuming that (a) losses at the pipe entry = $0.4\,v^2$, and (b) losses at the valve = $0.25\,v^2$.

Pipe diameter = 0.04 m, density of oil, $\rho = 770$ kg/m^3.

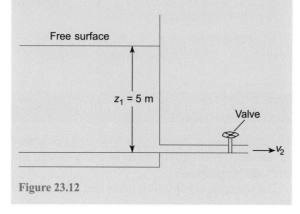

Figure 23.12

Let v_2 = velocity of oil through the outlet pipe.

From Bernoulli's equation:

$$\frac{P_1}{\rho} + \frac{v_1^2}{2} + gz_1$$

$$= \frac{P_2}{\rho} + \frac{v_2^2}{2} + gz_2 + 0.4\,v_2^2 + 0.25\,v_2^2$$

i.e. $\quad 0 + 0 + g(5\text{ m}) = 0 + \frac{v_2^2}{2} + 0 + 0.65\,v_2^2$

(where in the above, the following assumptions have been made: $P_1 = P_2$ = atmospheric pressure, and v_1 is negligible)

Hence, $\qquad 5\text{ m} \times 9.81\,\dfrac{\text{m}}{\text{s}^2} = (0.5 + 0.65)v_2^2$

Rearranging gives: $\quad 1.15\,v_2^2 = 49.05\,\dfrac{\text{m}^2}{\text{s}^2}$

Hence, $\qquad\qquad v_2^2 = \dfrac{49.05}{1.15}$

from which, $\qquad v_2 = \sqrt{\dfrac{49.05}{1.15}} = \mathbf{6.531}$ **m/s**

Cross-sectional area of pipe $= a_2 = \dfrac{\pi\,d_2^2}{4} = \dfrac{\pi \times 0.04^2}{4}$

$$= 0.001257\text{ m}^2$$

Mass rate of flow through the outlet pipe

$$= \rho\,a_2 v_2 = 770\,\frac{\text{kg}}{\text{m}^3} \times 1.257 \times 10^{-3}\text{m}^2 \times 6.531\,\frac{\text{m}}{\text{s}}$$

$$= \mathbf{6.321}\ \textbf{kg/s}$$

Flow through an orifice

Consider the flow of a liquid through a small orifice, as shown in Figures 23.13(a) and (b), where it can be seen that the vena contracta (VC) lies just to the right of the orifice. The cross-sectional area of the fluid is the smallest here and its decrease in area from the orifice is measured by the coefficient of contraction (C_c).

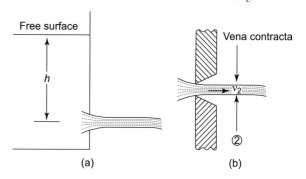

(a) (b)

Figure 23.13

Due to friction losses there will be a loss in velocity at the orifice; this is measured by the coefficient of velocity, namely C_v, so that:

$$C_d = C_V \times C_c = \textbf{the coefficient of discharge.}$$

Let a = area of orifice.

Due to the vena contracta the equivalent cross-sectional area = $C_c a$

Now the theoretical velocity at section 2 = $v_2 = \sqrt{2gh}$,

but due to friction losses, $v_2 = C_v \sqrt{2gh}$

and due to contraction, $v_2 = C_c C_v \sqrt{2gh}$

Hence $\qquad$ discharge = $C_c a \times C_v \sqrt{2gh}$

But $\qquad\qquad C_d = C_v C_c$

Therefore, $\qquad$ **discharge** $= C_d \times a\sqrt{2gh}$

Now try the following Practise Exercise

1. If in the storage tank of worked Problem 2 on page 268, Figure 23.12, $z_1 = 8$ m, determine the mass rate of flow from the outlet pipe.
   [7.995 kg/s]

2. If in the storage tank of worked Problem 2, page 268, Figure 23.12, $z_1 = 10$ m, determine the mass rate of flow from the outlet pipe.
   [8.939 kg/s]

3. If in Figure 23.13, $h = 6$ m, $C_c = 0.8$, $C_v = 0.7$, determine the values of C_d and actual v_2.   [$C_d = 0.56$, $v_2 = 6.08$ m/s]

4. If in Figure 23.13, $h = 10$ m, $C_c = 0.75$, $C_v = 0.65$, and the cross-sectional area is 1.5×10^{-3} m², determine the discharge and the actual velocity v_2.
   [$C_d = 0.488$, 6.83 m/s, 7.90 kg/s]

23.15   Impact of a jet on a stationary plate

The impact of a jet on a plate is of importance in a number of engineering problems, including the determination of pressures on buildings subjected to gusts of wind.

Consider the jet of fluid acting on the flat plate of Figure 23.14, where it can be seen that the velocity of the fluid is turned through 90°, or change of velocity = v.

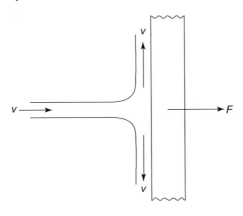

Figure 23.14

Now, momentum = mv and as v is constant, the change of momentum = $\dfrac{dm}{dt} \times v$

However, $\dfrac{dm}{dt}$ = mass rate of flow = $\rho a v$

Therefore, change of momentum = $\rho a v \times v = \rho a v^2$ but from Newton's second law of motion (see pages 157 and 163),

$$F = \text{rate of change of momentum}$$

i.e. $$F = \rho a v^2$$

where F = resulting normal force on the flat plate.

$$\textbf{Pressure} = \frac{\text{force}}{\text{area}} = \frac{\rho a v^2}{a} = \rho v^2$$

For wide surfaces, such as garden fences, the pressure can be calculated by the above formula, but **for tall buildings and trees**, civil engineers normally assume that:

$$\textbf{Pressure } p = \textbf{0.5 } \rho v^2$$

This is because the flow of fluid is similar to the plan view shown in Figure 23.15, where the change of momentum is much less.

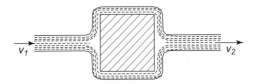

Figure 23.15

> **Problem 3.** Determine the wind pressure on a slim, tall building due to a gale of 100 km/h. Take density of air, $\rho = 1.2$ kg/m³.

For a tall building, pressure $p = 0.5 \, \rho v^2$

Velocity, $v = 100 \dfrac{\text{km}}{\text{h}} \times \dfrac{1000\,\text{m}}{\text{km}} \times \dfrac{1\,\text{h}}{3600\,\text{s}} = 27.78$ m/s

Hence,

wind pressure, $p = 0.5 \times 1.2 \dfrac{\text{kg}}{\text{m}^3} \times \left(27.78 \dfrac{\text{m}}{\text{s}}\right)^2$

$$= \textbf{462.96 N/m}^2 = \textbf{0.00463 bar.}$$

> **Problem 4.** What would be the wind pressure of Problem 3, if the gale were acting on a very wide and flat surface?

For a very wide surface,

pressure, $p = \rho v^2 = 1.2 \dfrac{\text{kg}}{\text{m}^3} \times \left(27.78 \dfrac{\text{m}}{\text{s}}\right)^2$

$$= \textbf{926.1 N/m}^2 = \textbf{0.00926 bar}$$

(or less than 1/100th of atmospheric pressure!)

Now try the following Practise Exercises

Practise Exercise 126   **Further problems on the impact of jets on flat surfaces**

1. A hurricane of velocity 220 km/h blows perpendicularly on to a very wide flat surface. Determine the wind pressure that acts on this surface due to this hurricane, when the density of air, $\rho = 1.2$ kg/m³.   [0.0448 bar]

2. What is the wind pressure for Problem 1 on a slim, tall building?   [0.0224 bar]

3. A tornado with a velocity of 320 km/h blows perpendicularly on to a very wide surface. Determine the wind pressure that acts on this surface due to this tornado, when the density of air, $\rho = 1.23$ kg/m³.   [0.0972 bar]

4. What is the wind pressure for Problem 3 on a slim, tall building?   [0.0486 bar]

5. If atmospheric pressure were 1.014 bar, what fraction of atmospheric pressure would be the wind pressure calculated in Problem 4?

$$\left[0.0479 \approx \frac{1}{21}\right]$$

Practise Exercise 127   **Short-answer questions on the measurement of fluid flow**

In the flowmeters listed 1 to 10, state typical practical applications of each.

1. Orifice plate.

2. Venturimeter.

3. Float and tapered-tube meter.

4. Electromagnetic flowmeter.

5. Pitot-static tube.

6. Hot-wire anemometer.

7. Turbine flowmeter.

8. Deflecting vane flowmeter.

9. Flow nozzles.

10. Rotary vane positive displacement meter.

11. Write down the relationship between the coefficients C_c, C_v and C_d

12. Write down the formula for the pressure due to a wind acting perpendicularly on a tall slender building.

Practise Exercise 128   **Multiple-choice questions on the measurement of fluid flow**

(Answers on page 298)

1. The term 'flow rate' usually refers to:
   (a) mass flow rate
   (b) velocity of flow
   (c) volumetric flow rate

2. The most suitable method for measuring the velocity of high-speed gas flow in a duct is:
   (a) venturimeter
   (b) orifice plate
   (c) Pitot-static tube
   (d) float and tapered-tube meter

3. Which of the following statements is false? When a fluid moves through a restriction in a pipe, the fluid
   (a) accelerates and the pressure increases
   (b) decelerates and the pressure decreases
   (c) decelerates and the pressure increases
   (d) accelerates and the pressure decreases

4. With an orifice plate in a pipeline the vena contracta is situated:
   (a) downstream at the position of minimum cross-sectional area of flow
   (b) upstream at the position of minimum cross-sectional area of flow
   (c) downstream at the position of maximum cross-sectional area of flow
   (d) upstream at the position of maximum cross-sectional area of flow

In Questions 5 to 14, select the most appropriate device for the particular requirements from the following list:

(a) orifice plate
(b) turbine flowmeter
(c) flow nozzle
(d) pitometer
(e) venturimeter
(f) cup anemometer
(g) electromagnetic flowmeter
(h) pitot-static tube
(i) float and tapered-tube meter
(j) hot-wire anemometer
(k) deflecting vane flowmeter.

5. Easy to install, reasonably inexpensive, for high-velocity flows.

6. To measure the flow rate of gas, incorporating a Wheatstone bridge circuit.

7. Very low flow rate of corrosive liquid in a chemical process.

8. To detect leakages from water mains.

9. To determine the flow rate of liquid metals without impeding its flow.

10. To measure the velocity of wind.

11. Constant flow rate, large bore pipe, in the general process industry.

12. To make a preliminary test of flow rate in order to specify permanent flow measuring equipment.

13. To determine the flow rate of fluid very accurately with low pressure loss.

14. To measure the flow rate of air in a ventilating duct.

15. For a certain wind velocity, what fraction of the pressure would act on a tall slender building in comparison with a very wide surface?

(a) 0.01   (b) 0
(c) 0.5   (d) 0.99

16. For a wind speed of 190 km/h, what fraction (approximate) of atmospheric pressure will this be, when blowing perpendicularly to a very wide surface?

(a) 2.5   (b) 0.5
(c) 1/30   (d) 0

Chapter 24

Ideal gas laws

The relationships that exist between pressure, volume and temperature in a gas are given in a set of laws called the gas laws, the most fundamental being those of Boyle's, Charles', and the pressure law, together with Dalton's law of partial pressures and the characteristic gas equation. These laws are used for all sorts of practical applications, including for designing pressure vessels, in the form of circular cylinders and spheres, which are used for storing and transporting gases. An example of this is the pressure in car tyres, which can increase due to a temperature increase, and can decrease due to a temperature decrease. Other examples are large and medium size gas storage cylinders and domestic spray cans, which can explode if they are heated. In the case of domestic spray cans, these can explode dangerously in a domestic situation if they are left on a window sill where the sunshine acting on them causes them to heat up or, if they are thrown on to a fire. In these cases, the consequence can be disastrous, so don't throw your 'full' spray can on to a fire; you may very sadly and deeply regret it! Another example of a gas storage vessel is that used by your 'local' gas companies, which supply natural gas (methane) to domestic properties, businesses, etc.

At the end of this chapter you should be able to:

- state and perform calculations involving Boyle's law
- understand the term isothermal
- state and perform calculations involving Charles' law
- understand the term isobaric
- state and perform calculations involving the pressure law
- state and perform calculations on Dalton's law of partial pressures

Mechanical Engineering Principles, Bird and Ross, ISBN 9780415517850

- state and perform calculations on the characteristic gas equation
- understand the term STP

24.1 Boyle's law

Boyle's law states:

the volume V of a fixed mass of gas is inversely proportional to its absolute pressure p at constant temperature

i.e. $\qquad p \propto \dfrac{1}{V}$

or $\qquad p = \dfrac{k}{V}$

or $\qquad \boldsymbol{pV = k}$ at constant temperature, where

$\quad p$ = absolute pressure in pascals (Pa),

$\quad V$ = volume in m³, and k = a constant.

Changes that occur at constant temperature are called **isothermal** changes. When a fixed mass of gas at constant temperature changes from pressure p_1 and volume V_1 to pressure p_2 and volume V_1 then:

$$p_1 V_1 = p_2 V_2$$

Problem 1. A gas occupies a volume of 0.10 m³ at a pressure of 1.8 MPa. Determine (a) the pressure if the volume is changed to 0.06 m³ at constant temperature, and (b) the volume if the pressure is changed to 2.4 MPa at constant temperature.

(a) Since the change occurs at constant temperature (i.e. an isothermal change), Boyle's law applies, i.e. $p_1V_1 = p_2V_2$

where $p_1 = 1.8$ MPa, $V_1 = 0.10$ m^3 and
$V_2 = 0.06$ m^3.

Hence, $(1.8)(0.10) = p_2(0.06)$

from which, **pressure** $p_2 = \dfrac{1.8 \times 0.10}{0.06} = \mathbf{3\ MPa}$

(b) $p_1V_1 = p_2V_2$ where $p_1 = 1.8$ MPa, $V_1 = 0.10$ m^3
and $p_2 = 2.4$ MPa.

Hence, $(1.8)(0.10) = (2.4)V_2$

from which, **volume** $V_2 = \dfrac{1.8 \times 0.10}{2.4} = \mathbf{0.075\ m^3}$

Problem 2. In an isothermal process, a mass of gas has its volume reduced from 3200 mm^3 to 2000 mm^3. If the initial pressure of the gas is 110 kPa, determine the final pressure.

Since the process is isothermal, it takes place at constant temperature and hence Boyle's law applies, i.e. $p_1V_1 = p_2V_2$, where $p_2 = 110$ kPa, $V_1 = 3200$ mm^3 and $V_2 = 2000$ mm^3.

Hence, $(110)(3200) = p_2(2000)$

from which, **final pressure,** $p_2 = \dfrac{110 \times 3200}{2000}$

$= \mathbf{176\ kPa}$

Problem 3. Some gas occupies a volume of 1.5 m^3 in a cylinder at a pressure of 250 kPa. A piston, sliding in the cylinder, compresses the gas isothermally until the volume is 0.5 m^3. If the area of the piston is 300 cm^2, calculate the force on the piston when the gas is compressed.

An isothermal process means constant temperature and thus Boyle's law applies, i.e. $p_1V_1 = p_2V_2$

where $V_1 = 1.5$ m^3, $V_2 = 0.5$ m^3 and $p_1 = 250$ kPa.

Hence, $(250)(1.5) = p_2(0.5)$

from which, pressure, $p_2 = \dfrac{250 \times 1.5}{0.5}$

$= 750$ kPa

$\text{Pressure} = \dfrac{\text{force}}{\text{area}}$, from which, force = pressure × area.

Hence, **force on the piston**

$= (750 \times 10^3 \text{ Pa})(300 \times 10^{-4} \text{ m}^2)$

$= \mathbf{22.5\ kN}$

Now try the following Practise Exercise

Practise Exercise 129 Further problems on Boyle's law

1. The pressure of a mass of gas is increased from 150 kPa to 750 kPa at constant temperature. Determine the final volume of the gas, if its initial volume is 1.5 m^3. [0.3 m^3]

2. In an isothermal process, a mass of gas has its volume reduced from 50 cm^3 to 32 cm^3. If the initial pressure of the gas is 80 kPa, determine its final pressure. [125 kPa]

3. The piston of an air compressor compresses air to $\dfrac{1}{4}$ of its original volume during its stroke. Determine the final pressure of the air if the original pressure is 100 kPa, assuming an isothermal change. [400 kPa]

4. A quantity of gas in a cylinder occupies a volume of 2 m^3 at a pressure of 300 kPa. A piston slides in the cylinder and compresses the gas, according to Boyle's law, until the volume is 0.5 m^3. If the area of the piston is 0.02 m^2, calculate the force on the piston when the gas is compressed. [24 kN]

24.2 Charles' law

Charles' law states:

for a given mass of gas at constant pressure, the volume V is directly proportional to its thermodynamic temperature T

i.e. $V \propto T$

or $V = kT$

or $\dfrac{V}{T} = k$ at constant pressure, where

T = thermodynamic temperature in Kelvin (K).

A process that takes place at constant pressure is called an **isobaric** process.

The relationship between the Celsius scale of temperature and the thermodynamic or absolute scale is given by:

kelvin = degrees Celsius + 273

i.e. $K = °C + 273$

or $°C = K − 273$ (as stated in Chapter 20).

If a given mass of gas at a constant pressure occupies a volume V_1 at a temperature T_1 and a volume V_2 at temperature T_2, then

$$\frac{V_1}{T_1} = \frac{V_2}{T_2}$$

Problem 4.   A gas occupies a volume of 1.2 litres at 20°C. Determine the volume it occupies at 130°C if the pressure is kept constant.

Since the change occurs at constant pressure (i.e. an isobaric process), Charles' law applies,

i.e. $\dfrac{V_1}{T_1} = \dfrac{V_2}{T_2}$

where $V_1 = 1.2$ litre, $T_1 = 20°C = (20 + 273)K = 293$ K

and $T_2 = (130 + 273)K = 403$ K.

Hence, $\dfrac{1.2}{293} = \dfrac{V_2}{403}$

from which, **volume at 130°C, $V_2 = \dfrac{(1.2)(403)}{293}$**

$$= \textbf{1.65 litres}$$

Problem 5.   Gas at a temperature of 150°C has its volume reduced by one-third in an isobaric process. Calculate the final temperature of the gas.

Since the process is isobaric it takes place at constant pressure and hence Charles' law applies,

i.e. $\dfrac{V_1}{T_1} = \dfrac{V_2}{T_2}$

where $T_1 = (150 + 273)K = 423$ K

and $V_2 = \dfrac{2}{3} V_1$

Hence $\dfrac{V_1}{423} = \dfrac{\frac{2}{3} V_1}{T_2}$

from which, **final temperature,**

$$T_2 = \frac{2}{3}(423) = \textbf{282 K}$$

or $(282 − 273)°C$ i.e. **9°C**

Now try the following Practise Exercise

Practise Exercise 130   Further problems on Charles' law

1. Some gas initially at 16°C is heated to 96°C at constant pressure. If the initial volume of the gas is 0.8 m³, determine the final volume of the gas.   [1.02 m³]

2. A gas is contained in a vessel of volume 0.02 m³ at a pressure of 300 kPa and a temperature of 15°C. The gas is passed into a vessel of volume 0.015 m³. Determine to what temperature the gas must be cooled for the pressure to remain the same.   [−57°C]

3. In an isobaric process gas at a temperature of 120°C has its volume reduced by a sixth. Determine the final temperature of the gas.   [54.5°C]

24.3   The pressure law

The **pressure law** states:

the pressure p of a fixed mass of gas is directly proportional to its thermodynamic temperature T at constant volume.

i.e. $p \propto T$ or $p = kT$ or $\dfrac{p}{T} = k$

When a fixed mass of gas at constant volume changes from pressure p_1 and temperature T_1, to pressure p_2 and temperature T_2 then:

$$\frac{p_1}{T_1} = \frac{p_2}{T_2}$$

Problem 6.   Gas initially at a temperature of 17°C and pressure 150 kPa is heated at constant volume until its temperature is 124°C. Determine the final pressure of the gas, assuming no loss of gas.

Since the gas is at constant volume, the pressure law applies, i.e. $\dfrac{p_1}{T_1} = \dfrac{p_2}{T_2}$

where $T_1 = (17 + 273)K = 290$ K,

             $T_2 = (124 + 273)K = 397$ K

and $p_1 = 150$ kPa

Hence $\dfrac{150}{290} = \dfrac{p_2}{397}$ from which,

final pressure, $p_2 = \dfrac{(150)(397)}{290}$

$= \textbf{205.3 kPa}$

Now try the following Practise Exercise

Practise Exercise 131 A further problem on the pressure law

1. Gas, initially at a temperature of 27°C and pressure 100 kPa, is heated at constant volume until its temperature is 150°C. Assuming no loss of gas, determine the final pressure of the gas. [141 kPa]

24.4 Dalton's law of partial pressure

Dalton's law of partial pressure states:

the total pressure of a mixture of gases occupying a given volume is equal to the sum of the pressures of each gas, considered separately, at constant temperature.

The pressure of each constituent gas when occupying a fixed volume alone is known as the **partial pressure** of that gas.

An **ideal gas** is one that completely obeys the gas laws given in Sections 24.1 to 24.4. In practice no gas is an ideal gas, although air is very close to being one. For calculation purposes the difference between an ideal and an actual gas is very small.

Problem 7. A gas R in a container exerts a pressure of 200 kPa at a temperature of 18°C. Gas Q is added to the container and the pressure increases to 320 kPa at the same temperature. Determine the pressure that gas Q alone exerts at the same temperature.

Initial pressure, $p_R = 200$ kPa, and the pressure of gases R and Q together, $p = p_R + p_Q = 320$ kPa

By Dalton's law of partial pressure, **the pressure of gas Q alone is**

$$p_Q = p - p_R = 320 - 200 = \textbf{120 kPa}.$$

Now try the following Practise Exercise

Practise Exercise 132 A further problem on Dalton's law of partial pressure

1. A gas A in a container exerts a pressure of 120 kPa at a temperature of 20°C. Gas B is added to the container and the pressure increases to 300 kPa at the same temperature. Determine the pressure that gas B alone exerts at the same temperature. [180 kPa]

24.5 Characteristic gas equation

Frequently, when a gas is undergoing some change, the pressure, temperature and volume all vary simultaneously. Provided there is no change in the mass of a gas, the above gas laws can be combined, giving

$$\dfrac{p_1 V_1}{T_1} = \dfrac{p_2 V_2}{T_2} = k \qquad \text{where k is a constant.}$$

For an ideal gas, constant $k = mR$, where m is the mass of the gas in kg, and R is the **characteristic gas constant**, i.e. $\dfrac{pV}{T} = mR$

or $\qquad pV = mRT$

This is called the **characteristic gas equation**. In this equation, p = absolute pressure in pascals,

V = volume in m³, m = mass in kg,

R = characteristic gas constant in J/(kg K),

and $\quad T$ = thermodynamic temperature in Kelvin.

Some typical values of the characteristic gas constant R include:
air, 287 J/(kg K), hydrogen 4160 J/(kg K), oxygen 260 J/(kg K) and carbon dioxide 184 J/(kg K).

Standard temperature and pressure (i.e. **STP**) refers to a temperature of 0°C, i.e. 273 K, and normal atmospheric pressure of 101.325 kPa.

24.6 Worked problems on the characteristic gas equation

Problem 8. A gas occupies a volume of 2.0 m³ when at a pressure of 100 kPa and a temperature of

120°C. Determine the volume of the gas at 15°C if the pressure is increased to 250 kPa.

Using the combined gas law: $\dfrac{p_1 V_1}{T_1} = \dfrac{p_2 V_2}{T_2}$

where $V_1 = 2.0$ m^3, $p_1 = 100$ kPa, $p_2 = 250$ kPa,

$T_1 = (120 + 273)K = 393$ K and

$T_2 = (15 + 273)$ K $= 288$ K, gives:

$$\frac{(100)(2.0)}{393} = \frac{(250)V_2}{288}$$

from which, **volume at 15°C,**

$$V_2 = \frac{(100)(2.0)(288)}{(393)(250)} = \mathbf{0.586 \ m^3}$$

Problem 9. 20000 mm^3 of air initially at a pressure of 600 kPa and temperature 180°C is expanded to a volume of 70000 mm^3 at a pressure of 120 kPa. Determine the final temperature of the air, assuming no losses during the process.

Using the combined gas law: $\dfrac{p_1 V_1}{T_1} = \dfrac{p_2 V_2}{T_2}$

where $V_1 = 20000$ mm^3,

$V_2 = 70000$ mm^3, $p_1 = 600$ kPa,

$p_2 = 120$ kPa, and

$T_1 = (180 + 273)$ K $= 453$ K

Hence $\dfrac{(600)(20000)}{453} = \dfrac{(120)(70000)}{T_2}$

from which, **final temperature,**

$$T_2 = \frac{(120)(70000)(453)}{(600)(20000)} = \mathbf{317 \ K} \ \text{ or } \ \mathbf{44°C}$$

Problem 10. Some air at a temperature of 40°C and pressure 4 bar occupies a volume of 0.05 m^3. Determine the mass of the air assuming the characteristic gas constant for air to be 287 J/(kg K).

From above, $pV = mRT$,

where $p = 4$ bar $= 4 \times 10^5$ Pa

(since 1 bar $= 10^5$ Pa – see Chapter 22),

$V = 0.05$ m^3,

$T = (40 + 273)$ K $= 313$ K,

and $R = 287$ J/(kg K).

Hence $(4 \times 10^5)(0.05) = m(287)(313)$

from which, **mass of air, m** $= \dfrac{(4 \times 10^5)(0.05)}{(287)(313)}$

$$= \mathbf{0.223 \ kg} \ \text{ or } \ \mathbf{223 \ g}$$

Problem 11. A cylinder of helium has a volume of 600 cm^3. The cylinder contains 200 g of helium at a temperature of 25°C. Determine the pressure of the helium if the characteristic gas constant for helium is 2080 J/(kg K).

From the characteristic gas equation, $pV = mRT$,

$V = 600$ cm$^3 = 600 \times 10^{-6}$ m^3, $m = 200$ g $= 0.2$ kg,

$T = (25 + 273)$ K $= 298$ K and $R = 2080$ J/(kg K)

Hence $(p)(600 \times 10^{-6}) = (0.2)(2080)(298)$

from which, **pressure, p** $= \dfrac{(0.2)(2080)(298)}{(600 \times 10^{-6})}$

$$= 206613333 \ \text{Pa}$$

$$= \mathbf{206.6 \ MPa}$$

Problem 12. A spherical vessel has a diameter of 1.2 m and contains oxygen at a pressure of 2 bar and a temperature of –20°C. Determine the mass of oxygen in the vessel. Take the characteristic gas constant for oxygen to be 0.260 kJ/(kg K).

From the characteristic gas equation, $pV = mRT$,

$V =$ volume of spherical vessel

$$= \frac{4}{3}\pi r^3 = \frac{4}{3}\pi \left(\frac{1.2}{2}\right)^3 = 0.905 \ \text{m}^3,$$

$p = 2$ bar $= 2 \times 10^5$ Pa,

$T = (-20 + 273)$ K $= 253$ K

and $R = 0.260$ kJ/(kg K) $= 260$ J/(kg K)

Hence, $(2 \times 10^5)(0.905) = m(260)(253)$

from which, **mass of oxygen, m** $= \dfrac{(2 \times 10^5)(0.905)}{(260)(253)}$

$$= \mathbf{2.75 \ kg}$$

Problem 13. Determine the characteristic gas constant of a gas which has a specific volume of 0.5 m^3/kg at a temperature of 20°C and pressure 150 kPa.

From the characteristic gas equation, $pV = mRT$

from which, $R = \dfrac{pV}{mT}$ where $p = 150 \times 10^3$ Pa,

$T = (20 + 273)$ K $= 293$ K and

specific volume, $V/m = 0.5$ m^3/kg.

Hence the **characteristic gas constant**,

$$R = \left(\frac{p}{T}\right)\left(\frac{V}{m}\right) = \left(\frac{150 \times 10^3}{293}\right)(0.5)$$

$$= \mathbf{256 \ J/(kg \ K)}$$

Now try the following Practise Exercise

Practise Exercise 133 Further problems on the characteristic gas equation

1. A gas occupies a volume of 1.20 m^3 when at a pressure of 120 kPa and a temperature of 90°C. Determine the volume of the gas at 20°C if the pressure is increased to 320 kPa. [0.363 m^3]

2. A given mass of air occupies a volume of 0.5 m^3 at a pressure of 500 kPa and a temperature of 20°C. Find the volume of the air at STP. [2.30 m^3]

3. A spherical vessel has a diameter of 2.0 m and contains hydrogen at a pressure of 300 kPa and a temperature of –30°C. Determine the mass of hydrogen in the vessel. Assume the characteristic gas constant R for hydrogen is 4160 J/(kg K). [1.24 kg]

4. A cylinder 200 mm in diameter and 1.5 m long contains oxygen at a pressure of 2 MPa and a temperature of 20°C. Determine the mass of oxygen in the cylinder. Assume the characteristic gas constant for oxygen is 260 J/(kg K). [1.24 kg]

5. A gas is pumped into an empty cylinder of volume 0.1 m^3 until the pressure is 5 MPa. The temperature of the gas is 40°C. If the cylinder mass increases by 5.32 kg when the gas has been added, determine the value of the characteristic gas constant. [300 J/(kg K)]

6. The mass of a gas is 1.2 kg and it occupies a volume of 13.45 m^3 at STP. Determine its characteristic gas constant. [4160 J/(kg K)]

7. 30 cm^3 of air initially at a pressure of 500 kPa and temperature 150°C is expanded to a volume of 100 cm^3 at a pressure of 200 kPa. Determine the final temperature of the air, assuming no losses during the process. [291°C]

8. A quantity of gas in a cylinder occupies a volume of 0.05 m^3 at a pressure of 400 kPa and a temperature of 27°C. It is compressed according to Boyle's law until its pressure is 1 MPa, and then expanded according to Charles' law until its volume is 0.03 m^3. Determine the final temperature of the gas. [177°C]

9. Some air at a temperature of 35°C and pressure 2 bar occupies a volume of 0.08 m^3. Determine the mass of the air assuming the characteristic gas constant for air to be 287 J/(kg K). (1 bar = 10^{5}Pa) [0.181 kg]

10. Determine the characteristic gas constant R of a gas that has a specific volume of 0.267 m^3/kg at a temperature of 17°C and pressure 200 kPa. [184 J/(kg K)]

24.7 Further worked problems on the characteristic gas equation

Problem 14. A vessel has a volume of 0.80 m^3 and contains a mixture of helium and hydrogen at a pressure of 450 kPa and a temperature of 17°C. If the mass of helium present is 0.40 kg determine (a) the partial pressure of each gas, and (b) the mass of hydrogen present. Assume the characteristic gas constant for helium to be 2080 J/(kg K) and for hydrogen 4160 J/(kg K).

(a) $V = 0.80$ m^3, $p = 450$ kPa, $T = (17 + 273)$K = 290 K, $m_{He} = 0.40$ kg, $R_{He} = 2080$ J/(kg K).

If p_{He} is the partial pressure of the helium, then using the characteristic gas equation,

$$p_{He}V = m_{He}R_{He}T \quad \text{gives:}$$
$$(p_{He})(0.80) = (0.40)(2080)(290)$$

from which, **the partial pressure of the helium,**

$$p_{He} = \frac{(0.40)(2080)(290)}{(0.80)} = \mathbf{301.6 \ kPa}$$

By Dalton's law of partial pressure the total pressure p is given by the sum of the partial pressures, i.e. $p = p_H + p_{He}$, from which,

the partial pressure of the hydrogen,
$$p_H = p - p_{He} = 450 - 301.6 = \textbf{148.4 kPa}$$

(b) From the characteristic gas equation,
$$p_H V = m_H R_H T$$
Hence, $(148.4 \times 10^3)(0.8) = m_H (4160)(290)$

from which, **mass of hydrogen,**
$$m_H = \frac{(148.4 \times 10^3)(0.8)}{(4160)(290)} = \textbf{0.098 kg} \quad \text{or} \quad \textbf{98 g}$$

Problem 15. A compressed air cylinder has a volume of 1.2 m³ and contains air at a pressure of 1 MPa and a temperature of 25°C. Air is released from the cylinder until the pressure falls to 300 kPa and the temperature is 15°C. Determine (a) the mass of air released from the container, and (b) the volume it would occupy at STP. Assume the characteristic gas constant for air to be 287 J/(kg K).

$V_1 = 1.2$ m³ $(= V_2)$, $p_1 = 1$ MPa $= 10^6$ Pa,

$T_1 = (25 + 273)K = 298$ K,

$T_2 = (15 + 273)K = 288$ K,

$p_2 = 300$ kPa $= 300 \times 10_3$ Pa

and $R = 287$ J/(kg K)

(a) Using the characteristic gas equation, $p_1 V_1 = m_1 R T_1$, to find the initial mass of air in the cylinder gives:
$$(10^6)(1.2) = m_1(287)(298)$$

from which,   mass $m_1 = \dfrac{(10^6)(1.2)}{(287)(298)} = 14.03$ kg

Similarly, using $p_2 V_2 = m_2 R T_2$ to find the final mass of air in the cylinder gives:
$$(300 \times 10^3)(1.2) = m_2(287)(288)$$

from which, mass $m_2 = \dfrac{(300 \times 10^3)(1.2)}{(287)(288)} = 4.36$ kg

Mass of air released from cylinder
$$= m_1 - m_2 = 14.03 - 4.36 = \textbf{9.67 kg.}$$

(b) At *STP*, $T = 273$ K and $p = 101.325$ kPa.
Using the characteristic gas equation $pV = mRT$

volume, $V = \dfrac{mRT}{p} = \dfrac{(9.67)(287)(273)}{101325} = \textbf{7.48 m}^3$

Problem 16. A vessel X contains gas at a pressure of 750 kPa at a temperature of 27°C. It is connected via a valve to vessel Y that is filled

with a similar gas at a pressure of 1.2 MPa and a temperature of 27°C. The volume of vessel X is 2.0 m³ and that of vessel Y is 3.0 m³. Determine the final pressure at 27°C when the valve is opened and the gases are allowed to mix. Assume R for the gas to be 300 J/(kg K).

For vessel X:

$p_X = 750 \times 10^3$ Pa, $T_X = (27 + 273)K = 300$ K,

$V_X = 2.0$ m³ and $R = 300$ J/(kg K)

From the characteristic gas equation, $p_X V_X = m_X R T_X$

Hence $(750 \times 10^3)(2.0) = m_X (300)(300)$

from which, mass of gas in vessel X,
$$m_X = \frac{(750 \times 10^3)(2.0)}{(300)(300)} = 16.67 \text{ kg.}$$

For vessel Y:

$p_Y = 1.2 \times 10^6$ Pa, $T_Y = (27 + 273)K = 300$ K,

$V_Y = 3.0$ m³ and $R = 300$ J/(kg K)

From the characteristic gas equation, $p_Y V_Y = m_Y R T_Y$

Hence $(1.2 \times 10^6)(3.0) = m_Y (300)(300)$

from which, mass of gas in vessel Y,
$$m_Y = \frac{(1.2 \times 10^6)(3.0)}{(300)(300)} = 40 \text{ kg}$$

When the valve is opened, mass of mixture,
$$m = m_X + m_Y = 16.67 + 40 = 56.67 \text{ kg.}$$

Total volume, $V = V_X + V_Y = 2.0 + 3.0 = 5.0$ m³,

$$R = 300 \text{ J/(kg K)}, \ T = 300 \text{ K.}$$

From the characteristic gas equation, $pV = mRT$

$$p(5.0) = (56.67)(300)(300)$$

from which, **final pressure,**

$$p = \frac{(56.67)(300)(300)}{5.0} = \textbf{1.02 MPa}$$

Now try the following Practise Exercises

Practise Exercise 134   Further questions on ideal gas laws

1. A vessel P contains gas at a pressure of 800 kPa at a temperature of 25°C. It is connected via a valve to vessel Q that is filled with similar gas at a pressure of 1.5 MPa

and a temperature of 25°C. The volume of vessel P is 1.5 m^3 and that of vessel R is 2.5 m^3. Determine the final pressure at 25°C when the valve is opened and the gases are allowed to mix. Assume R for the gas to be 297 J/(kg K).

[1.24 MPa]

2. A vessel contains 4 kg of air at a pressure of 600 kPa and a temperature of 40°C. The vessel is connected to another by a short pipe and the air exhausts into it. The final pressure in both vessels is 250 kPa and the temperature in both is 15°C. If the pressure in the second vessel before the air entered was zero, determine the volume of each vessel. Assume R for air is 287 J/(kg K).

[0.60 m^3, 0.72 m^3]

3. A vessel has a volume of 0.75 m^3 and contains a mixture of air and carbon dioxide at a pressure of 200 kPa and a temperature of 27°C. If the mass of air present is 0.5 kg determine (a) the partial pressure of each gas and (b) the mass of carbon dioxide. Assume the characteristic gas constant for air to be 287 J/(kg K) and for carbon dioxide 184 J/(kg K).

[(a) 57.4 kPa, 142.6 kPa (b) 1.94 kg]

4. A mass of gas occupies a volume of 0.02 m^3 when its pressure is 150 kPa and its temperature is 17°C. If the gas is compressed until its pressure is 500 kPa and its temperature is 57°C, determine (a) the volume it will occupy and (b) its mass, if the characteristic gas constant for the gas is 205 J/(kg K).

[(a) 0.0068 m^3 (b) 0.050 kg]

5. A compressed air cylinder has a volume of 0.6 m^3 and contains air at a pressure of 1.2 MPa absolute and a temperature of 37°C. After use the pressure is 800 kPa absolute and the temperature is 17°C. Calculate (a) the mass of air removed from the cylinder, and (b) the volume the mass of air removed would occupy at STP conditions. Take R for air as 287 J/(kg K) and atmospheric pressure as 100 kPa.

[(a) 2.33 kg (b) 1.83 m^3]

Practise Exercise 135   Short-answer questions on ideal gas laws

1. State Boyle's law.

2. State Charles' law.

3. State the Pressure law.

4. State Dalton's law of partial pressures.

5. State the relationship between the Celsius and the thermodynamic scale of temperature.

6. What is (a) an isothermal change, and (b) an isobaric change?

7. Define an ideal gas.

8. State the characteristic gas equation.

9. What is meant by STP?

Practise Exercise 136   Multiple-choice questions on ideal gas laws

(Answers on page 298)

1. Which of the following statements is false?
   (a) At constant temperature, Charles' law applies.
   (b) The pressure of a given mass of gas decreases as the volume is increased at constant temperature.
   (c) Isobaric changes are those which occur at constant pressure.
   (d) Boyle's law applies at constant temperature.

2. A gas occupies a volume of 4 m^3 at a pressure of 400 kPa. At constant temperature, the pressure is increased to 500 kPa. The new volume occupied by the gas is:
   (a) 5 m^3 (b) 0.3 m^3
   (c) 0.2 m^3 (d) 3.2 m^3

3. A gas at a temperature of 27°C occupies a volume of 5 m^3. The volume of the same mass of gas at the same pressure but at a temperature of 57°C is:
   (a) 10.56 m^3 (b) 5.50 m^3
   (c) 4.55 m^3 (d) 2.37 m^3

Part Four

4. Which of the following statements is false?

   (a) An ideal gas is one that completely obeys the gas laws.

   (b) Isothermal changes are those that occur at constant volume.

   (c) The volume of a gas increases when the temperature increases at constant pressure.

   (d) Changes that occur at constant pressure are called isobaric changes.

A gas has a volume of 0.4 m³ when its pressure is 250 kPa and its temperature is 400 K. Use this data in Questions 5 and 6.

5. The temperature when the pressure is increased to 400 kPa and the volume is increased to 0.8 m³ is:

   (a) 400 K   (b) 80 K
   (c) 1280 K   (d) 320 K

6. The pressure when the temperature is raised to 600 K and the volume is reduced to 0.2 m³ is:

   (a) 187.5 kPa   (b) 250 kPa
   (c) 333.3 kPa   (d) 750 kPa

7. A gas has a volume of 3 m³ at a temperature of 546 K and a pressure of 101.325 kPa. The volume it occupies at STP is:

   (a) 3 m³   (b) 1.5 m³
   (c) 6 m³

8. Which of the following statements is false?

   (a) A characteristic gas constant has units of J/(kg K).

   (b) STP conditions are 273 K and 101.325 kPa.

   (c) All gases are ideal gases.

   (d) An ideal gas is one that obeys the gas laws.

A mass of 5 kg of air is pumped into a container of volume 2.87 m³. The characteristic gas constant for air is 287 J/(kg K). Use this data in Questions 9 and 10.

9. The pressure when the temperature is 27°C is:

   (a) 1.6 kPa   (b) 6 kPa
   (c) 150 kPa   (d) 15 kPa

10. The temperature when the pressure is 200 kPa is:

   (a) 400°C   (b) 127°C
   (c) 127 K   (d) 283 K

The measurement of temperature

A change in temperature of a substance can often result in a change in one or more of its physical properties. Thus, although temperature cannot be measured directly, its effects can be measured. Some properties of substances used to determine changes in temperature include changes in dimensions, electrical resistance, state, type and volume of radiation and colour.

Temperature measuring devices available are many and varied. Those described in this chapter are those most often used in science and industry.

At the end of this chapter you should be able to:

- describe the construction, principle of operation and practical applications of the following temperature measuring devices:
    (a) liquid-in-glass thermometer (including advantages of mercury, and sources of error)
    (b) thermocouples (including advantages and sources of error)
    (c) resistance thermometer (including limitations and advantages of platinum coil)
    (d) thermistors
    (e) pyrometers (total radiation and optical types, including advantages and disadvantages)
- describe the principle of operation of
    (a) temperature indicating paints and crayons
    (b) bimetallic thermometers
    (c) mercury-in-steel thermometer
    (d) gas thermometer
- select the appropriate temperature measuring device for a particular application.

Mechanical Engineering Principles, Bird and Ross, ISBN 9780415517850

25.1 Liquid-in-glass thermometer

A **liquid-in-glass thermometer** uses the expansion of a liquid with increase in temperature as its principle of operation.

Construction

A typical liquid-in-glass thermometer is shown in Figure 25.1 and consists of a sealed stem of uniform small-bore tubing, called a capillary tube, made of glass, with a cylindrical glass bulb formed at one end. The bulb and part of the stem are filled with a liquid such as mercury or alcohol and the remaining part of the tube is evacuated. A temperature scale is formed by etching graduations on the stem. A safety reservoir is usually provided, into which the liquid can expand without bursting the glass if the temperature is raised beyond the upper limit of the scale.

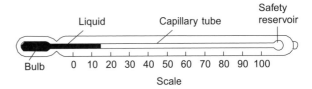

Figure 25.1

Principle of operation

The operation of a liquid-in-glass thermometer depends on the liquid expanding with increase in temperature and contracting with decrease in temperature. The position of the end of the column of liquid in the tube is a measure of the temperature of the liquid in the bulb – shown as 15°C in Figure 25.1, which is about room temperature. Two fixed points are needed to calibrate the thermometer, with the interval between these points being divided into 'degrees'. In the first thermometer, made by Celsius, the fixed points chosen were the temperature of melting ice (0°C) and that of boiling water at standard atmospheric pressure (100°C), in each case the blank stem being marked at the liquid level. The distance between these two points, called the fundamental interval, was divided into 100 equal parts, each equivalent to 1°C, thus forming the scale.

The **clinical thermometer**, with a limited scale around body temperature, the **maximum and/or minimum thermometer**, recording the maximum day temperature and minimum night temperature, and the **Beckman thermometer**, which is used only in accurate measurement of temperature change, and has no fixed points, are particular types of liquid-in-glass thermometer which all operate on the same principle.

Advantages

The liquid-in-glass thermometer is simple in construction, relatively inexpensive, easy to use and portable, and is the most widely used method of temperature measurement having industrial, chemical, clinical and meteorological applications.

Disadvantages

Liquid-in-glass thermometers tend to be fragile and hence easily broken, can only be used where the liquid column is visible, cannot be used for surface temperature measurements, cannot be read from a distance and are unsuitable for high temperature measurements.

Advantages of mercury

The use of mercury in a thermometer has many advantages, for mercury:

 (i)   is clearly visible,
 (ii) has a fairly uniform rate of expansion,
 (iii) is readily obtainable in the pure state,
 (iv) does not 'wet' the glass,
 (v)   is a good conductor of heat.

Mercury has a freezing point of –39°C and cannot be used in a thermometer below this temperature. Its boiling point is 357°C but before this temperature is reached some distillation of the mercury occurs if the space above the mercury is a vacuum. To prevent this, and to extend the upper temperature limits to over 500°C, an inert gas such as nitrogen under pressure is used to fill the remainder of the capillary tube. Alcohol, often dyed red to be seen in the capillary

tube, is considerably cheaper than mercury and has a freezing point of –113°C, which is considerably lower than for mercury. However it has a low boiling point at about 79°C.

Errors

Typical errors in liquid-in-glass thermometers may occur due to:

 (i)   the slow cooling rate of glass
 (ii) incorrect positioning of the thermometer
 (iii) a delay in the thermometer becoming steady (i.e. slow response time)
 (iv) non-uniformity of the bore of the capillary tube, which means that equal intervals marked on the stem do not correspond to equal temperature intervals.

25.2 Thermocouples

Thermocouples use the e.m.f. set up when the junction of two dissimilar metals is heated.

Principle of operation

At the junction between two different metals, say, copper and constantan, there exists a difference in electrical potential, which varies with the temperature of the junction. This is known as the 'thermo-electric effect'. If the circuit is completed with a second junction at a different temperature, a current will flow round the circuit. This principle is used in the thermocouple. Two different metal conductors having their ends twisted together are shown in Figure 25.2. If the two junctions are at different temperatures, a current I flows round the circuit.

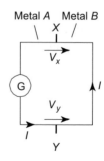

Figure 25.2

The deflection on the galvanometer G depends on the difference in temperature between junctions X and Y and is caused by the difference between voltages V_x and V_y. The higher temperature junction is usually called the 'hot junction' and the lower temperature junction the 'cold junction'. If the cold junction is kept at a constant known temperature, the galvanometer can be calibrated to indicate the temperature of the hot junction directly. The cold junction is then known as the reference junction.

In many instrumentation situations, the measuring instrument needs to be located far from the point at which the measurements are to be made. Extension leads are then used, usually made of the same material as the thermocouple but of smaller gauge. The reference junction is then effectively moved to their ends. The thermocouple is used by positioning the hot junction where the temperature is required. The meter will indicate the temperature of the hot junction only if the reference junction is at 0°C for:

(temperature of hot junction) = (temperature of the cold junction) + (temperature difference)

In a laboratory the reference junction is often placed in melting ice, but in industry it is often positioned in a thermostatically controlled oven or buried underground where the temperature is constant.

Construction

Thermocouple junctions are made by twisting together the ends of two wires of dissimilar metals before welding them. The construction of a typical copper-constantan thermocouple for industrial use is shown in Figure 25.3. Apart from the actual junction the two conductors used must be insulated electrically from each other with appropriate insulation and is shown in Figure 25.3 as twin-holed tubing. The wires and insulation are usually inserted into a sheath for protection from environments in which they might be damaged or corroded.

Applications

A copper-constantan thermocouple can measure temperature from –250°C up to about 400°C, and is used typically with boiler flue gases, food processing and with sub-zero temperature measurement. An iron-constantan thermocouple can measure temperature from –200°C to about 850°C, and is used typically in paper and pulp mills, re-heat and annealing furnaces and in chemical reactors. A chromel-alumel thermocouple can measure temperatures from –200°C to about 1100°C and is used typically with blast furnace gases, brick kilns and in glass manufacture.

For the measurement of temperatures above 1100°C radiation pyrometers are normally used. However, thermocouples are available made of platinum-platinum/rhodium, capable of measuring temperatures up to 1400°C, or tungsten-molybdenum which can measure up to 2600°C.

Advantages

A thermocouple:

(i)   has a very simple, relatively inexpensive construction
(ii)   can be made very small and compact
(iii) is robust
(iv)   is easily replaced if damaged
(v)   has a small response time
(vi)   can be used at a distance from the actual measuring instrument and is thus ideal for use with automatic and remote-control systems.

Sources of error

Sources of error in the thermocouple, which are difficult to overcome, include:

(i)   voltage drops in leads and junctions
(ii)   possible variations in the temperature of the cold junction
(iii) stray thermoelectric effects, which are caused by the addition of further metals into the 'ideal' two-metal thermocouple circuit.

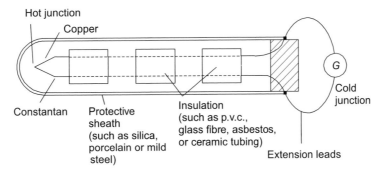

Figure 25.3

Additional leads are frequently necessary for extension leads or voltmeter terminal connections.

A thermocouple may be used with a battery- or mains-operated electronic thermometer instead of a millivoltmeter. These devices amplify the small e.m.f.'s from the thermocouple before feeding them to a multi-range voltmeter calibrated directly with temperature scales. These devices have great accuracy and are almost unaffected by voltage drops in the leads and junctions.

Problem 1. A chromel-alumel thermocouple generates an e.m.f. of 5 mV. Determine the temperature of the hot junction if the cold junction is at a temperature of 15°C and the sensitivity of the thermocouple is 0.04 mV/°C.

Temperature difference for 5 m $V = \dfrac{5\,\text{mV}}{0.04\,\text{mV/°C}}$

$$= 125°\text{C}.$$

Temperature at hot junction = temperature of cold junction + temperature difference

$$= 15°\text{C} + 125°\text{C} = \textbf{140°C}$$

Now try the following Practise Exercise

Practise Exercise 137   Further problem on the thermocouple

1. A platinum–platinum/rhodium thermocouple generates an e.m.f. of 7.5 mV. If the cold junction is at a temperature of 20°C, determine the temperature of the hot junction. Assume the sensitivity of the thermocouple to be 6 $\mu V/$°C   [1270°C]

25.3   Resistance thermometers

Resistance thermometers use the change in electrical resistance caused by temperature change.

Construction

Resistance thermometers are made in a variety of sizes, shapes and forms depending on the application for which they are designed. A typical resistance thermometer is shown diagrammatically in Figure 25.4. The most common metal used for the coil in such thermometers is platinum even though its sensitivity is not as high as other metals such as copper and nickel. However, platinum is a very stable metal and provides reproducible results in a resistance thermometer. A platinum resistance thermometer is often used as a calibrating device. Since platinum is expensive, connecting leads of another metal, usually copper, are used with the thermometer to connect it to a measuring circuit.

The platinum and the connecting leads are shown joined at A and B in Figure 25.4, although sometimes this junction may be made outside of the sheath. However, these leads often come into close contact with the heat source which can introduce errors into the measurements. These may be eliminated by including a pair of identical leads, called dummy leads, which experience the same temperature change as the extension leads.

Principle of operation

With most metals a rise in temperature causes an increase in electrical resistance, and since resistance can be measured accurately this property can be used to measure temperature. If the resistance of a length of wire at 0°C is R_0, and its resistance at θ°C is R_θ, then $R_\theta = R_0 (1 + \alpha\theta)$, where α is the temperature coefficient of resistance of the material.

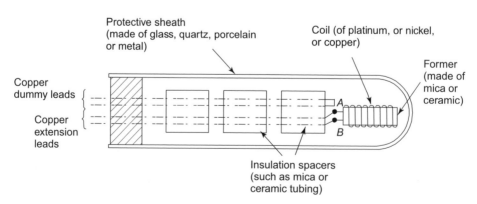

Protective sheath (made of glass, quartz, porcelain or metal)

Coil (of platinum, or nickel, or copper)

Former (made of mica or ceramic)

Copper dummy leads

Copper extension leads

A

B

Insulation spacers (such as mica or ceramic tubing)

Figure 25.4

Rearranging gives: **temperature,** $\theta = \dfrac{R_\theta - R_0}{\alpha R_0}$

Values of R_0 and α may be determined experimentally or obtained from existing data. Thus, if R_θ can be measured, temperature θ can be calculated. This is the principle of operation of a resistance thermometer. Although a sensitive ohmmeter can be used to measure R_θ, for more accurate determinations a **Wheatstone bridge circuit is used** as shown in Figure 25.5. This circuit compares an unknown resistance R_θ with others of known values, R_1 and R_2 being fixed values and R_3 being variable. Galvanometer G is a sensitive centre-zero microammeter. R_3 is varied until zero deflection is obtained on the galvanometer, i.e. no current flows through G and the bridge is said to be 'balanced'.

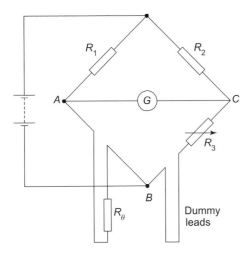

Figure 25.5

At balance: $\qquad\qquad R_2 R_\theta = R_1 R_3$

from which, $\qquad\qquad R_\theta = \dfrac{R_1 R_3}{R_2}$

and if R_1 and R_2 are of equal value, then $R_\theta = R_3$

A resistance thermometer may be connected between points A and B in Figure 25.5 and its resistance R_θ at any temperature θ accurately measured. Dummy leads included in arm BC help to eliminate errors caused by the extension leads which are normally necessary in such a thermometer.

Limitations

Resistance thermometers using a nickel coil are used mainly in the range –100°C to 300°C, whereas platinum resistance thermometers are capable of measuring

with greater accuracy temperatures in the range –200°C to about 800°C. This upper range may be extended to about 1500°C if high melting point materials are used for the sheath and coil construction.

Advantages and disadvantages of a platinum coil

Platinum is commonly used in resistance thermometers since it is chemically inert, i.e. un-reactive, resists corrosion and oxidation and has a high melting point of 1769°C. A disadvantage of platinum is its slow response to temperature variation.

Applications

Platinum resistance thermometers may be used as calibrating devices or in applications such as heat-treating and annealing processes and can be adapted easily for use with automatic recording or control systems. Resistance thermometers tend to be fragile and easily damaged especially when subjected to excessive vibration or shock.

> **Problem 2.** A platinum resistance thermometer has a resistance of 25 Ω at 0°C. When measuring the temperature of an annealing process a resistance value of 60 Ω is recorded. To what temperature does this correspond? Take the temperature coefficient of resistance of platinum as 0.0038/°C

$R_\theta = R_0 (1 + \alpha\theta)$, where $R_0 = 25\ \Omega$,

$R_\theta = 60\ \Omega$ and $\alpha = 0.0038/°C$.

Rearranging gives: **temperature,** $\theta = \dfrac{R_\theta - R_0}{\alpha R_0}$

$\qquad\qquad = \dfrac{60 - 25}{(0.0038)(25)} = \mathbf{368.4°C}$

Now try the following Practise Exercise

> **Exercise 138 Further problem on the resistance thermometer**
>
> 1. A platinum resistance thermometer has a resistance of 100 Ω at 0°C. When measuring the temperature of a heat process a resistance value of 177 Ω is measured using a Wheatstone bridge. Given that the temperature coefficient of resistance of platinum is 0.0038/°C, determine the temperature of the heat process, correct to the nearest degree.
> [203°C]

25.4 Thermistors

A thermistor is a semi-conducting material – such as mixtures of oxides of copper, manganese, cobalt, etc. – in the form of a fused bead connected to two leads. As its temperature is increased its resistance rapidly decreases. Typical resistance/temperature curves for a thermistor and common metals are shown in Figure 25.6. The resistance of a typical thermistor can vary from 400 Ω at 0°C to 100 Ω at 140°C.

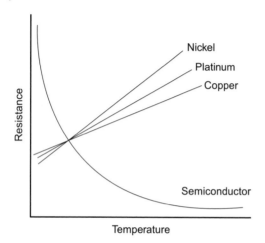

Figure 25.6

Advantages

The main advantages of a thermistor are its high sensitivity and small size. It provides an inexpensive method of measuring and detecting small changes in temperature.

25.5 Pyrometers

A pyrometer is a device for measuring very high temperatures and uses the principle that all substances emit radiant energy when hot, the rate of emission depending on their temperature. The measurement of thermal radiation is therefore a convenient method of determining the temperature of hot sources and is particularly useful in industrial processes. There are two main types of pyrometer, namely the total radiation pyrometer and the optical pyrometer.

Pyrometers are very convenient instruments since they can be used at a safe and comfortable distance from the hot source. Thus applications of pyrometers are found in measuring the temperature of molten metals, the interiors of furnaces or the interiors of volcanoes. Total radiation pyrometers can also be used in conjunction with devices which record and control temperature continuously.

Total radiation pyrometer

A typical arrangement of a total radiation pyrometer is shown in Figure 25.7. Radiant energy from a hot source, such as a furnace, is focused on to the hot junction of a thermocouple after reflection from a concave mirror. The temperature rise recorded by the thermocouple depends on the amount of radiant energy received, which in turn depends on the temperature of the hot source. The galvanometer G shown connected to the thermocouple records the current which results from the e.m.f. developed and may be calibrated to give a direct reading of the temperature of the hot source. The thermocouple is protected from direct radiation by a shield as shown and the hot source may be viewed through the sighting telescope. For greater sensitivity, a thermopile may be used, a thermopile being a number of thermocouples connected in series. Total radiation pyrometers are used to measure temperature in the range 700°C to 2000°C.

Optical pyrometers

When the temperature of an object is raised sufficiently, two visual effects occur; the object appears brighter

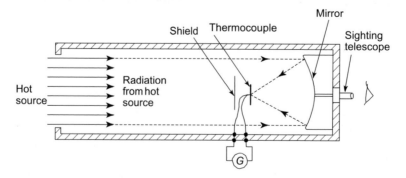

Figure 25.7

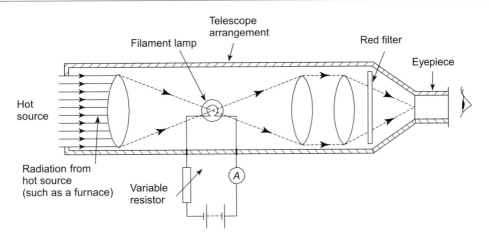

Figure 25.8

and there is a change in colour of the light emitted. These effects are used in the optical pyrometer where a comparison or matching is made between the brightness of the glowing hot source and the light from a filament of known temperature.

The most frequently used optical pyrometer is the disappearing filament pyrometer and a typical arrangement is shown in Figure 25.8. A filament lamp is built into a telescope arrangement which receives radiation from a hot source, an image of which is seen through an eyepiece. A red filter is incorporated as a protection to the eye.

The current flowing through the lamp is controlled by a variable resistor. As the current is increased, the temperature of the filament increases and its colour changes. When viewed through the eyepiece the filament of the lamp appears superimposed on the image of the radiant energy from the hot source. The current is varied until the filament glows as brightly as the background. It will then merge into the background and seem to disappear. The current required to achieve this is a measure of the temperature of the hot source and the ammeter can be calibrated to read the temperature directly. Optical pyrometers may be used to measure temperatures up to, and even in excess of, 3000°C.

Advantages of pyrometers
(i)   There is no practical limit to the temperature that a pyrometer can measure.
(ii) A pyrometer need not be brought directly into the hot zone and so is free from the effects of heat and chemical attack that can often cause other measuring devices to deteriorate in use.
(iii) Very fast rates of change of temperature can be followed by a pyrometer.

(iv) The temperature of moving bodies can be measured.
(v)   The lens system makes the pyrometer virtually independent of its distance from the source.

Disadvantages of pyrometers
(i)   A pyrometer is often more expensive than other temperature measuring devices.
(ii) A direct view of the heat process is necessary.
(iii) Manual adjustment is necessary.
(iv) A reasonable amount of skill and care is required in calibrating and using a pyrometer. For each new measuring situation the pyrometer must be re-calibrated.
(v)   The temperature of the surroundings may affect the reading of the pyrometer and such errors are difficult to eliminate.

25.6 Temperature indicating paints and crayons

Temperature indicating paints contain substances which change their colour when heated to certain temperatures. This change is usually due to chemical decomposition, such as loss of water, in which the change in colour of the paint after having reached the particular temperature will be a permanent one. However, in some types the original colour returns after cooling. Temperature indicating paints are used where the temperature of inaccessible parts of apparatus and machines is required. They are particularly useful in heat-treatment processes where the temperature of the component needs to be known before a quenching operation. There are several such paints available and most have only a small temperature range so that different paints have to be used for different temperatures. The usual range of temperatures covered by these paints is from about 30°C to 700°C.

Temperature sensitive crayons consist of fusible solids compressed into the form of a stick. The melting point of such crayons is used to determine when a given temperature has been reached. The crayons are simple to use but indicate a single temperature only, i.e. its melting point temperature. There are over 100 different crayons available, each covering a particular range of temperature. Crayons are available for temperatures within the range of 50°C to 1400°C. Such crayons are used in metallurgical applications such as preheating before welding, hardening, annealing or tempering, or in monitoring the temperature of critical parts of machines or for checking mould temperatures in the rubber and plastics industries.

25.7   Bimetallic thermometers

Bimetallic thermometers depend on the expansion of metal strips which operate an indicating pointer. Two thin metal strips of differing thermal expansion are welded or riveted together and the curvature of the bimetallic strip changes with temperature change. For greater sensitivity the strips may be coiled into a flat spiral or helix, one end being fixed and the other being made to rotate a pointer over a scale. Bimetallic thermometers are useful for alarm and over-temperature applications where extreme accuracy is not essential. If the whole is placed in a sheath, protection from corrosive environments is achieved but with a reduction in response characteristics. The normal upper limit of temperature measurement by this thermometer is about 200°C, although with special metals the range can be extended to about 400°C.

25.8   Mercury-in-steel thermometer

The **mercury-in-steel thermometer** is an extension of the principle of the mercury-in-glass thermometer. Mercury in a steel bulb expands via a small bore capillary tube into a pressure indicating device, say a Bourdon gauge, the position of the pointer indicating the amount of expansion and thus the temperature. The advantages of this instrument are that it is robust and, by increasing the length of the capillary tube, the gauge can be placed some distance from the bulb and can thus be used to monitor temperatures in positions which are inaccessible to the liquid-in-glass thermometer. Such thermometers may be used to measure temperatures up to 600°C.

25.9   Gas thermometers

The gas thermometer consists of a flexible U-tube of mercury connected by a capillary tube to a vessel containing gas. The change in the volume of a fixed mass of gas at constant pressure, or the change in pressure of a fixed mass of gas at constant volume, may be used to measure temperature. This thermometer is cumbersome and rarely used to measure temperature directly, but it is often used as a standard with which to calibrate other types of thermometer. With pure hydrogen the range of the instrument extends from –240°C to 1500°C and measurements can be made with extreme accuracy.

25.10   Choice of measuring devices

Problem 3.   State which device would be most suitable to measure the following:

(a)   metal in a furnace, in the range 50°C to 1600°C
(b)   the air in an office in the range 0°C to 40°C
(c)   boiler flue gas in the range 15°C to 300°C
(d)   a metal surface, where a visual indication is required when it reaches 425°C
(e)   materials in a high-temperature furnace in the range 2000°C to 2800°C
(f)   to calibrate a thermocouple in the range –100°C to 500°C
(g)   brick in a kiln up to 900°C
(h)   an inexpensive method for food processing applications in the range –25°C to –75°C.

(a)   **Radiation pyrometer**

(b)   **Mercury-in-glass thermometer**

(c)   **Copper-constantan thermocouple**

(d)   **Temperature sensitive crayon**

(e)   **Optical pyrometer**

(f)   **Platinum resistance thermometer or gas thermometer**

(g)   **Chromel-alumel thermocouple**

(h)   **Alcohol-in-glass thermometer.**

Now try the following Practise Exercises

Practise Exercise 139   Short-answer questions on the measurement of temperature

For each of the temperature measuring devices listed in 1 to 10, state very briefly its principle of operation and the range of temperatures that it is capable of measuring.

1. Mercury-in-glass thermomete.

2. Alcohol-in-glass thermometer.

3. Thermocouple.

4. Platinum resistance thermometer.

5. Total radiation pyrometer.

6. Optical pyrometer.

7. Temperature sensitive crayons.

8. Bimetallic thermometer.

9. Mercury-in-steel thermometer.

10. Gas thermometer.

Practise Exercise 140   Multiple-choice questions on the measurement of temperature

(Answers on page 298)

1. The most suitable device for measuring very small temperature changes is a

    (a) thermopile   (b) thermocouple
    (c) thermistor

2. When two wires of different metals are twisted together and heat applied to the junction, an e.m.f. is produced. This effect is used in a thermocouple to measure:

    (a) e.m.f.   (b) temperature
    (c) expansion   (d) heat

3. A cold junction of a thermocouple is at room temperature of 15°C. A voltmeter connected to the thermocouple circuit indicates 10 mV. If the voltmeter is calibrated as 20°C/mV, the temperature of the hot source is:

    (a) 185°C   (b) 200°C
    (c) 35°C   (d) 215°C

4. The e.m.f. generated by a copper-constantan thermometer is 15 mV. If the cold junction is at a temperature of 20°C, the temperature of the hot junction when the sensitivity of the thermocouple is 0.03 mV/°C is:

    (a) 480°C   (b) 520°C
    (c) 20.45°C   (d) 500°C

In Questions 5 to 12, select the most appropriate temperature measuring device from this list.

(a) copper-constantan thermocouple
(b) thermistor
(c) mercury-in-glass thermometer
(d) total radiation pyrometer
(e) platinum resistance thermometer
(f) gas thermometer
(g) temperature sensitive crayon
(h) alcohol-in-glass thermometer
(i) bimetallic thermometer
(j) mercury-in-steel thermometer
(k) optical pyrometer.

5. Over-temperature alarm at about 180°C.

6. Food processing plant in the range −250°C to +250°C.

7. Automatic recording system for a heat treating process in the range 90°C to 250°C.

8. Surface of molten metals in the range 1000°C to 1800°C.

9. To calibrate accurately a mercury-in-glass thermometer.

10. Furnace up to 3000°C.

11. Inexpensive method of measuring very small changes in temperature.

12. Metal surface where a visual indication is required when the temperature reaches 520°C.

This Revision Test covers the material contained in Chapters 22 to 25. *The marks for each question are shown in brackets at the end of each question.*

When required take the density of water to be 1000 kg/m^3 and gravitational acceleration as 9.81 m/s^2.

1. A circular piston exerts a pressure of 150 kPa on a fluid when the force applied to the piston is 0.5 kN. Calculate the diameter of the piston, correct to the nearest millimetre.   (6)

2. A tank contains water to a depth of 500 mm. Determine the water pressure

   (a) at a depth of 300 mm, and

   (b) at the base of the tank.   (6)

3. When the atmospheric pressure is 101 kPa, calculate the absolute pressure, to the nearest kilopascal, at a point on a submarine which is 50 m below the sea water surface. Assume that the density of sea water is 1030 kg/m^3.   (5)

4. A body weighs 2.85 N in air and 2.35 N when completely immersed in water. Determine

   (a) the volume of the body

   (b) the density of the body, and

   (c) the relative density of the body.   (9)

5. A submarine dives to a depth of 700 m. What is the gauge pressure on its surface if the density of seawater is 1020 kg/m^3 and $g = 9.81$ m/s^2.   (5)

6. State the most appropriate fluid flow measuring device for the following applications:

   (a) A high accuracy, permanent installation, in an oil pipeline.

   (b) For high velocity chemical flow, which does not suffer wear.

   (c) To detect leakage in water mains.

   (d) To measure petrol in petrol pumps.

   (e) To measure the speed of a viscous liquid.   (5)

7. A storage tank contains water to a depth of 7 m above an outlet pipe, as shown in Figure 23.12 on page 268. The system is in equilibrium until a valve in the outlet pipe is opened. Determine the initial mass rate of flow at the exit of the outlet pipe, assuming that losses at the pipe entry $= 0.3\, v^2$, and losses at the valve $= 0.2\, v^2$. The pipe diameter is 0.05 m and the water density, ρ, is 1000 kg/m^3.   (15)

8. Determine the wind pressure acting on a slender building due to a gale of 150 km/h that acts perpendicularly to the building. Take the density of air as 1.23 kg/m^3.   (5)

9. Some gas occupies a volume of 2.0 m^3 in a cylinder at a pressure of 200 kPa. A piston, sliding in the cylinder, compresses the gas isothermally until the volume is 0.80 m^3. If the area of the piston is 240 cm^2, calculate the force on the piston when the gas is compressed.   (5)

10. Gas at a temperature of 180°C has its volume reduced by a quarter in an isobaric process. Determine the final temperature of the gas.   (5)

11. Some air at a pressure of 3 bar and at a temperature of 60°C occupies a volume of 0.08 m^3. Calculate the mass of the air, correct to the nearest gram, assuming the characteristic gas constant for air is 287 J/(kg K).   (5)

12. A compressed air cylinder has a volume of 1.0 m^3 and contains air at a temperature of 24°C and a pressure of 1.2 MPa. Air is released from the cylinder until the pressure falls to 400 kPa and the temperature is 18°C. Calculate

    (a) the mass of air released from the container, and

    (b) the volume it would occupy at S.T.P. Assume the characteristic gas constant for air to be 287 J/(kg K).   (10)

13. A platinum resistance thermometer has a resistance of 24 Ω at 0°C. When measuring the temperature of an annealing process a resistance value of 68 Ω is recorded. To what temperature does this correspond ? Take the temperature coefficient of resistance of platinum as 0.0038/°C   (5)

14. State which device would be most suitable to measure the following:

    (a) materials in a high-temperature furnace in the range 1800°C to 3000°C.

    (b) the air in a factory in the range 0°C to 35°C.

    (c) an inexpensive method for food processing applications in the range –20°C to –80°C.

    (d) boiler flue gas in the range 15°C to 250°C.   (4)

Formula	Formula symbols	Units
Stress $= \dfrac{\text{applied force}}{\text{cross} - \text{sectional area}}$	$\sigma = \dfrac{F}{A}$	Pa
Strain $= \dfrac{\text{change in length}}{\text{original length}}$	$\varepsilon = \dfrac{x}{L}$	
Young's modulus of elasticity $= \dfrac{\text{stress}}{\text{strain}}$	$E = \dfrac{\sigma}{\varepsilon}$	Pa
Stiffness $= \dfrac{\text{force}}{\text{extension}}$	$k = \dfrac{F}{\delta}$	N/m
Modulus of rigidity $= \dfrac{\text{shear stress}}{\text{shear strain}}$	$G = \dfrac{\tau}{\gamma}$	Pa
Thermal strain = coefficient of linear expansion $\times$ temperature rise	$\varepsilon = \alpha T$	
Thermal stress in compound bar	$\sigma_1 = \dfrac{(\alpha_1 - \alpha_2) E_1 E_2 A_2 T}{(A_1 E_1 + A_2 E_2)}$	Pa
Ultimate tensile strength $= \dfrac{\text{maximum load}}{\text{original cross sectional area}}$		Pa
Moment = force $\times$ perpendicular distance	$M = Fd$	N m
$\dfrac{\text{stress}}{\text{distance from neutral axis}} = \dfrac{\text{bending moment}}{\text{second moment of area}}$ $= \dfrac{\text{Young's modulus}}{\text{radius of curvature}}$	$\dfrac{\sigma}{y} = \dfrac{M}{I} = \dfrac{E}{R}$	N/m^3
Torque = force $\times$ perpendicular distance	$T = Fd$	N m
Power = torque $\times$ angular velocity	$P = T\omega = 2\pi n T$	W
Horsepower	1 hp = 745.7 W	
Torque = moment of inertia $\times$ angular acceleration	$T = I\alpha$	N m
$\dfrac{\text{shear stress}}{\text{radius}} = \dfrac{\text{torque}}{\text{polar second moment of area}}$ $= \dfrac{(\text{rigidity})(\text{angle of twist})}{\text{length}}$	$\dfrac{\tau}{r} = \dfrac{T}{J} = \dfrac{G\theta}{L}$	N/m^3
Average velocity $= \dfrac{\text{distance travelled}}{\text{time taken}}$	$v = \dfrac{s}{t}$	m/s

Formula	Formula symbols	Units
Acceleration $= \dfrac{\text{change in velocity}}{\text{time taken}}$	$a = \dfrac{v - u}{t}$	m/s^2
Linear velocity	$v = \omega r$	m/s
Angular velocity	$\omega = \dfrac{\theta}{t} = 2\pi n$	rad/s
Linear acceleration	$a = r\alpha$	m/s^2
Relationships between initial velocity u, final velocity v, displacement s, time t and constant acceleration a	$\begin{cases} v_2 = v_1 + at \\ s = ut + \dfrac{1}{2}at^2 \\ v^2 = u^2 + 2as \end{cases}$	m/s m (m/s)^2
Relationships between initial angular velocity ω_1, final angular velocity ω_2, angle θ, time t and angular acceleration α	$\begin{cases} \omega_2 = \omega_1 + \alpha t \\ \theta = \omega_1 t + \dfrac{1}{2}\alpha t^2 \\ \omega_2^2 = \omega_1^2 + 2\alpha\theta \end{cases}$	rad/s rad (rad/s)^2
Momentum = mass × velocity		kg m/s
Impulse = applied force × time = change in momentum		kg m/s
Force = mass × acceleration	$F = ma$	N
Weight = mass × gravitational field	$W = mg$	N
Centripetal acceleration	$a = \dfrac{v^2}{r}$	m/s^2
Centripetal force	$F = \dfrac{mv^2}{r}$	N
Density $= \dfrac{\text{mass}}{\text{volume}}$	$\rho = \dfrac{m}{V}$	kg/m^3
Work done = force × distance moved	$W = Fs$	J
Efficiency $= \dfrac{\text{useful output energy}}{\text{input energy}}$		
Power $= \dfrac{\text{energy used (or work done)}}{\text{time taken}} =$ force × velocity	$P = \dfrac{E}{t} = Fv$	W
Potential energy = weight × change in height kinetic energy $= \dfrac{1}{2} \times$ mass × (speed)^2	$E_p = mgh$ $E_k = \dfrac{1}{2}mv^2$	J J

Formula	Formula symbols	Units
kinetic energy of rotation $= \dfrac{1}{2} \times$ moment of inertia $\times$ (angular velocity)^2	$E_k = \dfrac{1}{2} I\omega^2$	J
Frictional force = coefficient of friction $\times$ normal force	$F = \mu N$	N
Angle of repose, θ, on an inclined plane	$\tan\theta = \mu$	
Efficiency of screw jack	$\eta = \dfrac{\tan\theta}{\tan(\lambda + \theta)}$	
SHM periodic time $T = 2\pi\sqrt{\dfrac{\text{displacement}}{\text{acceleration}}}$	$T = 2\pi\sqrt{\dfrac{y}{a}}$	s
$T = 2\pi\sqrt{\dfrac{\text{mass}}{\text{stiffness}}}$	$T = 2\pi\sqrt{\dfrac{m}{k}}$	s
simple pendulum	$T = 2\pi\sqrt{\dfrac{L}{g}}$	s
compound pendulum	$T = 2\pi\sqrt{\dfrac{(k_G{}^2 + h^2)}{gh}}$	s
Force ratio $= \dfrac{\text{load}}{\text{effort}}$		
Movement ratio $= \dfrac{\text{distance moved by effort}}{\text{distance moved by load}}$		
Efficiency $= \dfrac{\text{force ratio}}{\text{movement ratio}}$		
Kelvin temperature = degrees Celsius + 273		
Quantity of heat energy = mass $\times$ specific heat capacity $\times$ change in temperature	$Q = mc(t_2 - t_1)$	J
New length = original length + expansion	$L_2 = L_1\,[1 + \alpha(t_2 - t_1)]$	m
New surface area = original surface area + increase in area	$A_2 = A_1\,[1 + \beta(t_2 - t_1)]$	m^2
New volume = original volume + increase in volume	$V_2 = V_1\,[1 + \gamma(t_2 - t_1)]$	m^3
Pressure $= \dfrac{\text{force}}{\text{area}}$ $=$ density $\times$ gravitational acceleration $\times$ height	$p = \dfrac{F}{A}$ $p = \rho g h$ $1\ \text{bar} = 10^5\ \text{Pa}$	Pa

Formula	Formula symbols	Units
Absolute pressure = gauge pressure + atmospheric pressure		
Metacentric height, GM	$GM = \dfrac{Px}{W} \cot \theta$	m
Bernoulli's equation	$\dfrac{P_1}{\rho} + \dfrac{v_1^{\,2}}{2} + gz_1$ $= \dfrac{P_2}{\rho} + \dfrac{v_2^{\,2}}{2} + g(z_2 + h_f)$	
Coefficient of discharge	$C_d = C_v \times C_c$	
Characteristic gas equation	$\dfrac{p_1 V_1}{T_1} = \dfrac{p_2 V_2}{T_2} = k$ $pV = mRT$	

Circular segment

In Figure F1, shaded area = $\dfrac{R^2}{2}(\alpha - \sin \alpha)$

Figure F1

Summary of standard results of the second moments of areas of regular sections

Shape	Position of axis	Second moment of area, I	Radius of gyration, k
Rectangle length d breadth b	(1) Coinciding with b	$\dfrac{bd^3}{3}$	$\dfrac{d}{\sqrt{3}}$
	(2) Coinciding with d	$\dfrac{db^3}{3}$	$\dfrac{b}{\sqrt{3}}$
	(3) Through centroid, parallel to b	$\dfrac{bd^3}{12}$	$\dfrac{d}{\sqrt{12}}$
	(4) Through centroid, parallel to d	$\dfrac{db^3}{12}$	$\dfrac{b}{\sqrt{12}}$
Triangle Perpendicular height h base b	(1) Coinciding with b	$\dfrac{bh^3}{12}$	$\dfrac{h}{\sqrt{6}}$
	(2) Through centroid, parallel to base	$\dfrac{bh^3}{36}$	$\dfrac{h}{\sqrt{18}}$
	(3) Through vertex, parallel to base	$\dfrac{bh^3}{4}$	$\dfrac{h}{\sqrt{2}}$
Circle radius r diameter d	(1) Through centre perpendicular to plane (i.e. polar axis)	$\dfrac{\pi r^4}{2}$ or $\dfrac{\pi d^4}{32}$	$\dfrac{r}{\sqrt{2}}$
	(2) Coinciding with diameter	$\dfrac{\pi r^4}{4}$ or $\dfrac{\pi d^4}{64}$	$\dfrac{r}{2}$
	(3) About a tangent	$\dfrac{5\pi r^4}{4}$ or $\dfrac{5\pi d^4}{64}$	$\dfrac{\sqrt{5}}{2}r$
Semicircle radius r	Coinciding with diameter	$\dfrac{\pi r^4}{8}$	$\dfrac{r}{2}$

Greek alphabet

Letter	Upper Case	Lower Case
Alpha	A	α
Beta	B	β
Gamma	Γ	γ
Delta	Δ	δ
Epsilon	E	ε
Zeta	Z	ζ
Eta	H	η
Theta	Θ	θ
Iota	I	ι
Kappa	K	κ
Lambda	Λ	λ
Mu	M	μ
Nu	N	ν
Xi	Ξ	ξ
Omicron	O	o
Pi	Π	π
Rho	P	ρ
Sigma	Σ	σ
Tau	T	τ
Upsilon	Y	υ
Phi	Φ	ϕ
Chi	X	χ
Psi	Ψ	ψ
Omega	Ω	ω

Answers to multiple-choice questions

Chapter 1
(Exercise 10, Page 15)

1. (b)	**2.** (d)	**3.** (a)	**4.** (d)	**5.** (a)
6. (b)	**7.** (c)	**8.** (a)	**9.** (b)	**10.** (c)
11. (b)	**12.** (d)	**13.** (a)	**14.** (c)	**15.** (a)
16. (c)	**17.** (d)	**18.** (b)	**19.** (a)	**20.** (c)

Chapter 2
(Exercise 17, Page 37)

1. (c)	**2.** (c)	**3.** (a)	**4.** (b)	**5.** (c)
6. (c)	**7.** (b)	**8.** (d)	**9.** (b)	**10.** (c)
11. (f)	**12.** (h)	**13.** (d)	**14.** (b)	**15.** (a)

Chapter 3
(Exercise 21, Page 45)

1. (f)	**2.** (d)	**3.** (g)	**4.** (b)

Chapter 4
(Exercise 29, Page 58)

1. (b)	**2.** (a)	**3.** (b)	**4.** (d)	**5.** (b)
6. (c)	**7.** (b)	**8.** (b)	**9.** (c)	**10.** (d)
11. (c)	**12.** (d)	**13.** (d)	**14.** (a)	

Chapter 5
(Exercise 35, Page 70)

1. (a)	**2.** (c)	**3.** (a)	**4.** (d)	**5.** (a)
6. (d)	**7.** (c)	**8.** (a)	**9.** (d)	**10.** (c)
11. (c)				

Chapter 6
(Exercise 40, Page 85)

1. (b)	**2.** (a)	**3.** (c)	**4.** (c)	**5.** (b)
6. (a)				

Chapter 7
(Exercise 44, Page 99)

1. (b)	**2.** (c)	**3.** (c)	**4.** (a)	**5.** (c)
6. (b)				

Chapter 8
(Exercise 50, Page 117)

1. (c)	**2.** (b)	**3.** (d)	**4.** (a)	**5.** (b)
6. (c)	**7.** (a)	**8.** (c)	**9.** (d)	**10.** (a)

Chapter 9
(Exercise 53, Page 125)

1. (b)	**2.** (b)	**3.** (c)

Chapter 10
(Exercise 59, Page 135)

1. (d)	**2.** (b)	**3.** (c)	**4.** (a)	**5.** (c)
6. (d)	**7.** (a)	**8.** (b)	**9.** (c)	**10.** (d)
11. (a)	**12.** (c)			

Chapter 11
(Exercise 62, Page 141)

1. (a)	**2.** (b)	**3.** (a)	**4.** (c)	**5.** (b)
6. (a)				

Chapter 12
(Exercise 68, Page 152)

1. (b)	**2.** (c)	**3.** (a)	**4.** (c)	**5.** (a)
6. (d)	**7.** (c)	**8.** (b)	**9.** (d)	**10.** (c)
11. (b)	**12.** (d)	**13.** (a)		

Chapter 13
(Exercise 72, Page 160)

1. (d)	**2.** (b)	**3.** (f)	**4.** (c)	**5.** (a)
6. (c)	**7.** (a)	**8.** (g)	**9.** (f)	**10.** (f)
11. (b)	**12**. (e)			

Chapter 14
(Exercise 77, Page 169)

1. (c)	**2.** (b)	**3.** (a)	**4.** (d)	**5.** (a)
6. (b)	**7.** (b)	**8.** (a)	**9.** (a)	**10.** (d)
11. (d)	**12.** (c)	**13.** (b)		

Chapter 15
(Exercise 84, Page 182)

1. (b)	**2.** (c)	**3.** (c)	**4.** (a)	**5.** (d)
6. (c)	**7.** (a)	**8.** (d)	**9.** (c)	**10.** (b)
11. (b)	**12.** (a)	**13.** (d)	**14.** (a)	**15.** (d)
16. (c)				

Chapter 16
(Exercise 89, Page 194)

1. (c)	**2.** (c)	**3.** (f)	**4.** (e)	**5.** (i)
6. (c)	**7.** (h)	**8.** (b)	**9.** (d)	**10.** (a)
11. (b)				

Chapter 17
(Exercise 94, Page 203)

1. (b)	**2.** (d)	**3.** (a)	**4.** (b)	**5.** (b)

Chapter 18
(Exercise 98, Page 210)

1. (b)	**2.** (c)	**3.** (b)	**4.** (a)

Chapter 19
(Exercise 105, Page 220)

1. (b)	**2.** (f)	**3.** (c)	**4.** (d)	**5.** (b)
6. (a)	**7.** (b)	**8.** (d)	**9.** (c)	**10.** (d)
11. (d)	**12.** (b)			

Chapter 20
(Exercise 111, Page 233)

1. (d)	**2.** (b)	**3.** (a)	**4.** (c)	**5.** (b)
6. (b)	**7.** (b)	**8.** (a)	**9.** (c)	**10.** (b)
11. (d)	**12.** (c)	**13.** (d)		

Chapter 21
(Exercise 115, Page 241)

1. (b)	**2.** (c)	**3.** (a)	**4.** (d)	**5.** (b)
6. (c)	**7.** (c)	**8.** (a)	**9.** (c)	**10.** (b)

Chapter 22
(Exercise 123, Page 259)

1. (b)	**2.** (d)	**3.** (a)	**4.** (a)	**5.** (c)
6. (d)	**7.** (b)	**8.** (c)	**9.** (c)	**10.** (d)
11. (d)	**12.** (d)	**13.** (c)	**14.** (b)	**15.** (c)
16. (a)	**17.** (b)	**18.** (f)	**19.** (a)	**20.** (b)
21. (c)	**22.** (c)	**23.** (a)	**24.** (c)	

Chapter 23
(Exercise 128, Page 270)

1. (c)	**2.** (c)	**3.** (d)	**4.** (a)	**5.** (c)
6. (j)	**7.** (i)	**8.** (d)	**9.** (g)	**10.** (f)
11. (a)	**12.** (h)	**13.** (e)	**14.** (k)	**15.** (c)
16. (c)				

Chapter 24
(Exercise 136, Page 279)

1. (a)	**2.** (d)	**3.** (b)	**4.** (b)	**5.** (c)
6. (d)	**7.** (b)	**8.** (c)	**9.** (c)	**10.** (b)

Chapter 25
(Exercise 140, Page 289)

1. (c)	**2.** (b)	**3.** (d)	**4.** (b)	**5.** (i)
6. (a)	**7.** (e)	**8.** (d)	**9.** (e or f)	**10.** (k)
11. (b)	**12.** (g)			

Index

Part Four